Electric Mobility Systems

Electric Mobility Systems

Fundamentals, Infrastructure, and Future Trends

Marcos Tostado-Véliz

Francisco Jurado

ELSEVIER

Elsevier
Radarweg 29, PO Box 211, 1000 AE Amsterdam, Netherlands
125 London Wall, London EC2Y 5AS, United Kingdom
50 Hampshire Street, 5th Floor, Cambridge, MA 02139, United States

ISBN: 978-0-443-49397-3

For information on all Elsevier publications
visit our website at https://www.elsevier.com/books-and-journals

Publisher: Priscilla Braglia
Acquisitions Editor: Ambika Kapoor
Editorial Project Manager: Arushi Rawat
Production Project Manager: Fahmida Sultana
Cover Designer: Bakyalakshmi S

Typeset by VTeX

To our families, friends, and colleagues.

With love to Fany

Contents

About the authors

Marcos Tostado-Véliz
Department of Electrical Engineering, University of Jaén, Linares, 23700, Spain

Was born in Plasencia in 1987, he received the B.Sc. degree in Electrical Engineering from University of Seville in 2016, graduating first in his class, and the M.Sc. degree in Electrical Engineering in 2017. He obtained the Ph.D. degree from University of Jaén in 2020, receiving the Extraordinary Doctorate Award. Since 2017, he has developed his academic career through a competitive pre-doctoral and post-doctoral research contract, and he currently serves as Associate Professor and Secretary of the Department of Electrical Engineering at the University of Jaén. Since 2018, he has been a member of the INYTE Research Group (TEP152).

His research focuses on power flow analysis, optimal energy management in smart grids, and energy community modeling. He has co-supervised five Ph.D. theses, co-led funded research projects, and authored more than 200 high-impact journal papers, accumulating over 6000 citations and an h-index above 40. He has also been recognized among the world's top 2% most influential researchers by Stanford University for three consecutive years. In addition, he has taught several undergraduate and graduate courses related to electrical circuits, electrical machines, and electric vehicles.

Francisco Jurado
Department of Electrical Engineering, University of Jaén, Linares, 23700, Spain

Was born in Linares. He received the Ph.D. degree in 1999 from National University of Distance Education (UNED). He is currently a Senior Member of Institute of Electrical and Electronics Engineers. His research activities have been devoted to several topics, including power systems and renewable energy.

He is author and co-author of more than 590 papers published in journals indexed in the Journal Citation Reports (JCR), around 280 papers in prestigious international conference proceedings, 11 teaching books, and 3 patents. He has participated in numerous research projects funded by the Spanish Ministry of Science and Education and the

European Commission. His scientific contributions have led to his inclusion among the world's most influential researchers according to the rankings published by Stanford University since 2017. In addition, he received the Electronics and Electrical Engineering in Spain Leader Award in 2024, 2025, and 2026 from Research.com.

Foreword

The electrification of transportation represents one of the most significant technological transitions of the twenty-first century. Driven by the urgent need to reduce greenhouse gas emissions, improve energy efficiency, and increase the sustainability of modern societies, electric mobility has rapidly evolved from a niche research field into a strategic pillar of future energy and transportation systems. This transformation involves not only the vehicles themselves, but also the complex interaction among energy storage technologies, charging infrastructures, intelligent energy management systems, renewable energy integration, and emerging digital solutions.

In this context, Electric Mobility Systems: Fundamentals, Infrastructure, and Future Trends provides a timely, comprehensive, and well-structured overview of the most relevant aspects of modern electric mobility. The book successfully combines fundamental concepts with current technological developments and future perspectives, offering readers a broad understanding of the challenges and opportunities associated with the ongoing electrification of transportation.

One of the main strengths of this work lies in its multidisciplinary approach. The authors do not limit the discussion to electric passenger cars, but instead extend the analysis to unmanned aerial vehicles, maritime applications, rail transportation, and electric aviation. In addition, the book addresses essential enabling technologies such as batteries, hydrogen systems, charging infrastructures, Vehicle-to-Everything paradigms, and smart charging strategies. This integrated perspective is particularly valuable in a field where technological convergence and system-level interactions are becoming increasingly important.

Another remarkable aspect of this book is its balance between accessibility and technical rigor. The material is presented in a clear and systematic manner, making the book suitable not only for researchers and professionals working in electric mobility and power systems, but also for undergraduate and graduate students seeking a solid introduction to the field. At the same time, the inclusion of recent technological trends, operational challenges, and infrastructure considerations ensures the relevance of the book for more advanced readers.

The authors have extensive expertise in electrical engineering, energy systems, and smart grid technologies, which is clearly reflected throughout the book. Their research experience and academic background allow them to present the subject from both scientific and practical perspectives, contributing to the overall coherence and quality of the work.

I am confident that this book will become a valuable reference for researchers, engineers, students, and professionals interested in electric mobility and sustainable transportation systems. Its broad scope, technical quality, and contemporary perspective make it a significant contribution to the growing body of literature in this important field.

Pierluigi Siano
Department of Management and Innovation Systems, University of Salerno, Fisciano, Italy

Preface

The rapid evolution of electric mobility is transforming the transportation sector and redefining the relationship between energy systems, infrastructure, and society. In recent years, electric vehicles have progressed from a promising technological alternative to a central element in global strategies aimed at reducing greenhouse gas emissions, improving energy efficiency, and promoting sustainable development. This transition is being driven not only by advances in vehicle technologies, but also by the increasing maturity of charging infrastructures, energy storage systems, intelligent control strategies, and regulatory frameworks.

Within this context, the objective of *Electric Mobility Systems: Fundamentals, Infrastructure, and Future Trends* is to provide a comprehensive and accessible overview of the principal technologies, concepts, and challenges associated with modern electric mobility. Rather than focusing exclusively on a single domain or application, this book adopts a broad perspective that integrates historical, technological, infrastructural, and operational aspects of electric transportation systems. The intention is to offer readers a structured reference work capable of serving both as an introductory text and as a technical resource for more advanced study.

The first part of the book introduces the foundations of electric mobility. Chapter 1 reviews the historical evolution of electric vehicles, highlighting the principal technological milestones and the external drivers that have influenced their development. Chapters 2 and 3 focus specifically on electric cars, presenting their main typologies and providing an overview of their current status and market penetration worldwide. Chapters 4 and 5 extend the discussion beyond passenger vehicles by addressing unmanned aerial vehicles, maritime transport, rail transportation, and electric aviation, thereby offering a broader perspective on the diversity of electric mobility applications.

The second part of the book examines the enabling technologies that support electric mobility systems. Chapter 6 analyses charging technologies and infrastructure, including charging modes, standards, and deployment trends. Chapters 7 through 9 are devoted to energy storage systems, covering batteries, hydrogen technologies, supercapacitors, and flywheel storage systems. These chapters discuss not only the operating principles and technical characteristics of each technology, but also their practical limitations, industrial maturity, and future prospects.

Finally, the last part of the book addresses advanced operational and energy management concepts. Chapter 10 introduces the Vehicle-to-Everything (V2X) paradigm and its different operating modes, while Chapter 11 focuses on smart charging strategies and coordinated energy management approaches for modern charging infrastructures. Together, these chapters illustrate the growing integration between electric mobility and intelligent energy systems, emphasizing the importance of optimization, flexibility, and digitalization in future transportation networks.

This book is primarily intended for undergraduate and graduate students, researchers, engineers, and professionals interested in electric mobility, smart grids, renewable energy integration, and sustainable transportation systems. Although several chapters include technical discussions and engineering concepts, the book has been written with the objective of remaining accessible to readers from different academic and professional backgrounds.

The authors would like to express their gratitude to colleagues, students, collaborators, and institutions that have contributed, directly or indirectly, to the development of this work. Special appreciation is extended to the research community whose continuous efforts are accelerating the transition toward cleaner and more sustainable mobility systems worldwide. Finally, the authors are deeply grateful to their families for their constant encouragement, patience, and unconditional support throughout the preparation of this book.

Marcos Tostado-Véliz
Francisco Jurado

History of electric vehicles

1

1.1 The birth and early prototypes

The origin of the electric vehicle (EV) is often attributed to Robert Anderson, a Scottish inventor active during the nineteenth century who, sometime between the early and mid-1830s, developed one of the earliest electrically propelled carriages. Although the exact technical details of Anderson's invention are not fully documented, his work represents one of the first known attempts to apply electric propulsion to road transportation. More broadly, however, the emergence of EVs—and electric mobility as a technological concept—can be traced slightly earlier, to the late 1820s, when several inventors across Europe began experimenting with the principles of electromagnetic motion and early electric motors.

Among the most notable pioneers was Ányos Jedlik (1800–1895), a Hungarian engineer and Benedictine priest. Around 1827, Jedlik developed one of the earliest experimental electric motors, marking a significant milestone in the technological foundations of electric propulsion. His work later evolved into a motor design incorporating the three essential components of a modern direct-current (DC) electric machine: the stator, the rotor, and the commutator. These elements remain central to many electric motor architectures used today. Some of Jedlik's original experimental devices have been preserved and can still be observed in operation at the Museum of Applied Arts Budapest (see Fig. 1.1), illustrating the ingenuity of these early developments and their importance in the history of electrical engineering.

At roughly the same time that Jedlik was conducting his research, other innovators were exploring the application of electric propulsion to transportation systems. Many

Figure 1.1 Jedlik's "lightning-magnetic self-rotor". (Image courtesy of Wikimmedia Commons.)

Electric Mobility Systems. https://doi.org/10.1016/B978-0-44-349397-3.00009-4

of these early prototypes were small experimental devices rather than practical vehicles, yet they played a crucial role in demonstrating the feasibility of electrically driven motion. One notable example is the work of Thomas Davenport (1802–1851), a blacksmith from Vermont (United States). In the 1830s, Davenport developed and patented one of the first practical electric motors and used it to power a small vehicle that moved along a circular rail track. Although simple in design, this apparatus represented one of the earliest demonstrations of electric propulsion applied to transportation technology.

While Anderson's electric carriage suggested that road vehicles powered by electricity were technically feasible, the first large-scale experiments with electric propulsion in real transportation environments took place in railway systems. In this context, the Scottish chemist and inventor Robert Davidson (1804–1894) made a pioneering contribution by developing one of the earliest full-scale electric locomotives powered by galvanic cells. Davidson constructed the Galvani locomotive in 1837, and the invention was later presented to the Royal Scottish Society of Arts in 1841. The Galvani locomotive represented a remarkable engineering achievement for its time. Weighing approximately 7100 kg, it was capable of transporting a load of around 6100 kg at a speed of about 6.4 km/h over a distance of roughly 2.4 km. In 1842, the locomotive was tested on the railway line between Edinburgh and Glasgow, becoming one of the earliest full-scale EVs demonstrated on an operational transportation infrastructure.

1.2 Combustion vs. Electricity: parallel developments

The final decades of the nineteenth century witnessed rapid technological progress in both major paradigms of mechanized mobility: internal combustion engines (ICEs) and electric propulsion. Important advances in internal combustion technology occurred during this period, particularly in Central Europe. One of the early pioneers was Siegfried Marcus (1831–1898), who in 1864 built an experimental gasoline-powered vehicle equipped with a single-cylinder ICE. This early prototype lacked a clutch mechanism, which meant that the rear wheels had to be lifted off the ground in order to start the engine. Although this design was still far from a practical automobile, it demonstrated the potential of liquid fuels as a source of mechanical power.

A major milestone occurred in 1886, when Carl Benz (1844–1929) developed the Benz Patent-Motorwagen, widely regarded as the first vehicle specifically designed to be powered by an ICE (see Fig. 1.2). The vehicle was equipped with a gasoline-powered engine integrated into a lightweight three-wheeled chassis. Its historical significance was further reinforced in 1888, when Bertha Benz (1849–1944), the inventor's wife, undertook the first long-distance automobile journey without her husband's prior knowledge. She drove from Mannheim to Pforzheim and back—a round trip of roughly 180 km—demonstrating the practical viability of the automobile and marking a defining moment in the history of road transportation.

Several other innovators contributed to the rapid evolution of internal combustion technology. In 1876, Nicolaus Otto (1832–1891) developed the first successful four-

Figure 1.2 The Benz Patent-Motorwagen developed by Carl Friedrich known as the first automobile to complete a long-distance trip. (Image courtesy of Wikimmedia Commons.)

Figure 1.3 The Reitwagen developed by Daimler and Maybach known as the first motorcycle. (Image courtesy of Wikimmedia Commons.)

stroke gasoline engine, commonly known as the Otto cycle, which became the fundamental operating principle for most ICEs developed thereafter. Otto also co-founded one of the earliest engine manufacturing companies together with Eugen Langen (1833–1895). Later, Gottlieb Daimler (1834–1900) and Wilhelm Maybach (1846–1929) introduced numerous improvements to combustion engines, including early carburetor designs. In 1885 they also developed the Reitwagen (see Fig. 1.3), often considered the first motorcycle in history.

Parallel to these developments, important innovations were taking place in the field of electric mobility. One of the most remarkable figures in this area was

Ferdinand Porsche (1875–1951). Although largely self-taught as an engineer, Porsche made several groundbreaking contributions to vehicle technology. In 1899 he joined the Viennese company Lohner-Werke, where he developed an innovative wheel-hub electric motor that integrated traction, transmission, and braking components within the wheel hub itself. This technology was incorporated into the Lohner-Porsche Electromobile, presented at the Paris Exposition Universelle 1900. The vehicle employed two electric motors rated at approximately 2.5 HP each and could reach a maximum speed of about 37 km/h.

Porsche's work attracted the attention of a British entrepreneur named E.W. Hart, who commissioned a vehicle capable of operating both on electricity and gasoline while transporting four passengers. The request reflected two key technological limitations of the time: the lack of charging infrastructure and the considerable weight of early battery systems. Porsche responded by developing what is widely regarded as one of the first hybrid vehicles in history, known as the Semper Vivus ("always alive"). Technically, the Semper Vivus functioned as a series hybrid or range-extended EV, combining wheel-hub electric motors powered by batteries with a gasoline engine connected to a generator that produced electricity to recharge the batteries.

To understand how such vehicles became possible, it is important to consider earlier developments in electrical engineering. The evolution of electric mobility was closely linked to advances in electric generators and energy storage technologies. In the early 1830s, the French instrument maker Hippolyte Pixii (1808–1835) built one of the first electric generators based on the principle of electromagnetic induction discovered by Michael Faraday (1791–1867). Later improvements led to the development of practical industrial generators during the mid-nineteenth century. Important contributions in this area were made by figures such as Charles Wheatstone (1802–1875), Werner von Siemens (1816–1892), and Samuel Alfred Varley (1832–1921), whose work helped establish the foundations of large-scale electrical power generation.

At nearly the same time, major progress was achieved in electrical energy storage. In 1859, the French physicist Gaston Planté (1834–1889) invented the lead–acid rechargeable battery, a crucial breakthrough that significantly improved the ability to store electrical energy. Planté's design reduced the internal resistance that had limited the performance of earlier electrochemical cells. However, the initial design required a lengthy formation process to prepare the electrodes. In 1881, the French chemical engineer Camille Alphonse Faure (1840–1898) introduced an important improvement by adding lead grids filled with lead-oxide paste, greatly increasing the battery's capacity and enabling large-scale industrial production.

Advances in both electricity generation and energy storage stimulated further experimentation with electrically powered vehicles. In 1867, the Austrian inventor Franz Kravogl presented an early two-wheeled EV at the Paris Universal Exhibition, although the design remained largely experimental. During the 1880s, the French electrical engineer Gustave Trouvé (1839–1902) carried out several notable experiments in electric mobility. In 1880 he electrified a tricycle originally manufactured by Starley using an electric motor developed by Siemens. Trouvé also made important contri-

Figure 1.4 Andreas Flocken and his wife aboard the Flocken Elektrowagen, developed in 1888. (Image courtesy of Wikimmedia Commons.)

butions in other areas, including the invention of the outboard electric motor and the development of one of the first electrically powered boats, capable of navigating rivers at speeds of approximately 3.6 km/h upstream and 9 km/h downstream.

Electric rail transport also advanced significantly during this period. Around 1879, Werner von Siemens demonstrated one of the first practical electric railways in Berlin. The system used a locomotive powered by a 2.2 kW electric motor pulling three carriages and reaching speeds of approximately 13 km/h. Siemens later developed one of the first electric tramways, inaugurated in 1881 in Lichterfelde, Germany. Similar electrified transport systems soon appeared elsewhere, including the Volk's Electric Railway in 1883 and the electric tram system of the Richmond Union Passenger Railway in 1888.

The closing years of the nineteenth century witnessed two major milestones in the development of electric automobiles. In 1888, the German engineer Andreas Flocken (1845–1913) introduced what is often considered the first four-wheeled electric car (EC), the Flocken Elektrowagen (see Fig. 1.4). The vehicle weighed roughly 400 kg, with the battery system accounting for about 100 kg, and could reach speeds of approximately 15 km/h using an electric motor rated at around 0.7 kW.

Just over a decade later, EVs achieved a spectacular demonstration of performance. In 1899, the Belgian engineer and racing driver Camille Jenatzy (1868–1913) set a new land-speed record in Paris by driving an electrically powered vehicle called La Jamais Contente (see Fig. 1.5). With its torpedo-shaped body made of lightweight alloy, the vehicle became the first automobile in history to exceed 100 km/h, reaching a speed of approximately 105 km/h. This achievement highlighted the remarkable performance potential of electric propulsion at a time when the technological competition between electric and combustion vehicles was still far from being decided.

Figure 1.5 The La Jamais Contente, exhibited at the Compiègne Automobile Museum, was the first vehicle to reach 100 km/h. (Image courtesy of Wikimedia Commons.)

1.3 The manure crisis

By the end of the nineteenth century, urban transportation in most of the world's major cities relied heavily on horse-drawn vehicles. Before the widespread adoption of mechanized mobility technologies such as the ICE and the EV, horses constituted the backbone of urban logistics, passenger transport, and goods distribution. In cities such as London, it is estimated that more than 300,000 horses were present on the streets every day, pulling carriages, omnibuses, taxis, and freight wagons.

The extensive use of horses created a serious public health and environmental challenge. Each horse could produce roughly 10 kg of manure and about one liter of urine per day. When multiplied by the number of animals operating daily in large metropolitan areas, these quantities generated enormous volumes of organic waste that accumulated in streets and public spaces. The decomposition of this waste released gases—primarily methane and ammonia—contributing to unpleasant odors and raising growing concerns about urban sanitation and the spread of disease. These problems were exacerbated by the rapid industrialization taking place in many cities during the same period, which was already contributing to deteriorating air quality.

In addition to the sanitary implications, municipal authorities faced increasing difficulties regarding the collection, storage, and disposal of animal waste. Contemporary observers warned that, if urban populations and horse-based transportation systems continued to grow at the same rate, cities could eventually face overwhelming waste management challenges. Some period commentators even suggested that, within a few decades, horse manure could accumulate in the streets to alarming levels, illustrating the magnitude of the perceived crisis.

These concerns became particularly visible during an international urban planning conference held in New York City in 1889. The meeting is often remembered as one of the earliest occasions in which the problem of urban transportation was discussed not only from a logistical perspective but also in relation to public health, sanitation, and the overall well-being of city residents. Although the conference did not produce concrete policy solutions—and reportedly ended earlier than planned due to disagreements among participants—it represented an important step in the professionalization of urban planning and in the recognition that transportation systems could have significant environmental and societal impacts.

Interestingly, the so-called "manure crisis" was ultimately resolved not through direct waste-management policies but through technological transformation. The rapid development and adoption of mechanized transportation—including ICE-powered vehicles and, to a lesser extent at the time, early EC and other forms of EV—gradually displaced horse-drawn transport in urban environments. As motorized vehicles became more practical and economically viable during the early twentieth century, horses were progressively removed from city streets, thereby eliminating one of the most pressing urban sanitation problems of the nineteenth century.

1.4 The golden years

During the late nineteenth and early twentieth centuries, motorized vehicles gradually replaced horse-drawn transport as the dominant form of urban mobility. At that time, several technological alternatives coexisted, including steam-powered vehicles, ICE vehicles powered by gasoline, and early EVs. Initially, steam-powered automobiles—representing a more mature technology—held a significant share of the market. In the United States, for example, steam vehicles accounted for roughly 40% of automobile sales between 1900 and 1905, compared with approximately 38% for EVs and 22% for gasoline-powered ICE vehicles. Despite this initial distribution, ECs quickly gained popularity, particularly in large urban areas where their operational advantages were most evident. By the end of the nineteenth century, electric taxis were already operating in cities such as London, which became an important center for early electric mobility innovation. Engineers such as Walter Bersey (1874–1950) and Thomas Parker (1843–1915) developed several functional EVs, including taxis, delivery vans, and buses, while also contributing to improvements in electrical energy storage technologies.

At the time, gasoline-powered ICE vehicles were still relatively immature and presented several practical disadvantages. Early gasoline cars were difficult to operate, requiring complex manual gear shifting and a physically demanding hand-crank mechanism to start the engine. In contrast, ECs offered a number of appealing features: they were quiet, easy to drive, mechanically simpler, and produced no local exhaust emissions. Their main limitation, however, was limited driving range due to the restricted energy density of early batteries. This drawback was compounded by the limited availability of electrical infrastructure, which made vehicle recharging difficult in many areas. Nevertheless, road networks were still relatively sparse and often in poor condition, meaning that most travel needs were local rather than long-distance. In such contexts, EVs represented a highly practical solution for urban mobility.

Steam-powered vehicles gradually disappeared from urban environments because of their operational complexity, long start-up times, and maintenance requirements.

Their decline contributed to the growing prominence of EVs in the early automotive market. During this period, the United States emerged as a major center for EV manufacturing, with companies such as Detroit Electric, Studebaker, Riker Electric Vehicle Company, and Anthony Electric Company producing various electric automobiles. Meanwhile, European production developed unevenly. France emerged as a leading automobile manufacturing nation, with companies such as Peugeot and Renault becoming major exporters before the outbreak of World War I. In contrast, production volumes remained more limited in countries such as Germany and the United Kingdom during this early stage of industrial development.

Electrification also began to influence public transportation systems. One of the earliest electric tramway demonstrations was developed by Siemens & Halske during the Berlin Industrial Exposition 1879. This prototype later evolved into operational tram systems such as the line inaugurated in Lichterfelde in 1881, which could transport several passengers using electric traction. Other experimental electrified transport systems soon followed, including early trolleybus concepts such as the Electromote, tested in Halensee.

Electric buses also appeared during the early twentieth century. In London, electric buses operated as part of the services provided by the London Electrobus Company. Later, in 1915, Thomas Edison (1847–1931) collaborated in the development of the Edison–Baker electric car, an EC produced by Baker Motor Vehicle Company and equipped with Edison's nickel–iron battery technology. These vehicles were sold in several cities, including York, West Bromwich, and Loughborough.

As the twentieth century progressed, electrical infrastructure expanded significantly, with power grids gradually reaching a growing number of households. This development made it easier for owners of EVs to recharge their vehicles at home. Early charging equipment was typically bulky and rudimentary, often requiring users to manually regulate charging voltage and current. Nevertheless, several early examples of collective and public charging infrastructure also appeared. During the 1910s, for instance, more than 700 public charging stations reportedly existed in the United States. In cities such as New York City, drivers could leave their ECs overnight in garages where batteries were recharged. Companies such as General Electric even developed early charging networks, including the Electrant system.

Another innovative approach involved battery swapping. This concept allowed depleted batteries to be replaced with fully charged units in a matter of minutes, thereby avoiding the long charging times—often several hours—required by conventional charging systems. One of the earliest implementations of battery swapping was developed by the Hartford Electric Light Company, which operated such systems for electric trucks between 1910 and 1924. Similar initiatives were also implemented in cities such as Chicago for vehicles produced by the Milburn Wagon Company.

Despite the promising growth of EVs during this period, several developments in the early twentieth century led to the rapid dominance of gasoline-powered ICE vehicles. One decisive factor was the introduction of the Ford Model T in 1908 by Henry Ford (see Fig. 1.6). Mass production techniques dramatically reduced the cost of gasoline vehicles. By 1912, a Model T could be purchased for around $650, while

Figure 1.6 The Ford Model T represented a turning point in automotive history, displacing the EC as the preferred choice for American drivers. (Image courtesy of Wikimedia Commons.)

electric automobiles often cost approximately $1750. This large price difference made gasoline vehicles far more accessible to the average consumer.

Another key technological advance occurred in 1912 when Charles F. Kettering introduced the electric starter motor for gasoline vehicles. This innovation eliminated the need for the cumbersome and sometimes dangerous hand-crank starting mechanism, significantly improving the usability of ICE vehicles.

During the 1920s, additional structural changes further strengthened the dominance of gasoline automobiles. The rapid expansion and improvement of the road network in the United States encouraged longer-distance travel between cities. For such journeys, gasoline-powered vehicles offered greater driving range and the convenience of quick refueling, advantages that EVs could not yet match. At the same time, major oil discoveries in Texas significantly increased the supply of crude oil and reduced gasoline prices, making ICE vehicles even more economically attractive.

Finally, broader economic and geopolitical developments also influenced transportation technologies. The increasing militarization that preceded World War II accelerated the development of gasoline-based propulsion technologies, which were generally considered more suitable for military applications due to their greater range and refueling flexibility. Meanwhile, changes in urban transport systems also occurred. In many cities, electric tram networks were gradually replaced by buses, which offered greater operational flexibility and did not require fixed infrastructure such as overhead catenary lines. In some countries—including the United States, Argentina, and Spain—tram networks were largely dismantled during the twentieth century, while in other regions, particularly in parts of Central Europe and Japan, tram systems continued to operate and evolve.

1.5 The long decline

The continuous reduction in gasoline prices during the early twentieth century, combined with rapid technological improvements in the ICE, led to a dramatic decline in the production of EVs during the 1910s. Most manufacturers that had previously experimented with electric propulsion gradually abandoned the technology in favor of gasoline-powered vehicles, which were becoming increasingly affordable and practical. Only a few isolated projects attempted to maintain electric automobile development. One of the most notable examples was the Henney Kilowatt, developed in the late 1950s by Henney Motor Company in collaboration with National Union Electric Company. This small EC was produced in configurations operating at 36 V and 72 V. It could reach speeds close to 96 km/h, although its practical driving range remained limited to roughly one hour of operation under typical conditions.

In this context, the use of EVs was largely confined to applications in which driving range was not a critical factor. During the 1920s, companies such as Clark Equipment Company and Yale & Towne Manufacturing Company developed electrically powered industrial forklifts and warehouse vehicles, which proved highly suitable for indoor logistics operations. In Europe—particularly in the United Kingdom—EVs continued to be used for specific urban services such as milk delivery. These "milk floats" became a familiar feature of urban streets and represented a significant proportion of the EVs in operation throughout much of the twentieth century. Similarly, electric golf carts began to appear during the 1950s, when companies such as Lektro started producing electrically powered vehicles designed for use in airports and golf courses.

A brief resurgence of interest in EVs emerged in the early 1970s following major geopolitical disruptions in global energy markets. In 1973, members of the Organization of the Petroleum Exporting Countries imposed an oil embargo on several countries—including the United States, Japan, Netherlands, United Kingdom, Canada, and others—because of their support for Israel during the Yom Kippur War. This event triggered what became known as the 1973 Oil Crisis. Within a few months, the price of crude oil increased by approximately 300%, rising from about $3 to $12 per barrel.

The crisis prompted affected countries to intensify efforts to reduce dependence on imported oil. In the United States, policies such as the creation of the Strategic Petroleum Reserve and the introduction of fuel-efficiency regulations were implemented. One important initiative was the establishment of the Corporate Average Fuel Economy standards, which required automobile manufacturers to improve the fuel efficiency of their vehicles. At the same time, the U.S. government increased funding for research into alternative propulsion technologies, including EVs and hybrid vehicles. In 1976, the U.S. Congress authorized the United States Department of Energy to support research and development programs focused on electric mobility.

These initiatives led several manufacturers to explore electric propulsion once again. As a result, the 1970s witnessed the introduction of several experimental EC models. One notable example was the CitiCar, produced between 1974 and 1977 by Sebring-Vanguard (see Fig. 1.7). The vehicle used battery configurations ranging from six to eight units operating at 36 V or 48 V. It could reach speeds slightly

Figure 1.7 The CitiCar was one of the few EV models developed by the American industry between the 1950s and 1970s. (Image courtesy of Wikimedia Commons.)

above 60 km/h and had a driving range of approximately 60–65 km. During the same period, the government of Japan launched ambitious research programs aimed at promoting the development of electric and energy-efficient vehicles, initiatives that would later contribute to the leadership of Japanese companies in advanced vehicle technologies.

Despite the general decline of EVs during much of the twentieth century, some remarkable milestones in electric mobility still occurred. In 1971, the Lunar Roving Vehicle became the first EV to operate beyond Earth when it was deployed during the Apollo 15 (see Fig. 1.8). The vehicle allowed astronauts to travel several kilometers across the lunar surface and demonstrated the reliability of electric propulsion in extreme environments.

Electric propulsion also began to appear in early unmanned aerial vehicles (UAVs). Although the conceptual origins of unmanned flight date back centuries—including experimental designs proposed by Leonardo da Vinci—practical UAV development accelerated during the twentieth century, particularly in connection with military research programs. Early UAVs developed during World War I and World War II were typically powered by ICE engines. However, by the late 1950s researchers began experimenting with electrically powered radio-controlled aircraft using DC motors and nickel–cadmium batteries.

During the 1970s, solar-powered UAV concepts also began to emerge. One pioneering project was the AstroFlight Sunrise, designed in 1970 by engineers at AstroFlight Incorporated. The aircraft completed its first successful flight in 1974 from Bicycle Lake, reaching an altitude of approximately 500 feet. After the original prototype was lost during its 28th flight due to turbulence, it was replaced by an improved version known as Sunrise II.

Figure 1.8 The Apollo 15 Lunar Roving Vehicle became the first EV to leave planet Earth. (Image courtesy of Wikimedia Commons.)

Another early solar-powered aircraft was the Solar Riser, developed in the late 1970s by Larry Mauro (see Fig. 1.9). The aircraft was derived from the Easy Riser glider and incorporated nickel–cadmium batteries that could be recharged using a 350 W solar panel array. The system powered a DC motor rated at approximately 2.6 kW. Although the Solar Riser could only sustain flight for a few minutes, it represented an important technological demonstration of solar-powered aviation. Its first flight took place on April 29, 1979, departing from Flabob Airport in California.

While the mid-twentieth century represented a relatively quiet period for EVs, electric propulsion continued to achieve significant milestones in rail transport. In 1939, the Italian electric train ElettroTreno ETR 200 set a railway speed record of approximately 204 km/h. This achievement occurred within the broader electrification of major railway lines in Italy, particularly the corridor connecting cities such as Milan, Bologna, Florence, Rome, and Naples. The project was developed with the participation of the Politecnico di Torino, which played an important role in advancing high-speed rail technology.

A few decades later, electric rail transport reached another historic milestone when Japan inaugurated the first Shinkansen high-speed railway line in 1964. Often referred to as the "bullet train," the Shinkansen was designed to address rapidly growing passenger demand between the country's largest metropolitan areas. The success of this

Figure 1.9 The Solar Riser, devised by Larry Mauro in 1979 as a modification of the UFM Easy Riser. (Image courtesy of Wikimedia Commons.)

electrified high-speed rail system demonstrated the long-term viability and efficiency of electric propulsion in large-scale transportation infrastructure, even during a period when EVs themselves remained largely absent from the automotive market.

1.6 The EV1 and its demise

The 1990s marked a period of growing environmental awareness, particularly regarding the consequences of large-scale fossil fuel consumption. Among the most pressing concerns were the severe levels of air pollution observed in many urban areas and the associated public health risks. In response to these concerns, several governmental agencies in the United States introduced regulatory initiatives aimed at reducing emissions produced by vehicles powered by the ICE.

One of the most influential policies of this period was the Zero Emission Vehicle (ZEV) mandate introduced by the California Air Resources Board (CARB) in 1990. This regulation represented one of the earliest attempts to accelerate the deployment of zero-emission transportation technologies. The ZEV mandate required automobile manufacturers selling vehicles in California to ensure that a minimum percentage of their total sales corresponded to vehicles producing zero tailpipe emissions. Initially, the regulation established a target of 2% zero-emission vehicles by 1998, increasing to 5% by 2001 and 10% by 2003. Financial incentives were also introduced to encourage consumers to adopt zero-emission vehicles. A key feature of the ZEV mandate was that it encouraged the development of alternative propulsion technologies beyond the conventional ICE. In particular, battery-powered EVs emerged as one of the most promising technological pathways. This policy therefore represented a significant shift from previous regulatory approaches, which had typically focused on incremental improvements in fuel efficiency and emissions control for conventional vehicles.

Within this regulatory context, General Motors (GM) launched the PreView program in 1994. The objective of this initiative was to lease approximately fifty prototype EVs, known as the Impact, for short evaluation periods of one or two weeks. The public response was far greater than expected. Instead of the roughly eighty inquiries anticipated by the company, more than 10,000 requests were reportedly received, forcing the temporary suspension of the program's telephone lines.

Encouraged by this unexpectedly strong interest, GM proceeded with the development of the EV1, a two-seat fully EC introduced in 1996. The first generation of the EV1 incorporated 16.5 kWh lead–acid batteries. Combined with its exceptionally low aerodynamic drag coefficient and lightweight structure—achieved partly through the use of an aluminum chassis—the vehicle offered a driving range of approximately 113 km. Unlike many earlier experimental EVs, the EV1 was designed from the outset as a dedicated electric platform rather than as a conversion of a conventional ICE vehicle.

Despite these technological advances, GM adopted a cautious commercialization strategy. The EV1 was offered exclusively through leasing agreements and only in the states of California and Arizona. Approximately 1100 units were produced in total. Charging infrastructure remained limited, with an estimated 300 public charging stations installed primarily in the regions where the vehicle was available. Leasing costs ranged between $399 and $459 per month, typically including both a residential charging unit and access to public charging infrastructure.

The leasing contract allowed a maximum mileage of 30,000 miles and did not include a purchase option. Nevertheless, feedback from drivers was largely positive. The EV1 was widely praised for its quiet operation, smooth acceleration, and ease of driving. The vehicle could accelerate from 0 to 100 km/h in roughly nine seconds and did not require a conventional multi-gear transmission. For many users, its range was sufficient to meet the needs of daily commuting.

In 1999, GM introduced a second generation of the EV1 equipped with nickel–metal hydride batteries, which increased the driving range to more than 200 km under favorable conditions. This version was marketed primarily in California, where regulatory incentives and infrastructure availability created more favorable conditions for EV adoption. Despite the positive reception among many drivers, the EV market remained relatively small during this period. Meanwhile, the regulatory framework that had originally stimulated EV development gradually evolved. Modifications to the ZEV mandate introduced additional flexibility for manufacturers by allowing alternative technologies, such as hybrid-electric vehicles (HEVs) and natural gas vehicles, to contribute toward compliance targets. These policy adjustments, combined with limited consumer demand at the time, reduced the pressure on manufacturers to deploy large numbers of battery electric vehicles (BEVs).

In 2001, General Motors announced the termination of the EV1 program. Vehicle recalls began in 2002, and by 2003 most EV1 units had been recovered by the manufacturer. The decision generated considerable debate among EV enthusiasts and early adopters, many of whom had expressed strong satisfaction with the vehicle. Some drivers even offered to extend their leasing agreements at their own expense in order to continue using the vehicles. Nevertheless, GM proceeded with the recall of the fleet. The end of the EV1 program remains one of the most widely discussed

Figure 1.10 The Th!nk City is an example of the neighborhood EVs developed at the beginning of the twenty-first century. (Image courtesy of Wikimedia Commons.)

episodes in the modern history of electric mobility. The events surrounding the project are examined in detail in the documentary Who Killed the Electric Car? directed by Chris Paine, which explores the technological, economic, and policy factors that influenced the program's outcome.

Although other manufacturers also experimented with EV technologies during the 1990s—such as Honda with its EV+ program—none achieved the same level of visibility as the EV1. Following the cancellation of several early EV initiatives, large-scale commercial development of passenger ECs slowed considerably during the early years of the twenty-first century. Most ECs available at the time were neighborhood cars or microcars designed primarily for short-distance urban mobility. One notable example from this period was the Th!nk City (see Fig. 1.10), a small BEV originally developed in Norway and later marketed in several European countries, including Sweden, Denmark, the United Kingdom, Germany, Italy, and the Netherlands. Despite the limited presence of full-size ECs in the early 2000s, global deployment of small urban EVs was not negligible. By the beginning of the decade, it was estimated that approximately 479,000 such vehicles were in operation worldwide, with the United States, Europe, and China representing the largest markets.

While passenger ECs experienced limited commercial expansion during this period, important technological advances in other areas of electric mobility were simultaneously taking place. One significant development was the increasing adoption of brushless DC motors in small electric propulsion systems. Unlike conventional DC motors, brushless machines use electronic commutation instead of mechanical brushes. This design reduces friction and mechanical wear, improves efficiency, lowers noise levels, and significantly increases operational lifetime. These advantages made brushless motors particularly suitable for compact electric propulsion systems.

Table 1.1 Main disadvantages of lead- and nickel-based storage technologies for electric mobility applications.

Lead-based batteries	Nickel-based batteries
Low energy density	Memory effect
Peukert effect: higher power demand reduces effective capacity	High self-discharge rates
Limited lifetime	Significant heat generation during rapid charge–discharge cycles capacity
Waste management challenges (lead toxicity)	Intermediate energy density

The widespread adoption of brushless DC motors contributed significantly to the rapid development of UAVs. During the 1990s, UAV technology advanced considerably, including the emergence of small tactical UAV systems. A representative example is the AeroVironment FQM-151 Pointer, widely considered one of the first tactical UAVs deployed by the United States military, which acquired approximately fifty units in 1990. The system was powered by lithium batteries driving a 0.3 kW electric motor, enabling a flight endurance of roughly one hour, a maximum speed of approximately 73 km/h, and an operational altitude of around 300 meters.

At the same time, several forms of electrified urban transport were experiencing renewed interest. In particular, electric tram systems began to re-emerge in a number of European cities. Although the roots of this revival can be traced back to the energy concerns raised during the 1973 oil crisis, large-scale deployment occurred mainly from the 1980s onward. France played a particularly important role in this process, developing modern tram networks in cities such as Nantes (1985), Strasbourg (1994), Rouen (1994), Bordeaux (2003), Nice (2007), and Toulouse (2010). The success of these systems demonstrated the continued relevance of electrified public transportation and encouraged many other cities to adopt similar solutions.

1.7 The lithium revolution

Until the early twenty-first century, electric mobility relied almost entirely on lead–acid and nickel-based batteries. Although these technologies played an important role in the early development of the EV, they presented significant limitations for large-scale mobility applications. The main drawbacks of these storage technologies are summarized in Table 1.1.

Because of these limitations, the battery technologies available throughout most of the twentieth century represented a major bottleneck for the widespread adoption of EVs. This constraint was particularly critical for applications that could not rely on external power supply infrastructure, such as catenary systems used in electric rail transport. As a result, the limited energy density and performance of early batteries significantly restricted the development of ECs and UAVs.

The foundations of modern lithium-based batteries, however, had been established several decades earlier. The first major step toward lithium battery technology occurred during the energy crisis triggered by the 1973 oil shock, when concerns over energy security stimulated research into alternative energy storage technologies. During this period, the American chemist Michael Stanley Whittingham (1941–) developed one of the first functional lithium battery prototypes while working at Exxon. His design used titanium disulfide (TiS_2) as the cathode material and metallic lithium as the anode. Although the concept demonstrated promising energy density, early designs faced significant safety challenges due to the formation of lithium dendrites, which could cause internal short circuits.

Important improvements were achieved during the 1980s through the work of several researchers. In particular, John B. Goodenough (1922–2023) discovered that lithium cobalt oxide ($LiCoO_2$) could be used as a cathode material capable of delivering significantly higher energy densities. Shortly afterward, the Japanese chemist Akira Yoshino (1948–) introduced a major safety improvement by replacing the metallic lithium anode with a carbon-based material derived from petroleum coke. This carbon material enabled the reversible intercalation of lithium ions, eliminating many of the safety risks associated with metallic lithium electrodes. These combined discoveries laid the foundations for the modern lithium-ion battery, a breakthrough that later earned Whittingham, Goodenough, and Yoshino the 2019 Nobel Prize in Chemistry.

The 1990s marked the beginning of the commercial expansion of lithium-ion batteries. A key milestone occurred when the Japanese company Sony introduced lithium-ion batteries in consumer electronics such as camcorders and portable devices. The primary motivation for adopting this technology was its significantly higher energy density compared with lead-based batteries, which were generally too heavy for portable applications. At that time, however, lithium-ion batteries were still extremely expensive. In the early 1990s, battery costs were estimated to exceed $4,000 per kWh, compared with roughly $100 per kWh for conventional lead–acid batteries. As a result, lithium-based technologies initially remained confined to high-value electronic devices where weight reduction was critical.

Another important factor influencing the development of lithium technologies was the structure of the global lithium supply chain. During the early 1990s, a significant portion of lithium production originated from hard-rock mining, particularly in the United States. These extraction methods were relatively costly, and the limited market demand for lithium prevented the emergence of significant economies of scale. This situation began to change in the late 1990s with the expansion of lithium production in Chile, particularly from the brine deposits of the Salar de Atacama. In these deposits, lithium could be extracted by pumping brine to the surface and concentrating it through solar evaporation, a process considerably cheaper than traditional hard-rock mining. The entry of Chile into the global lithium market significantly increased supply and contributed to a gradual reduction in battery costs. As illustrated in Fig. 1.11, the cost of lithium-ion batteries began to decline steadily toward the beginning of the twenty-first century, making the technology progressively more

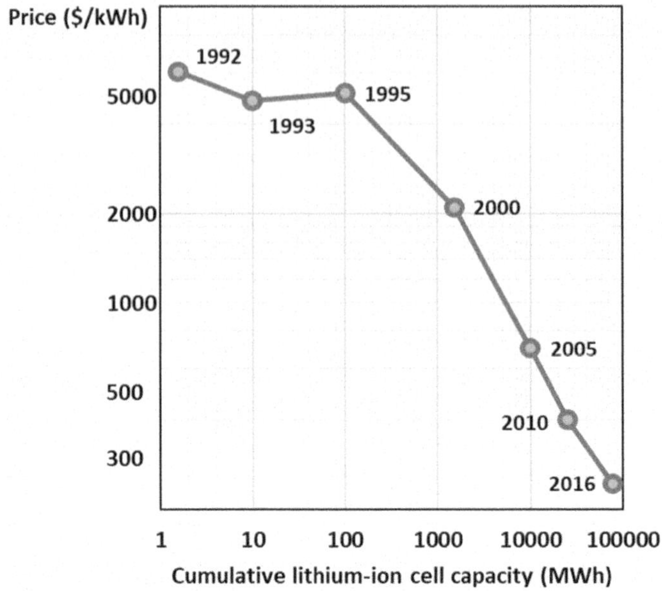

Figure 1.11 Historical decline in lithium-ion battery pack costs between 1992 and 2016. Continuous improvements in manufacturing processes, economies of scale, and technological innovation have reduced battery costs by more than an order of magnitude. This decline has been a key factor enabling the large-scale deployment of EVs.

competitive. In contrast, several higher-cost mining operations in the United States became economically unviable and were eventually closed.

At the beginning of the twenty-first century, lithium-ion batteries benefited from growing economies of scale as their use expanded in consumer electronics such as laptop computers and mobile devices. Increased production volumes, improvements in manufacturing processes, and continuous technological innovation progressively reduced costs while improving performance. These developments created the technological conditions necessary for the re-emergence of EVs as a viable transportation technology. A key milestone in this process occurred in 2008, when the Tesla Roadster became the first production EC to employ large-scale lithium-ion battery packs, achieving a driving range of approximately 300 km. The vehicle demonstrated that lithium-ion technology could provide sufficient energy density to support long-range electric mobility, overcoming one of the principal limitations that had historically constrained EV development.

The global financial crisis of 2008 also indirectly accelerated investment in energy technologies, including lithium batteries. Governments and investors increasingly recognized the strategic importance of advanced energy storage technologies for both energy security and the transition toward lower-carbon energy systems. As a result, research and industrial investment in lithium battery technology expanded rapidly during the following decade.

The impact of lithium batteries was not limited to passenger EVs. Their high energy density and relatively low weight also played a critical role in the rapid development of UAV technologies. Earlier battery technologies were often too heavy to allow small aerial platforms to achieve sustained flight. Lithium batteries, however, enabled multirotor UAVs to generate sufficient thrust to lift their own weight while carrying sensors, cameras, and communication equipment. These technological improvements facilitated the emergence of lightweight consumer UAVs such as the Parrot AR.Drone and the DJI Phantom, quadcopters weighing roughly one kilogram including their battery packs. These platforms represented important milestones in the popularization of UAV technology, as their relatively low cost and compact size made them accessible to a broad range of users beyond military and research applications.

The rapid progress of lithium-ion batteries therefore marked a turning point in the history of electric mobility. By significantly improving the energy density, efficiency, and practicality of electric propulsion systems, lithium technology removed one of the most persistent barriers to the widespread adoption of EVs and other forms of electric transportation.

1.8 2010s: the definitive breakthrough?

The 2010s represented a decisive turning point in the history of the EV, marking a period of rapid expansion in both sales and technological maturity. One of the most significant milestones occurred in 2010, when the Nissan Leaf was introduced to the market. The Leaf became the first mass-produced five-door battery EC launched by a major automotive manufacturer since the EV1 program of the 1990s. Its release demonstrated that large-scale commercialization of EVs had become technically and economically viable. Following the introduction of the Leaf, a wide range of manufacturers entered the EV market between 2010 and 2013. Companies such as BMW, Renault, Honda, Toyota, Mercedes-Benz, Fiat, Volkswagen, and Coda Automotive released their own electric models during this period. In 2011, the Mitsubishi i-MiEV became the first EV to surpass 10,000 units sold, a milestone that was soon eclipsed by the Nissan Leaf, which reached 50,000 cumulative sales by 2013.

The EV market experienced another major technological milestone with the introduction of Tesla's Model S in 2012, followed shortly afterward by the Model X. Tesla's strategy of combining long driving range with high-performance electric drivetrains helped reshape public perception of EVs, demonstrating that electric propulsion could compete directly with ICE vehicles in both performance and practicality. In 2013, the Tesla Model S became the best-selling car in Norway, accounting for 5.1% of total car sales in the country. This event marked the first time an EC became the best-selling model in any national automotive market.

By 2014, approximately 500,000 plug-in vehicles were registered worldwide, with the majority located in the United States. Global EV adoption continued to accelerate, surpassing one million vehicles in 2015. During this same year, China emerged as the second-largest EV market in the world, second only to the United States in terms of annual sales. Another key milestone occurred in 2016, when Tesla unveiled the Model 3, a BEV offering a driving range of approximately 345 km. The model generated unprecedented consumer interest, registering more than 115,000 reservations prior to its commercial launch. That same year, global sales of BEVs surpassed one million units annually for the first time.

Some countries played a particularly important role in accelerating EV adoption. Norway, for example, implemented a combination of fiscal incentives, tax exemptions, and infrastructure investments that significantly favored EV adoption. By 2016, plug-in vehicles accounted for approximately 5% of the total vehicle fleet, while combined EV and plug-in hybrid sales reached 51.4% of new vehicle registrations, making Norway the first country where electrified vehicles outsold conventional ICE cars. By 2018, the global EV fleet had grown to approximately five million vehicles. Despite this impressive growth, EVs still represented a relatively small fraction of the global automotive fleet. By the end of the decade, roughly one out of every 250 vehicles worldwide was electric, indicating that the transition toward electrified mobility was still in its early stages.

A key driver of EV adoption during the 2010s was the rapid expansion of charging infrastructure. At the beginning of the twenty-first century, charging infrastructure was extremely limited due to the low penetration of EVs. Until then, the EV1 had represented the most visible attempt to introduce electric mobility on a commercial scale, but its deployment had been geographically limited to California and Arizona. The EV1 used a proprietary inductive charging system located at the front of the vehicle, which limited interoperability with other charging technologies. Early attempts to develop alternative infrastructure models were not always successful. One notable example was the Better Place battery-swapping system, which aimed to replace depleted batteries with fully charged ones at specialized stations. Despite significant investment, the project was ultimately abandoned in 2013, highlighting the challenges associated with establishing new infrastructure standards.

At the beginning of the 2010s, the global charging network consisted of roughly 47,000 slow chargers and approximately 1900 fast chargers, with the largest concentrations located in Japan and the United States. During the decade, governments and private companies invested heavily in expanding this infrastructure. According to the Global EV Outlook 2020, by the end of 2019 there were approximately 7.3 million charging points worldwide, of which 6.5 million were private residential chargers. At the same time, important progress was made in the standardization of charging technologies, leading to the widespread adoption of fast-charging standards such as Combined Charging System (CCS) and CHAdeMO.

While electric passenger cars received most of the public attention during this period, the 2010s also witnessed significant electrification across several other mobility sectors. One particularly notable example was the rapid expansion of UAVs. During this decade, UAVs transitioned from being primarily military or experimental platforms to becoming widely available consumer products. A key milestone occurred in 2013 with the introduction of the DJI Phantom, a compact and relatively affordable UAV priced at approximately $700. The Phantom was among the first consumer drones sold as a ready-to-fly platform, significantly lowering the barrier to entry for aerial photography and recreational drone flight. The commercial UAV market expanded rapidly following this introduction. While the global UAV market was valued at roughly $700 million in 2014–2015, it grew dramatically over the following years, reaching approximately $22.5 billion within the next decade. In 2013, another influential consumer UAV—the Hubsan X4 H107D—was introduced at a price of around $150. This small quadcopter helped popularize indoor drone flight and first-person-view (FPV) operation, both of which later became widespread in recreational drone communities. Between 2014 and 2016, several new UAV models were introduced in response to the growing popularity of aerial photography. These included the DJI Phantom 3 Professional and the 3DR Solo in 2015, followed by the DJI Phantom 4 and the Yuneec Typhoon H in 2016. Toward the end of the decade, manufacturers began prioritizing portability and compact design. Notable examples include the Parrot Bebop 2 and the DJI Mavic Mini, both weighing less than one kilogram at takeoff and capable of flight times of up to 30 minutes, with maximum speeds approaching 60 km/h.

The electrification trend also extended to urban micro-mobility systems. In 2017, Travis VanderZanden launched Bird, a shared electric scooter service initially deployed in California. By 2024, Bird operated in more than 350 cities worldwide, contributing to the emergence of a global shared micro-mobility industry spanning more than 20 countries. Electric scooters and shared mobility platforms rapidly became an important component of short-distance urban transportation systems. Public transportation systems also experienced significant electrification during the 2010s. In 2010, the Chinese company BYD introduced the K9 electric bus, one of the first mass-produced battery electric buses. The deployment of electric buses expanded rapidly in China, culminating in 2017, when the city of Shenzhen became the first city in the world to operate a fully electrified bus fleet. Since then, many other cities have begun similar transitions toward electric public transportation. Additional forms of electric mobility also emerged during this decade. In 2014, the city of Madrid launched BiciMAD, a public electric bicycle sharing system designed to facilitate sustainable urban mobility. Similar e-bike sharing systems have since been implemented in numerous cities worldwide.

Electrification also began to expand in maritime transportation during the 2010s. One of the most widely cited milestones occurred in 2015, when the MF Ampere entered service in Norway as the first fully electric ferry in regular commercial operation. Operated by the company Norled, the vessel uses two 450 kW electric motors powered by a 1 MWh lithium-ion battery system, enabling zero-emission crossings on short fjord routes. The success of the MF Ampere demonstrated the technical and economic feasibility of electrifying short-distance maritime transport and inspired similar projects in several countries. Following this example, additional electric ferries were introduced in Scandinavia and Northern Europe, including vessels operating in Denmark and Finland. Hybrid-electric ferries also began to appear in other regions, combining battery systems with conventional propulsion to reduce fuel consumption and emissions in coastal transport. In aviation, the 2010s also witnessed growing interest in electrically powered aircraft, although the technology remained largely at the experimental or prototype stage. Several research initiatives focused on small aircraft for short-distance regional transport and pilot training. Among the most notable examples was the Eviation Alice, unveiled in 2019, a fully electric regional aircraft designed to carry nine passengers. Other important developments included the Pipistrel Alpha Electro, one of the first commercially available electric training aircraft, and experimental projects supported by major aerospace companies, such as Airbus' E-Fan demonstrator. Although the limited energy density of batteries still represented a major technological barrier for large commercial aircraft, these projects demonstrated the potential of electric propulsion for short-haul aviation and specialized applications.

Finally, the railway sector, which has historically been one of the most electrified forms of transportation, continued to innovate during this decade. In regions where railway electrification via overhead catenary systems was already widespread, technological progress focused mainly on improving efficiency, regenerative braking systems, and onboard energy management. However, an important breakthrough occurred with the introduction of alternative propulsion technologies designed to decarbonize non-electrified rail lines. A prominent example was the Alstom Coradia iLint, which became the first hydrogen fuel cell train to enter commercial service in Germany in 2018. By generating electricity onboard using hydrogen fuel cells, the train eliminates the need for continuous overhead electrification infrastructure, offering a promising solution for regional railway lines where the installation of catenary systems would be economically prohibitive. In parallel, several railway operators began testing battery-electric trains capable of operating on partially electrified routes, further expanding the range of technological options available for low-emission rail transport.

Further reading

Readers interested in a broader historical and societal perspective on electric mobility may consult the following accessible resources.

- **Wikipedia – History of the Electric Vehicle**. https://en.wikipedia.org/wiki/History_of_the_electric_vehicle
- **U.S. Department of Energy – The History of the Electric Car**. https://www.energy.gov/timeline/timeline-history-electric-car
- **Documentary: *Who Killed the Electric Car?*** (2006). Directed by Chris Paine, this documentary investigates the rise and disappearance of early modern electric vehicles in the 1990s, focusing particularly on the General Motors EV1 program.
- **The New York Times – "The Rebirth of the Electric Car"**. https://www.nytimes.com/2007/09/20/technology/circuits/20pogue-email.html
- **Global EV Outlook**. https://www.iea.org/reports/global-ev-outlook-2025
- **The Complete History of Drone Technology**. https://www.zenatech.com/history-of-drone-technology/
- **Wikipedia – History of rail transport**. https://en.wikipedia.org/wiki/History_of_rail_transport

Types of electric cars

2

2.1 Battery electric cars (BECs)

BECs[1] are vehicles purely propelled by electrical energy and, therefore, they do not include an ICE. As a result, they are inherently emission free at the point of use, which is why they are sometimes referred to as zero emission vehicles. Since no combustion process is involved, these vehicles do not require a conventional fuel tank. Instead, they incorporate a battery system that stores electrical energy and supplies it to the electric motor responsible for providing traction to the wheels.

In BECs, the battery system is rechargeable from external sources, typically through plug in charging infrastructure or, in some configurations, through an onboard generation system (see Section 2.2). In addition, the battery can be partially recharged during operation by means of regenerative braking (see Section 2.1.6), which allows part of the vehicle kinetic energy to be recovered and converted back into electrical energy.

These vehicles typically operate at medium or high voltage levels (on the order of 800 V), where the traction battery system is designed to function. Alongside this high voltage system, BECs also include a low voltage auxiliary battery system (typically 12 V), which is dedicated to supplying power to auxiliary loads such as control units, lighting systems, infotainment, and other onboard electronics. The traction battery is commonly composed of lithium-ion cell packs due to their high energy density and favorable performance characteristics, whereas lead-acid batteries are still frequently used in the low voltage subsystem due to their cost effectiveness, robustness, and well established technology.

2.1.1 Solar-powered BECs

Solar cars can be understood as a particular category of BECs that incorporate an onboard photovoltaic (PV) generation system. Through the integration of solar panels, these vehicles are capable of supplying electrical energy, either partially or in favorable conditions almost entirely, to the traction system. This additional energy source can contribute to extending the driving range or reducing the dependency on external charging infrastructure.

In most designs, solar cars include a battery system that plays a key role in regulating the power supplied to the electric motor (acting as an energy buffer) and in storing excess energy generated during periods of high solar irradiance. This battery system also enables the storage of energy recovered through regenerative braking (see Section 2.1.6). Furthermore, many solar vehicles retain the capability of plug

[1] In the literature, the term electric vehicle is often referred for electric cars. In this book, however, a distinction is made between passenger cars and other types of electric vehicles. For the sake of terminological consistency, the term electric car is therefore used when referring specifically to electric passenger cars, rather than the broader designation of electric vehicle.

Electric Mobility Systems. https://doi.org/10.1016/B978-0-44-349397-3.00010-0

in charging in order to complement the PV system and ensure reliable operation under low solar availability conditions.

Despite their conceptual appeal, solar cars have achieved very limited market penetration, with their use largely restricted to experimental prototypes, technology demonstrators, and competitive events. One notable exception is the *Lightyear 0* (see Fig. 2.1), which became the first solar car to reach limited production in 2021, with an initial price of $250,000. This vehicle integrated PV panels on its roof and hood, providing a total active surface of approximately 5 m^2 and incorporating 70 PV cells. Under favorable conditions, the system was capable of delivering up to 70 km of additional driving range, contributing to a total range of up to 625 km.

Figure 2.1 The Lightyear 0 was a car that incorporated solar panels integrated into its bodywork launched in December 2022. Its production was halted one month later. (Image courtesy of Wikimedia Commons.)

The first units were delivered in December 2022. However, production was halted in early 2023 due to financial difficulties [1], after only a very limited number of vehicles (on the order of a dozen units) had been manufactured. This outcome highlights the current economic and technological challenges associated with large scale deployment of solar powered passenger vehicles.

2.1.2 Piezoelectric cars

Piezoelectric materials are widely used in conventional combustion vehicles, particularly in fuel injection systems where precise actuation is required. Beyond this established application, these materials have also been investigated as a potential means of electrical energy generation in ECs. The underlying principle relies on the ability of piezoelectric materials to generate an electric charge when subjected to mechanical deformation.

In the context of vehicles, piezoelectric elements are typically integrated into components that experience continuous mechanical stress, such as the suspension system. As the vehicle moves, vibrations and dynamic loads induce deformation in the piezoelectric materials, which in turn produce electrical energy. This concept enables the harvesting of otherwise dissipated mechanical energy during normal vehicle operation.

However, the practical implementation of piezoelectric energy harvesting in ECs faces significant limitations. The conversion efficiency of such systems is relatively low, the associated costs are high, and the amount of generated energy is generally insufficient to meaningfully contribute to the traction power demand. For these reasons, their application has remained largely confined to experimental studies and prototype developments.

In most demonstrated cases, the harvested energy is used to supply low power auxiliary systems, such as monitoring devices or embedded sensors, rather than being directly utilized for propulsion. Although some experimental designs have explored the possibility of storing the generated energy in the main battery system, the overall contribution remains marginal. Consequently, piezoelectric generation is not currently considered a viable primary or even secondary energy source for traction in practical EC applications [2].

2.1.3 Components and design of BECs

Following the framework proposed in [3], the architecture of a BEC can be structured into three main subsystems, namely the propulsion subsystem, the energy source, and an auxiliary subsystem, as illustrated in Fig. 2.2. The primary control inputs to the propulsion subsystem are provided by the accelerator and brake pedals, which determine the activation and deactivation of the elements within the electronic controller.

The electronic controller acts as the main actuator of the system and integrates power electronics devices that govern the operation of the power converter. This converter is responsible for adapting the electrical power supplied by the energy source to the requirements of the electric motor. For instance, if the energy source delivers high voltage DC, the converter can transform it into alternating current (AC) at an appropriate voltage level when required by the motor.

In addition to power adaptation, the converter also regulates the energy flow between the energy source and the motor. It is important to note that bidirectional power flow is possible due to regenerative braking (see Section 2.1.6), which allows energy to flow from the motor back to the power converter and subsequently to the energy source (typically the battery system). Consequently, the energy source must be receptive, meaning that it must be capable of both supplying and absorbing power. This requirement explains why most BECs incorporate energy storage systems. In contrast, generation systems such as PV panels or piezoelectric devices are not

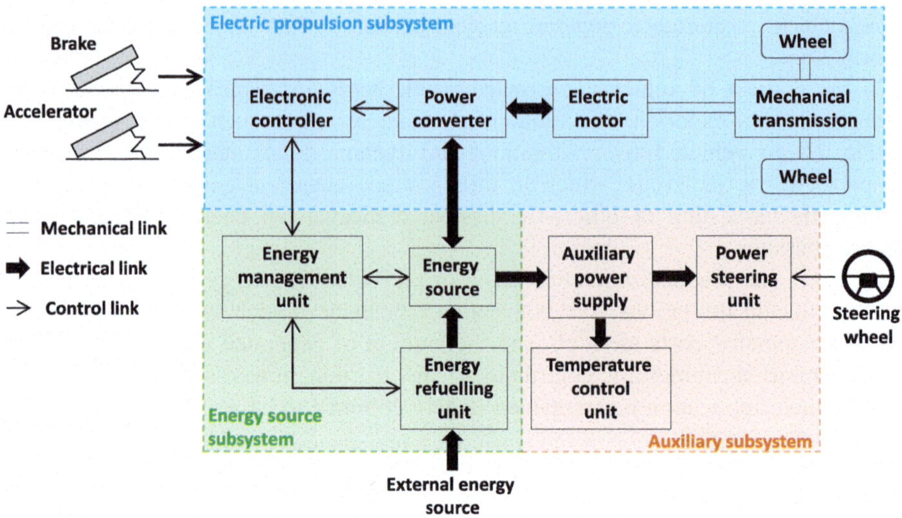

Figure 2.2 Main components of BECs and energy/information flows.

inherently receptive and therefore cannot directly accommodate reverse power flows. A similar limitation applies to certain battery technologies, which are consequently unsuitable for electric mobility applications.

The energy management unit (often referred to as the vehicle control unit) can be regarded as the central supervisory system of the vehicle. Its primary function is to monitor the state of the energy source and determine the direction and magnitude of power flows within the vehicle. The main objective of this unit is to manage these energy flows in real time in order to maximize overall efficiency while minimizing battery degradation.

To achieve this, the energy management unit interacts with multiple subsystems, including the energy source, the electronic controller, and the charging unit. The latter monitors the condition of both the internal energy source and the external supply, adapting the incoming power flow so that it can be effectively utilized by the onboard system. Additionally, the energy management unit coordinates with the auxiliary power source, which is independent of the main energy source and is typically dedicated to supplying auxiliary loads such as thermal management systems, control electronics, and other low voltage components (commonly through a 12 V battery system).

The interaction among these subsystems plays a crucial role in both the design and the techno economic evaluation of the vehicle. This interaction is often analyzed using a mnemonic arrangement known as the quality function matrix, shown in Fig. 2.3 for a typical BEC configuration. The design of a BEC inherently combines principles from electrical and mechanical engineering, while also incorporating aspects of chemistry, electronics, and structural design. Although the energy source is arguably the most critical component, a balanced and integrated design approach is required to ensure overall system performance.

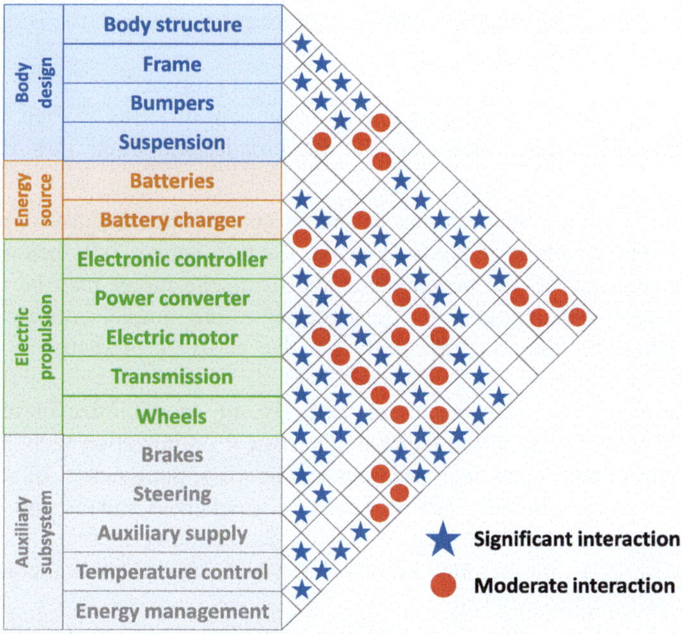

Figure 2.3 Qualification matrix for a BEC.

In contrast to internal combustion vehicles, driving range is a critical constraint in BECs, as these vehicles typically exhibit more limited autonomy and longer refueling times. Therefore, optimizing energy flows is essential. The energy management unit relies on a wide range of sensor inputs distributed throughout the vehicle, including measurements of internal and external temperature, voltage and current levels in the energy source and motor, as well as vehicle speed and acceleration.

Among its main functions, the energy management unit is responsible for:

- Optimizing internal energy flows in order to maximize efficiency while reducing component degradation.
- Monitoring the energy source and estimating the remaining driving range.
- Evaluating driving patterns and suggesting more efficient driving modes when appropriate.
- Controlling auxiliary systems such as thermal management and lighting based on environmental conditions.
- Recording and analyzing the operating conditions of the energy source, maintaining historical data for diagnostics and optimization.
- Detecting and diagnosing abnormal operating conditions or component faults.

It is important to distinguish the energy management unit from the battery management system (BMS). The former operates as a centralized supervisory controller for the entire vehicle, whereas the BMS is specifically dedicated to the battery pack. Thus, a hierarchical control structure exists in which the BMS operates as a subordinate control loop.

Different energy management strategies have been proposed for BECs, all sharing the common objective of optimizing energy usage, but differing in their underlying methodologies [4]. These strategies can be broadly classified into three main categories:

- Rule based strategies: these approaches define control rules that govern power flow decisions. For example, if the SoC of the traction battery falls below a certain threshold (e.g., 20%), the power demanded from the motor may be reduced to prevent accelerated degradation. These strategies are simple, intuitive, and easy to implement, but they are inherently reactive and lack predictive optimization capabilities.
- Optimization based strategies: these methods aim to minimize or maximize a predefined objective function using mathematical optimization techniques. This includes both exact (analytical) methods and heuristic approaches. Exact methods guarantee global optimality but may require specialized solvers and significant computational effort, which can limit their applicability in real time contexts. Heuristic methods, on the other hand, offer reduced computational complexity at the expense of optimality [5].
- Learning based or artificial intelligence strategies: these more advanced approaches leverage machine learning techniques, such as neural networks or deep learning models, to infer optimal control actions from data. They combine elements of adaptability and predictive capability, but their performance strongly depends on the availability of sufficiently large and representative training datasets.

It should be noted that the strategies described above focus on internal vehicle energy management. Additional strategies exist that address external interactions, such as optimizing charging processes or coordinating with the electrical grid.

Another important aspect in the design of BECs is the interaction between subsystems based on different engineering domains (electrical, mechanical, electronic, among others). These interactions are often complex and require a multidisciplinary approach. In this context, system level simulation becomes essential, enabling the analysis of component behavior under realistic operating conditions and supporting design optimization. Several software tools are available for multiphysics simulation of BECs, including Ansys,[2] GT Suite,[3] and CarMaker.[4]

Simulation and optimization are key elements in BEC design, as engineers must address trade offs between conflicting objectives that impact both technical performance and economic viability. Some of the most relevant considerations include [3]:

[2] https://www.ansys.com/technology-trends/electrification.

[3] https://staging.gtisoft.com/electric-powertrain-simulation-solutions/.

[4] https://www.ipg-automotive.com/solutions/product-portfolio/carmaker.

- The interaction between subsystems must be thoroughly analyzed to optimize overall vehicle performance, although this often leads to complex models requiring advanced computational tools.
- A balance must be achieved between model accuracy and simulation usability, which directly influences the choice of modeling approach and software tools.
- The system voltage level represents a critical design trade off. Higher voltage typically requires a greater number of cells connected in series, increasing cost and weight. However, it also reduces current levels (and therefore conduction losses), while improving acceleration and driving range. As a result, the system voltage must be carefully optimized as a compromise solution.

2.1.4 BEC architectures

BECs exhibit a significantly more flexible internal configuration than ICE vehicles. This flexibility arises from the fact that many of the interconnections between components are implemented electrically through wiring, rather than through rigid mechanical or hydraulic linkages. As a result, energy transfer within the vehicle is not constrained by mechanical layouts, allowing designers to decouple subsystems more effectively and to optimize their placement according to criteria such as efficiency, packaging, thermal management, and safety.

From a system engineering perspective, this inherent flexibility enables a more modular approach to vehicle design. Subsystems can be designed, analyzed, and optimized independently before being integrated into the overall architecture. This modularity also facilitates scalability and platform sharing across different vehicle models, reducing development costs and increasing design versatility. Consequently, the subdivision of a BEC into functional subsystems is both more natural and more convenient for modeling, simulation, and control design.

A typical subsystem level decomposition is shown in Fig. 2.2, where the internal structure of the BEC is divided into three main subsystems, namely the electric propulsion subsystem, the energy source subsystem, and an auxiliary subsystem. The main components within each subsystem, as well as their roles within the overall vehicle architecture, are summarized as follows:

- The electric propulsion subsystem includes electrical components such as the electronic controller, the power converter, and the electric motor, as well as mechanical elements such as the transmission and the wheels. The primary function of this subsystem is to process the driver input signals, typically obtained from the accelerator and brake pedals, and to deliver the appropriate torque and speed to the wheels. To achieve this, electrical energy is drawn from the energy source subsystem and delivered to the DC bus. The power converter, usually implemented as a voltage source inverter, conditions this energy and converts it into an AC waveform suitable for driving the electric motor. The motor then transforms the electrical energy into mechanical rotational energy, which is transmitted to the wheels through the drivetrain. In addition to propulsion, this subsystem is also responsible for implementing regenerative braking, requiring bidirectional energy flow capability and precise torque control.

- The energy source subsystem comprises the energy management unit, the primary energy source (typically the battery system), and the charging unit. This subsystem is responsible for managing energy flows throughout the vehicle, determining when, how much, and in which direction energy should be transferred. The energy management unit plays a central role in coordinating the operation of the vehicle, ensuring that energy is supplied efficiently while maintaining the operational limits of the battery system. This includes enforcing constraints related to SoC, temperature, and current limits. The charging unit enables interaction with external energy sources, allowing the battery to be replenished either through AC or DC charging. Overall, the objective of this subsystem is to maximize system efficiency, extend battery lifetime, and guarantee safe operation under all conditions.
- The auxiliary subsystem includes additional components required for vehicle operation, such as thermal management systems, sensing and monitoring devices, and low voltage electrical loads. Thermal management is particularly critical in BECs, as both the battery system and the power electronics are sensitive to temperature variations, which can significantly affect performance and lifetime. This subsystem may also incorporate an independent auxiliary energy source if it differs from the main energy source of the vehicle, as is commonly the case with low voltage battery systems (e.g., 12 V). These auxiliary systems ensure the proper operation of lighting, infotainment, control units, and safety related functions, even when the main traction system is inactive.

The inherent flexibility in BEC design enables multiple possible drivetrain architectures, as illustrated in Fig. 2.4. These configurations differ in terms of component arrangement, mechanical complexity, efficiency, and performance. The most representative configurations are discussed below:

- In Fig. 2.4a, a first configuration is presented, which closely resembles the conventional architecture of an internal combustion vehicle. This layout typically adopts a front motor and front wheel drive arrangement. It includes an electric motor coupled to the wheel axle through a clutch, a gearbox, and a differential. The presence of a gearbox and clutch allows the adjustment of the transmission ratio, enabling the control of torque delivered to the wheels. Short gear ratios provide high torque at low speeds, which is beneficial for vehicle launch and acceleration, while long gear ratios enable efficient operation at higher speeds. The differential allows the wheels on the same axle to rotate at different speeds, improving vehicle handling during cornering. This configuration is commonly found in converted internal combustion vehicles, where existing mechanical components are reused.
- In the configuration of Fig. 2.4a, the gearbox and clutch can be replaced by a fixed gear, resulting in the architecture shown in Fig. 2.4b. This simplification reduces both weight and mechanical complexity, leading to improved reliability and lower maintenance requirements. In this case, the system consists of an electric motor, a fixed gear, and a differential. The wide torque speed operating range of electric motors makes it possible to eliminate multi-speed gearboxes in many applications.

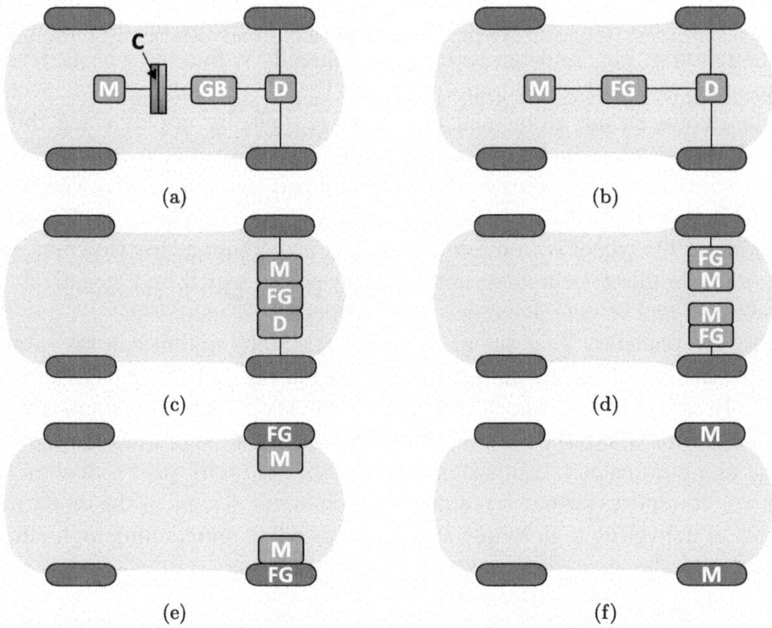

Figure 2.4 Different BEC configurations depending on the electric propulsion. C: clutch, D: differential, FG: fixed gearing, GB: gearbox, M: electric motor.

However, this configuration is not suitable when a thermal propulsion system is present, as in hybrid vehicles, since the absence of a gearbox and clutch prevents achieving the torque speed characteristics required by an ICE.

- A further development of the configuration in Fig. 2.4b is shown in Fig. 2.4c, where the main components are arranged along a common axle. This configuration reduces mechanical stress and friction losses by eliminating extended drivetrain elements such as long shafts and additional couplings. It also improves packaging efficiency, freeing up space that can be used for other components, particularly the battery system. This can contribute to increasing vehicle range. Additionally, this architecture is highly modular and can be integrated into different vehicle platforms with minimal structural modifications. This arrangement is commonly referred to as an e-axle and is widely adopted in modern ECs.
- The dual motor configuration (shown in Fig. 2.4d) eliminates the need for a mechanical differential, as each wheel (for example, in a front wheel drive configuration) is driven by an independent electric motor. These motors can operate at different speeds and deliver different torque levels, enabling advanced control strategies. This architecture significantly improves vehicle performance under low adhesion conditions, such as wet, icy, or uneven surfaces, by allowing independent torque control at each wheel. This capability can be used for traction control and torque vectoring, enhancing both safety and handling. Furthermore, the combined power of multiple motors increases overall performance, enabling higher acceleration and torque availability.

- In order to further reduce mechanical transmission elements, the motors in the dual configuration of Fig. 2.4d can be integrated directly within the wheels, leading to the so called in-wheel configuration shown in Fig. 2.4e. In this case, the mechanical coupling between the motor and the wheel is typically implemented through a planetary gear set, which reduces the motor speed to match the wheel speed. This configuration minimizes drivetrain complexity and allows for a very compact design. It also improves dynamic response, as torque is delivered directly from the motor to the wheel with minimal intermediate components. However, placing motors inside the wheels increases unsprung mass, which may negatively affect ride comfort and vehicle dynamics if not properly managed.
- Finally, the planetary gear set in the in-wheel configuration can be eliminated, resulting in a direct drive architecture as shown in Fig. 2.4f. In this case, the motor is directly coupled to the wheel, and therefore must operate at low rotational speeds consistent with wheel speed. This configuration further simplifies the mechanical design and eliminates transmission losses associated with gear reduction stages. However, it imposes stringent requirements on motor design, as the motor must be capable of delivering high torque at low speeds while maintaining high efficiency. Additionally, the increase in motor size and weight may pose challenges in terms of integration and vehicle dynamics.

2.1.5 Electric motor

The electric motor constitutes the primary propulsion device in any BEC. Its function is to convert electrical energy supplied by the primary energy source (typically the battery system) into mechanical torque, which is transmitted to the wheels and ultimately produces vehicle motion. Although electric motors are a mature technology with widespread industrial use throughout the twentieth century, their application in transportation has historically been limited. As a result, their integration into BECs requires a detailed understanding of the available technologies, as well as their respective advantages and limitations.

Modern BECs predominantly employ AC motors due to their high efficiency, particularly at medium and high speeds, and their robust operational characteristics. However, these motors require sophisticated control systems. Specifically, power electronic converters and advanced control algorithms are needed to regulate both torque and rotational speed with high precision.

Two main categories of AC motors are commonly used in electric mobility, namely induction motors (also referred to as asynchronous motors) and permanent magnet synchronous motors. Induction motors operate based on electromagnetic induction principles. They consist of two electrical circuits, a stator and a rotor, magnetically coupled through a ferromagnetic core and separated by a small air gap. The stator windings are typically embedded in slots located on the inner surface of the motor housing, while the rotor can be implemented either as a wound rotor or as a squirrel cage structure.

When the stator is supplied with alternating current, a rotating magnetic field is generated. This time varying magnetic field induces currents in the rotor conductors (which are effectively short circuited). The interaction between the induced rotor currents and the stator magnetic field produces electromagnetic torque, causing the rotor to rotate. A key characteristic of induction motors is that the rotor speed is slightly lower than the synchronous speed of the stator field, hence the term asynchronous.

In contrast, permanent magnet synchronous motors have gained significant prominence in recent years and are now widely adopted in BECs [6]. In these machines, the rotor incorporates permanent magnets instead of conductive windings. These magnets align with the rotating magnetic field produced by the stator, resulting in synchronous operation, meaning that the rotor rotates at the same speed as the stator magnetic field, which is determined by the frequency of the supplied AC.

Permanent magnet motors offer several advantages, most notably higher power density and reduced weight, making them particularly attractive for electric mobility applications. However, they rely on magnetic materials that are often expensive and subject to supply constraints. Additionally, there is a risk of demagnetization under extreme operating conditions (such as high inductive currents during fault events).

One of the most relevant aspects of electric motors in mobility applications is their torque speed characteristic (see Fig. 2.5), which differs significantly from that of ICEs and represents a major advantage of BECs. In general, traction motors exhibit two distinct operating regions, namely a constant torque region and a constant power region. At low speeds, the motor can deliver nearly constant torque, primarily limited by the maximum allowable current in the stator windings and by thermal constraints. This enables ECs to provide high torque from standstill, eliminating the need for complex multi-gear transmissions and significantly improving acceleration performance.

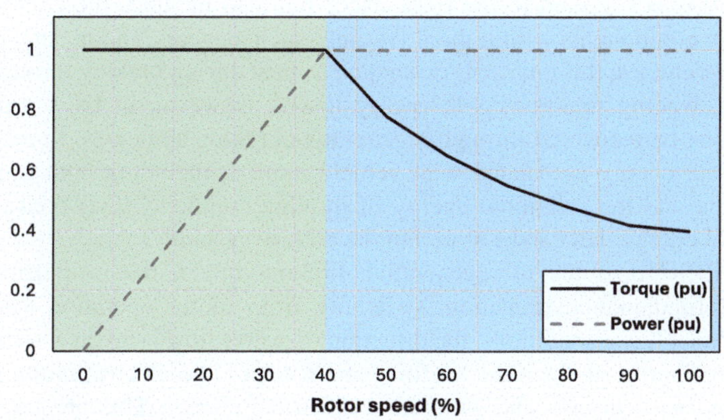

Figure 2.5 Typical Torque-speed curve of an induction motor with a base speed of 40%. The constant-toque region (green) is delimited within speeds 0–40%, limit beyond which the motor starts to work in the constant-power area (blue).

As the motor speed increases and reaches the base speed, the system transitions into the constant power region. In this regime, the inverter progressively reduces the magnetic flux (a technique known as field weakening) to prevent overvoltage conditions. As a consequence, the available torque decreases approximately inversely with speed, while the delivered mechanical power remains nearly constant over a wide operating range.

Although induction motors and permanent magnet synchronous motors currently dominate vehicle applications, alternative technologies have emerged in recent years with the aim of improving efficiency, reducing cost, or minimizing reliance on critical materials. Among these, synchronous reluctance motors have attracted attention due to their simple rotor construction and the absence of permanent magnets. Their operation is based on the tendency of magnetic flux to follow paths of minimum reluctance within the rotor structure. This results in lower material costs and reduced dependence on complex supply chains. However, these motors typically exhibit lower power density and greater sensitivity to operating conditions.

Another promising technology is the axial flux motor. Unlike conventional radial flux machines, where the magnetic flux flows radially, axial flux motors feature a magnetic flux path parallel to the axis of rotation. This configuration enables more compact and lightweight designs with high power density, which is particularly advantageous in applications where space and weight are critical constraints. Nevertheless, their more complex construction and the challenges associated with thermal management have so far limited their widespread industrial adoption.

2.1.6 Regenerative braking

An electric motor stores kinetic energy during rotation, primarily as a function of its moment of inertia and angular velocity. In general terms, the greater the rotor mass and rotational speed, the higher the kinetic energy accumulated in the system. This stored energy is not limited to the rotor itself, but also includes the contribution of other rotating components within the drivetrain, such as gears, shafts, and wheels. In conventional vehicles, this energy is dissipated as heat during braking through friction mechanisms, leading to irreversible energy losses. However, in ECs, a portion of this energy can be recovered through regenerative braking strategies. In this process, the mechanical energy associated with vehicle motion, including both translational kinetic energy and the rotational energy of moving components, is converted back into electrical energy that can be stored in the battery system.

When a BEC operates in regenerative braking mode, the propulsion system undergoes a fundamental transition, switching from motor operation to generator operation. Under these conditions, the direction of power flow is reversed. The wheels, driven by the inertia of the vehicle, force the rotor of the electric motor to rotate, inducing electrical currents and generating electrical energy. This process produces a resistive torque on the shaft, opposing the direction of rotation and contributing to vehicle deceleration. From a control perspective, this braking torque can be modulated continuously, allowing for smooth deceleration profiles and improved drivability.

From an electromagnetic standpoint, this operating mode involves a reversal in the direction of energy exchange between the magnetic field and the electrical system, while preserving the same underlying electromechanical conversion principles. The same machine that delivers torque during propulsion is therefore capable of converting mechanical energy into electrical energy during braking, highlighting the intrinsic bidirectional nature of electric machines. This dual functionality is one of the key advantages of electric propulsion systems compared to conventional internal combustion engines.

Regenerative braking is particularly effective in urban driving environments, which are characterized by frequent acceleration and deceleration cycles. In such conditions, a significant portion of the energy expended during acceleration can be recovered during braking phases, thereby improving overall vehicle efficiency and extending driving range. This effect is especially relevant in stop and go traffic scenarios, where repeated braking events create multiple opportunities for energy recovery. In contrast, during highway driving or steady speed operation, where braking events are less frequent and less intense, the impact of regenerative braking is considerably reduced.

The effective implementation of regenerative braking requires appropriate coordination among the different vehicle subsystems. In particular, power electronic converters must be capable of bidirectional operation, allowing energy to flow from the battery to the motor during propulsion and from the motor back to the battery during regeneration. These systems typically include inverters that operate as DC to AC converters in traction mode and as AC to DC converters in generation mode, ensuring precise control of voltages and currents. Additionally, the control system must coordinate regenerative braking with conventional friction braking, blending both mechanisms to achieve the desired deceleration while maximizing energy recovery and maintaining vehicle stability and safety.

Despite its advantages in terms of energy efficiency, regenerative braking is subject to several practical limitations. One of the main constraints arises from the energy storage system itself. Batteries have a limited capacity to absorb power, especially at high SoC. As the battery approaches its maximum charge level, the allowable charging current must be reduced to prevent overcharging and accelerated degradation, which directly limits the amount of recoverable energy. Thermal constraints also play a significant role, as high charging currents can increase battery temperature, requiring additional thermal management measures.

Another important limitation occurs at low vehicle speeds. As speed decreases, the ability of the electric motor to generate sufficient voltage for effective energy transfer to the battery is also reduced. This is particularly relevant in AC machines, where the generated voltage is proportional to the rotational speed. Consequently, regenerative braking becomes less effective in this operating region, and conventional friction braking must complement the deceleration process. Moreover, safety and comfort considerations may impose limits on the maximum regenerative braking torque, especially under low adhesion conditions such as wet or icy roads.

Additional factors that influence regenerative braking performance include the efficiency of the power electronics, the internal resistance of the battery, and the control strategy used to manage energy flows. In practice, not all the theoretically available kinetic energy can be recovered, as losses occur at each stage of the energy conversion and storage process.

Despite these limitations, regenerative braking can recover a substantial fraction of the energy otherwise lost during braking, reaching values of up to approximately 30% in typical urban driving cycles. This capability represents a key advantage of BECs, contributing not only to improved energy efficiency and extended driving range, but also to reduced wear of mechanical braking components, thereby lowering maintenance requirements and enhancing overall system sustainability.

2.1.7 Working modes

Depending on the direction of internal power flows and the interaction with external systems, BECs can operate under different modes, each of which involves specific configurations of the electrical, electronic, and control subsystems. The most representative operating modes are described below (see Fig. 2.6):

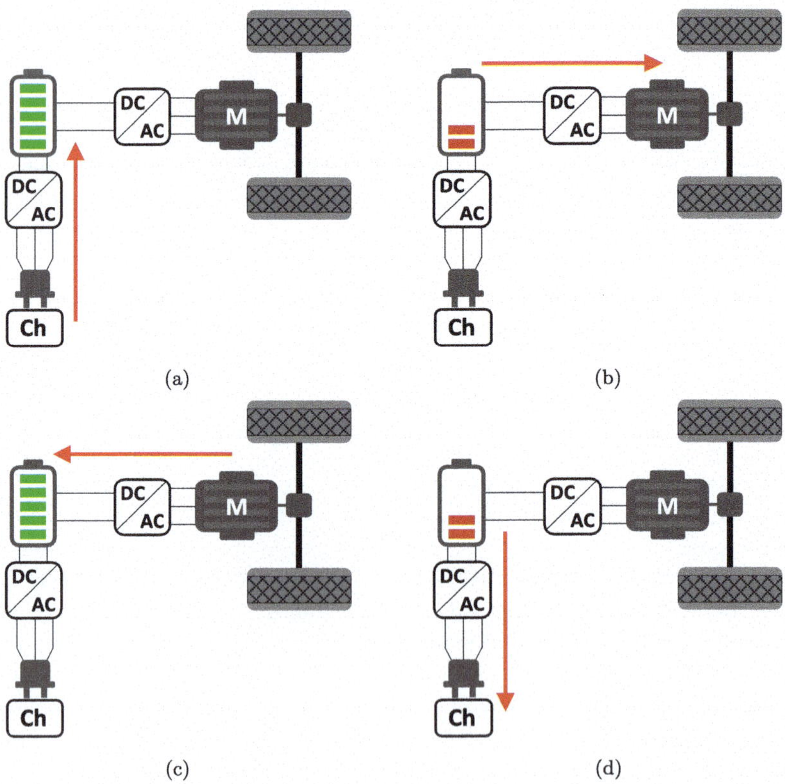

Figure 2.6 Representation of the different working modes of BECs. Ch: charger, M: electric motor.

- **External charging** (Fig. 2.6a). In this mode, the vehicle receives electrical energy from an external source in order to recharge its traction battery. The supplied energy can be in the form of AC, typically associated with low or medium power charging, or DC, as in fast charging stations. In the case of AC charging, the vehicle must include an AC/DC converter (onboard charger) that rectifies and conditions the incoming energy for storage, regulating parameters such as voltage, current, and charging profile according to the battery state. This process is typically coordinated by the BMS and the energy management unit, which ensure safe operation by enforcing limits related to temperature, state of charge, and allowable current rates. In contrast, during DC charging, the conversion process is carried out externally, allowing higher power levels to be delivered directly to the battery system and reducing charging time. This type of charging requires communication between the vehicle and the charging station to negotiate voltage and current levels, ensuring compatibility and safety. Although DC fast charging significantly reduces charging times, it may accelerate battery degradation if used frequently, due to higher thermal and electrochemical stress on the cells.
- **Traction mode** (Fig. 2.6b). In this operating mode, the energy required for vehicle propulsion is supplied exclusively by the energy storage system. The traction battery delivers current to the DC bus of the vehicle, from which an inverter converts it into AC to supply the electric motor in the case of AC machines. The inverter not only performs waveform conversion, but also regulates the frequency and amplitude of the signal, enabling precise control of motor torque and speed. This control is typically achieved through advanced modulation and control techniques, such as pulse width modulation and vector control, which allow the motor to operate efficiently across a wide range of speeds and loads. The traction mode also involves continuous monitoring of system variables, including current, voltage, and temperature, to ensure safe and efficient operation. The overall performance in this mode is strongly influenced by the efficiency of the power electronics and the motor, as well as by the characteristics of the battery system.
- **Regenerative braking** (Fig. 2.6c). In this mode, the direction of power flow is reversed with respect to traction mode. The electric motor operates as a generator, converting the mechanical energy of the vehicle into electrical energy that is transferred back to the energy storage system. The generated current, typically in AC form for AC machines, must be rectified by the inverter, which in this case operates as an AC/DC converter. The effectiveness of regenerative braking depends on several factors, including vehicle speed, battery SoC, and the limits imposed by the power electronics. At higher speeds, a greater amount of kinetic energy is available for recovery, whereas at low speeds the effectiveness decreases due to reduced voltage generation capability. In practice, regenerative braking is often combined with conventional friction braking in a coordinated manner to achieve the desired deceleration while maximizing energy recovery and maintaining vehicle stability.

- **Vehicle to X (V2X)** (Fig. 2.6d). In this mode, the vehicle acts as an energy source for external applications, extending its functionality beyond transportation. The concept of V2X encompasses several modalities, including supplying energy to a household, a building, or even the electrical grid. In these cases, the energy stored in the traction battery is delivered externally, typically in the form of AC. This requires the use of a DC/AC inverter to adapt the electrical characteristics of the output signal to the requirements of the load or the grid. From a system perspective, V2X operation introduces additional requirements in terms of control, communication, and grid compliance. For instance, when interfacing with the electrical grid, the inverter must meet specific standards related to voltage, frequency, and power quality. Moreover, bidirectional power flow capability must be ensured, and appropriate protection mechanisms must be implemented. Beyond technical considerations, V2X also enables new use cases such as demand response, peak shaving, and emergency backup power, thereby increasing the overall value proposition of ECs within the broader energy ecosystem.

2.2 Hybrid-electric cars (HECs)

A HEC is a vehicle whose propulsion can be provided by two distinct energy sources, one of which is electrical in nature and the other typically an ICE fueled by conventional fuels such as gasoline or diesel. In most configurations, these vehicles incorporate two propulsion systems: an electric motor supplied by a traction battery system, and a thermal engine connected to a fuel tank.

Both propulsion systems can operate either independently or jointly, depending on the vehicle architecture and the adopted energy management strategy. However, there are specific configurations in which the ICE does not directly contribute to traction, but instead operates exclusively as an electrical generator supplying energy to the onboard electrical system, as discussed in Section 2.2.1.

The combination of an electric propulsion system with a thermal one provides several advantages compared to their standalone operation. For instance, the electric motor is capable of delivering high torque at low speeds, whereas an ICE typically exhibits improved performance at higher speeds. Therefore, their combined operation allows the vehicle to achieve high torque over a wider speed range, enhancing overall drivability and efficiency [7].

Similarly to BECs, HECs incorporate two battery systems. The traction battery is commonly based on lithium-ion technology, although it typically operates at lower voltage levels than in BECs (on the order of 100 to 400 V). In addition, an auxiliary battery system is included, usually based on lead-acid technology and operating at 12 V, which supplies power to auxiliary loads and vehicle electronics.

2.2.1 Degree of hybridization (DoH)

In HECs, it is useful to define the degree of hybridization (DoH), which provides a quantitative measure of the level of electrification of the vehicle. Formally, the DoH is defined in (2.1) as the ratio between the rated power of the electric propulsion system and the total installed propulsion power of the vehicle. This metric captures the relative contribution of the electric drivetrain with respect to the overall propulsion capability

$$\text{DoH} = \frac{\text{Electric motor power}}{\text{Total power}}. \tag{2.1}$$

This power based definition is the most widely used for performance characterization, as it is directly related to the instantaneous capability of the vehicle to deliver torque and power through the electric system. Nevertheless, alternative formulations based on energy have also been proposed in the literature, particularly when the focus is placed on driving range or energy consumption rather than peak performance. According to the DoH value, HECs can be classified as shown in Table 2.1.

It is important to note that the DoH is an intrinsic design parameter derived from the nominal ratings of the propulsion components. As such, it does not directly reflect how the vehicle is actually operated in real world conditions. The effective degree of hybridization during operation depends strongly on the adopted energy management strategy, as well as on the driving profile, environmental conditions, and user behavior. In particular, vehicles with identical DoH values may exhibit significantly different fuel consumption and emission levels depending on the achieved electric driving share, commonly quantified through the Utility Factor. This highlights that DoH alone is not sufficient to fully characterize the operational efficiency of a HEC, but rather must be interpreted in conjunction with control strategies and usage patterns.

Although the DoH is defined in terms of power, its practical interpretation is also closely linked to the energy capacity of the onboard battery system. In general, higher DoH values are associated with larger battery capacities, as a greater degree of electrification requires not only higher power capability but also sufficient energy storage to sustain electric operation over meaningful driving distances. In this context, Fig. 2.7 conceptually illustrates the typical range of battery capacities associated with different vehicle categories as a function of DoH. As expected, increasing the DoH generally implies a larger and heavier battery system, which introduces additional design considerations related to cost, weight, thermal management, and packaging.

It is also important to recognize that increasing the DoH leads to a gradual shift in the dominant energy pathway within the vehicle. At low DoH values, the ICE remains the primary energy conversion device, with the electric system playing a secondary and supportive role. As the DoH increases, the electric drivetrain progressively assumes a more central role in propulsion, eventually becoming the dominant or even exclusive source of traction.

In this context, both PHECs and ERECs represent a qualitative shift in the electrification of mobility. In the case of PHECs, the availability of external charging infrastructure allows for significantly larger battery capacities compared to conventional hybrids. This enables extended operation in pure electric mode, often covering daily commuting distances without requiring the use of the ICE. As a result, the electric system can effectively act as the primary propulsion source under typical

Table 2.1 Classification of HECs according their DoH.

Classification	**DoH**	**Characteristics**
Micro hybrid	0–0.05	Traction is exclusively thermal. The electrical system is limited to auxiliary functions such as start stop and very limited energy recovery. No electric propulsion capability is available.
Mild hybrid	0.05–0.25	The electrical system can provide torque assistance to the ICE (torque assist), but pure electric traction is not possible under normal operating conditions. Regenerative braking is included, improving overall efficiency.
Full hybrid	0.25–0.5	The vehicle can be propelled either by the electric system or the ICE independently, and can recover energy through regenerative braking, but it does not support external charging (non plug in). The energy source is typically selected based on parameters such as SoC, torque demand, or speed.
Plug in hybrid (PHEC)	0.5–0.8	Incorporates larger battery capacity and external charging capability. It can operate in pure electric mode over significant distances. The thermal system acts as backup or support for long trips or high power demand.
Extended range (EREC)	0.8–1	Traction is exclusively electric. The ICE is not mechanically coupled to the wheels and operates only as a generator to extend driving range.
BEC	1	Purely electric vehicles powered exclusively by batteries.

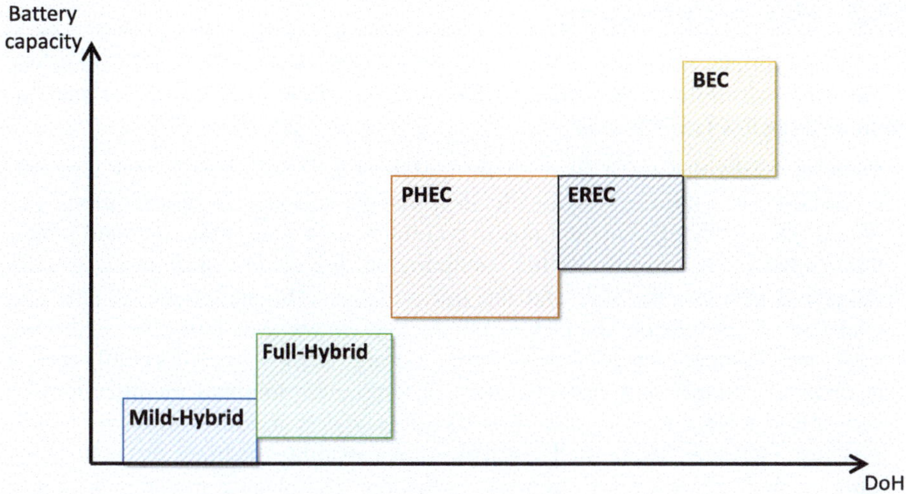

Figure 2.7 Conceptual view on battery capacity depending on the DoH.

Table 2.2 Functionalities of HECs.

Classification	Start-stop	Regenera-tive braking	Electric assist	Pure electric traction	External charging
Micro hybrid	✓	✓			
Mild hybrid	✓	✓	✓		
Full hybrid	✓	✓	✓	✓	
PHEC	✓	✓	✓	✓	✓
EREC	✓	✓	✓	✓	✓

usage conditions, while the ICE provides flexibility and range extension for long distance travel or in situations where charging infrastructure is not readily available.

On the other hand, ERECs follow a distinct design philosophy. These vehicles typically exhibit a DoH close to unity, reflecting the fact that the combustion engine is completely decoupled from the drivetrain. In this configuration, propulsion is entirely electric, and the ICE operates solely as an onboard generator that produces electrical energy to recharge the battery when needed. This decoupling allows the ICE to operate under steady and optimal conditions, potentially improving its efficiency compared to conventional operation. From a conceptual standpoint, ERECs can be interpreted as BECs equipped with an auxiliary fuel based generation system, effectively bridging the gap between fully electric vehicles and conventional hybrids.

As a synthesis, Table 2.2 summarizes the main functionalities enabled in the different hybrid vehicle categories discussed above, highlighting the progressive increase in electrification capabilities as the DoH increases.

2.2.2 HEC architectures

There are multiple ways to combine the propulsion systems in a HEC, leading to different drivetrain architectures. The most representative configurations are summarized below (see Fig. 2.8):

- **Parallel configuration with a single clutch** (Fig. 2.8a). This layout typically corresponds to a mild hybrid vehicle, in which the ICE and the electric motor can operate independently, but their power is combined on a common shaft and jointly transmitted to the wheels. In this configuration, the electric machine is usually integrated between the ICE and the transmission. Although both systems can contribute to propulsion, the lack of full decoupling prevents pure electric driving under normal operating conditions. From a design perspective, this configuration is relatively simple and cost effective, as it requires minimal modifications to conventional powertrain layouts. It is therefore widely used in entry level hybrid systems. However, its limited electrification capability restricts the achievable gains in fuel efficiency and emission reduction. The electric motor is primarily used for torque assist, start-stop functionality, and limited regenerative braking.
- **Parallel configuration with two clutches** (Fig. 2.8b). In this configuration, a clutch is incorporated in order to enable the decoupling of the ICE from the drivetrain, thereby allowing purely electric traction under appropriate operating conditions. This additional degree of freedom significantly enhances the flexibility of the overall propulsion system, as it enables a more versatile management of power flows between the thermal and electrical subsystems. In particular, the ability to mechanically isolate the ICE allows the vehicle to operate in fully electric mode when the power demand and the SoC of the traction battery permit it, thus reducing fuel consumption and local emissions. Moreover, this configuration facilitates more efficient energy recovery during transient events. For instance, during braking or deceleration phases, the ICE can be decoupled to eliminate parasitic losses associated with engine drag, thereby maximizing the amount of kinetic energy that can be converted into electrical energy through regenerative braking.
- **Parallel configuration with dual transmission** (Fig. 2.8c). In the previous configuration, the use of two clutches arranged in series may introduce packaging constraints, particularly in applications where space availability within the powertrain is limited. In order to mitigate these issues, an alternative configuration can be adopted in which the electric motor is connected to an independent transmission (which can be interpreted as a subsystem of the main drivetrain), rather than being directly coupled to the common crankshaft of the ICE. This arrangement provides greater flexibility in terms of mechanical integration, as it allows a more compact distribution of components within the vehicle layout. At the same time, it preserves the capability to decouple the thermal and electrical subsystems, enabling independent operation of each propulsion source under suitable conditions. However, this configuration also introduces additional complexity at the system level. In particular, the presence of multiple transmission

paths requires more advanced coordination between subsystems to ensure smooth operation, proper torque delivery, and acceptable drivability.

- **Parallel configuration with split axle** (Fig. 2.8d). In this configuration, the electric motor and the ICE are completely decoupled mechanically and are assigned to different axles. For example, the ICE may drive the front axle while the electric motor drives the rear axle. This arrangement enables true all wheel drive capability without requiring a mechanical link between axles. From an operational standpoint, this architecture provides high flexibility and allows for advanced control strategies, including torque vectoring and independent axle control. It is particularly attractive for vehicles aiming at higher DoH values. Additionally, the ICE can be operated independently to recharge the battery if required. However, this configuration increases system cost and complexity, as it effectively duplicates certain drivetrain components.
- **Series configuration** (Fig. 2.8e). This architecture corresponds to an EREC, in which the ICE is not mechanically connected to the wheels. Instead, it drives a generator that produces electrical energy, which is either stored in the battery or directly supplied to the electric motor for traction. One of the main advantages of this configuration is the decoupling between the ICE and the drivetrain, allowing the thermal engine to operate at its optimal efficiency point regardless of vehicle speed or load. Furthermore, the absence of a conventional mechanical transmission simplifies the drivetrain and can free up space for larger battery systems. However, the main limitation of this architecture is its lower overall efficiency, as energy must undergo multiple conversion stages (mechanical to electrical, electrical to chemical and back to electrical, and finally electrical to mechanical). These conversion losses can be significant, particularly during sustained operation. As a result, this configuration is typically more suitable for specific use cases where efficiency trade offs are acceptable in exchange for design simplicity or operational benefits.
- **Series parallel configuration** (Fig. 2.8f). This architecture combines features of both series and parallel configurations by incorporating a clutch that allows the ICE to be mechanically connected to the drivetrain when required. In addition, the ICE is coupled to a generator, enabling battery charging and electrical power generation. This hybrid arrangement allows multiple operating modes, including pure electric driving, parallel propulsion (ICE and electric motor working together), and series operation where the ICE generates electricity. By reducing the number of energy conversion stages compared to a pure series configuration, this architecture can achieve higher overall efficiency. On the other hand, the system becomes more complex, as it requires both mechanical and electrical coupling elements, as well as advanced control strategies to manage transitions between operating modes. Additionally, the need for a mechanical transmission reduces some of the packaging advantages observed in pure series configurations.
- **Power split configuration** (Fig. 2.8g). This architecture represents a more advanced implementation that combines the advantages of series and parallel

configurations through the use of a power splitting device, typically based on a planetary gear set. In this arrangement, the power generated by the ICE is divided into two paths: one portion is transmitted mechanically to the wheels, while the other is converted into electrical energy through a generator. The electrical energy can then be used to supply the traction motor or to charge the battery, enabling a high degree of flexibility in energy management. This configuration allows the ICE to operate closer to its optimal efficiency region while maintaining a direct mechanical path to the wheels, thus reducing conversion losses. Despite its advantages, the power split architecture is mechanically and control wise complex, requiring precise coordination between multiple components. It is commonly used in commercially successful hybrid vehicles due to its balance between efficiency and performance, although it typically involves higher development and manufacturing costs.

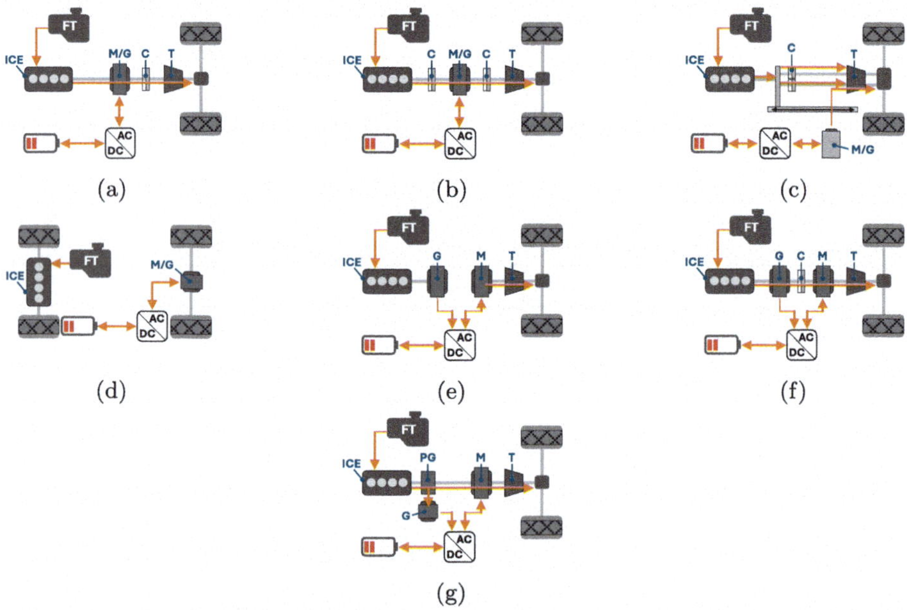

Figure 2.8 Different HEC configurations. C: clutch, FT: fuel tank, G: generator, M: electric motor, PG: planetary gear, T: transmission.

2.2.3 Operation

The operational behavior of a HEC can be systematically described through a set of characteristic operating modes, which are defined as a function of the instantaneous power demand, the SoC of the traction battery, and the prevailing operating conditions of the propulsion system. These modes capture the different ways in which power flows are managed between the ICE, the electric motor, and the energy storage system. Such a classification provides a structured framework for understanding the energy management strategy of hybrid vehicles. In this context, six principal operating modes can be identified [7]:

- **Start up**. If the SoC of the traction battery is sufficiently high, the electric motor is used to drive the ICE at a low rotational speed (approximately 1000 rpm), thereby enabling engine ignition without the need for a conventional starter. However, under certain conditions (such as low SoC, low ambient temperature, or the presence of system faults), the electric motor may not be able to perform this function. In such cases, a conventional starter motor is employed, powered by the auxiliary battery system. This dual starting capability enhances system reliability and ensures proper vehicle operation under a wide range of conditions.
- **Acceleration**. During acceleration phases, the propulsion system is required to deliver increased power to meet the driver demand. Under normal conditions, the electric motor can provide traction power either independently or in combination with the ICE, depending on the operating point and control strategy. When the SoC is relatively low (but still above a predefined minimum threshold), electric assistance is typically restricted and only allowed under demanding conditions (such as wide open throttle operation). If the SoC reaches its minimum allowable value, electric assistance is completely disabled, and the ICE becomes the sole source of traction power. This strategy prevents excessive battery depletion and ensures system durability.
- **Cruising**. In steady state driving conditions (characterized by moderate and relatively constant power demand), the energy management strategy aims to maintain an appropriate SoC level while minimizing fuel consumption. If the SoC falls below a predefined threshold, the ICE operates not only to supply the required traction power but also to recharge the traction battery. Conversely, when the SoC is sufficiently high, the electric motor can contribute to propulsion either independently or in combination with the ICE. This flexible operation allows the system to optimize overall efficiency depending on driving conditions.
- **Deceleration**. During deceleration events (i.e., when the fuel supply to the ICE is interrupted), the vehicle motion is sustained by the kinetic energy stored in the rotating wheels. In this mode, the electric motor can operate as a generator, converting mechanical energy into electrical energy through regenerative braking. This recovered energy is then stored in the traction battery, improving overall system efficiency and reducing energy losses associated with conventional braking.
- **Braking**. This mode is closely related to deceleration but corresponds to situations where higher braking forces are required. Under these conditions, regenerative braking is maximized in order to recover as much energy as possible. However, when the required braking power exceeds the capability of the electric system, conventional mechanical braking systems are engaged in combination with regenerative braking. This coordinated operation ensures adequate braking performance, reduces stopping distances, and improves driving comfort while still enabling partial energy recovery.
- **Idling**. In idling conditions (i.e., when the vehicle is stationary or operating at very low load), the ICE may remain running at low speed if the SoC of the traction

battery is below a desired level. This allows the system to maintain readiness for subsequent power demands and enables rapid torque delivery without requiring a transition through the start up mode. Additionally, this operating mode can be used to gradually recharge the battery when necessary, contributing to the overall energy management strategy.

2.2.4 Energy management

Energy management strategies in HECs differ substantially from those employed in BECs. Although, from a conceptual standpoint, both approaches rely on the formulation and solution of an optimization problem (typically using the techniques described in Section 2.1.3), the presence of multiple energy sources in HECs introduces an additional layer of complexity. In particular, the coexistence of a thermal system and an electrical system leads to higher dimensional optimization problems, increased coupling between variables, and a more intricate representation of system dynamics and constraints.

In most cases, the energy management strategy in a HEC is formulated with the objective of minimizing fuel consumption or, more generally, an equivalent cost function. This equivalent cost accounts for both the fuel consumption and the electrical energy usage, typically through an energy equivalence factor that enables a unified treatment of both energy domains. This formulation differs from the standard objective in BECs, which is primarily focused on the optimal use of electrical energy. Furthermore, HEC optimization problems often include binary decision variables, which are used to represent the on/off states of the different energy sources (thermal and electrical). This allows the control strategy to explicitly determine which source is active at each time instant, thereby capturing the hybrid nature of the propulsion system.

Another key distinction lies in the management of the SoC of the traction battery. In this regard, two main operational strategies can be identified [8].

On the one hand, in the Charge Sustaining mode (which is characteristic of conventional hybrid vehicles), the primary objective of the energy management system is to maintain the SoC within a relatively narrow band around a nominal value. In this operating regime, the battery does not act as a primary energy source but rather as an intermediate buffer that decouples instantaneous power generation from consumption. The ICE assumes the dominant role as the main energy provider, while the electric motor is used to enhance overall efficiency through functions such as acceleration assistance and energy recovery during braking. From a control perspective, this behavior is typically enforced by imposing a cyclic constraint on the SoC, such that its final value over a driving cycle remains approximately equal to its initial value.

On the other hand, in the Charge Depleting mode (which is typical of PHECs), the traction battery is used as the primary energy source over a significant portion of the driving cycle. In this case, the objective is not to maintain the SoC at a given level, but rather to allow its gradual depletion as the stored electrical energy (previously obtained through external charging) is consumed. This mode enables extended operation in purely electric conditions, thereby reducing the use of the ICE and, consequently, lowering fuel consumption and local emissions. From a control standpoint, the SoC is no longer treated as a strictly regulated variable but rather as an energy resource that must be optimally allocated over the driving horizon.

In PHECs, both operating modes coexist in a sequential manner. Typically, the system initially operates in Charge Depleting mode, prioritizing the use of electrical energy until the SoC reaches a predefined lower threshold. Beyond this point, the vehicle transitions to Charge Sustaining mode, adopting a behavior similar to that of a conventional hybrid vehicle. This combined strategy allows the system to exploit the benefits of electric propulsion over short distances, while preserving the flexibility and extended driving range provided by the thermal engine for longer trips.

2.3 Fuel-cell-electric cars (FCECs)

FCECs constitute one of the main technological alternatives within the electrification of transport. In contrast to BECs (where energy is stored electrochemically in batteries), FCECs generate electricity on board from a fuel (typically hydrogen) by means of a fuel cell (FC) system. This approach enables a partial decoupling between energy generation and electrical storage, thereby introducing a distinct energy architecture within the broader context of electric mobility.

From a functional perspective, a FCEC can be interpreted as an EC in which the primary energy source is not a battery, but an onboard electrochemical generation system. Nevertheless, in practical implementations, these vehicles also incorporate an auxiliary electrical storage system (typically a battery of lower capacity or a supercapacitor based system). The role of this auxiliary system is to manage power transients, absorb regenerative energy, and improve the overall efficiency of the system. Although this configuration differs conceptually from that of HECs, it introduces certain similarities in terms of energy flow management and coupling between energy sources.

In terms of architecture, FCECs typically adopt configurations in which traction is fully electric, in a manner analogous to BECs. The FC system operates as the main electricity generator, supplying power to the inverter and the electric traction motor, as well as recharging the auxiliary storage system when required. As a result, from the perspective of the powertrain, FCECs share several characteristics with BECs, including the absence of complex conventional transmission systems and the capability to implement energy recovery strategies during braking.

Another relevant aspect of this technology is its scalability. FCECs are particularly well suited for applications in which the energy density of electrical storage becomes a limiting factor (such as heavy duty vehicles, long distance transport, or high utilization fleets). In these contexts, the ability to store large amounts of energy in the form of fuel and to refuel the vehicle in short time intervals provides significant operational advantages. However, in smaller scale or urban applications, the benefits with respect to BECs may be less pronounced, particularly when considering the increased system complexity and the associated infrastructure requirements.

References

[1] S. Hanely, Lightyear files for bankruptcy, retrieved 2 Feb. 2023, https://cleantechnica.com/2023/02/02/lightyear-files-for-bankruptcy/.

[2] Y.L. Morangueira, J.C. de C. Pereira, Energy harvesting assessment with a coupled full car and piezoelectric model, Energy 210 (2020) 118668, https://doi.org/10.1016/j.energy.2020.118668.

[3] C. Chan, The state of the art of electric and hybrid vehicles, Proceedings of the IEEE 90 (2) (2002) 247–275, https://doi.org/10.1109/5.989873.

[4] J. Oladigbolu, A. Mujeeb, L. Li, Optimization and energy management strategies, challenges, advances, and prospects in electric vehicles and their charging infrastructures: a comprehensive review, Computers & Electrical Engineering 120 (2024) 109842, https://doi.org/10.1016/j.compeleceng.2024.109842.

[5] A. Kermansaravi, S.S. Refaat, M. Trabelsi, H. Vahedi, Ai-based energy management strategies for electric vehicles: challenges and future directions, Energy Reports 13 (2025) 5535–5550, https://doi.org/10.1016/j.egyr.2025.04.053.

[6] M. Takahashi, P. Jaswani, Electric motors for electric vehicles 2026–2036: Technologies, materials, markets, and forecasts, Market report, IDTechEx Research, 2024, https://www.idtechex.com.

[7] T. Denton, Electric and Hybrid Vehicles, Routledge, 2020.

[8] X. Lü, S. Li, X. He, C. Xie, S. He, Y. Xu, J. Fang, M. Zhang, X. Yang, Hybrid electric vehicles: a review of energy management strategies based on model predictive control, Journal of Energy Storage 56 (2022) 106112, https://doi.org/10.1016/j.est.2022.106112.

Electric cars, current status

3

3.1 Introduction

Among the different types of EVs that provide alternative solutions for electric mobility, ECs constitute, without doubt, the most relevant segment from the end-user perspective. This prominence is mainly due to their direct interaction with private consumers and their growing presence in everyday transportation systems. As a result, the analysis of EC market penetration offers a meaningful and representative indication of the actual level of adoption of electric mobility within society. In this sense, ECs can be regarded as a key indicator of the progress achieved in the transition toward more sustainable transport systems.

Despite the significant technological advances in electric mobility achieved throughout the nineteenth and twentieth centuries, the large-scale deployment of these technologies has only materialized in recent decades. Even so, their penetration in the twenty-first century has not followed a homogeneous trajectory. In particular, the adoption of ECs exhibits substantial variability depending on the geographical region under consideration. While some countries have experienced rapid growth and high levels of market penetration, others remain at relatively early stages of adoption. This uneven development reflects the influence of a wide range of factors that extend beyond purely technological considerations.

In this regard, the expansion of ECs is strongly influenced by elements such as public policies, economic incentives, the availability and accessibility of charging infrastructure, and the specific characteristics of local markets. Regulatory frameworks aimed at reducing emissions, financial support mechanisms for vehicle acquisition, and investments in charging networks have proven to be decisive in accelerating adoption. At the same time, aspects such as consumer behavior, vehicle cost, and energy prices also play a relevant role in shaping demand. Consequently, understanding the current status of EC deployment requires not only the analysis of aggregate figures, but also a careful examination of the context in which these figures are generated.

Within this context, several relevant questions naturally arise. It is of particular interest to determine the current level of penetration of ECs in the global market, to analyze how this penetration is distributed across different regions, and to identify the main factors that explain the observed differences. Addressing these questions is essential to obtain a comprehensive and realistic understanding of the state of electric mobility and its potential future evolution.

To provide a rigorous and up-to-date answer to these issues, this chapter is based on information obtained from widely recognized and authoritative sources. In particular, the analysis draws upon data from the Global EV Outlook,[1] published annually by the International Energy Agency (IEA), complemented by market analyzes

[1] https://www.iea.org/reports/global-ev-outlook-2025.

Electric Mobility Systems. https://doi.org/10.1016/B978-0-44-349397-3.00011-2

from Bloomberg[2] and reports produced by the European Automobile Manufacturers Association (ACEA).[3] These sources provide a robust and consistent foundation for assessing the evolution of ECs at both global and regional levels. Furthermore, they allow the incorporation of the most recent data available at the time of writing, ensuring that the discussion reflects the current state of the sector.

It is important to emphasize that the objective of this chapter is not limited to the presentation of quantitative data. While numerical indicators such as sales volumes, market shares, and fleet size are essential to characterize the evolution of ECs, they do not, by themselves, provide a complete understanding of the underlying dynamics. For this reason, the analysis adopts an interpretative perspective, aiming to identify trends, highlight structural patterns, and explain the factors that have driven recent developments. This approach enables a more comprehensive understanding of the processes shaping the deployment of electric mobility.

From this perspective, ECs are examined within a broader framework that encompasses not only technological aspects, but also their economic, regulatory, and societal dimensions. The transition toward electric mobility is closely linked to wider transformations in the energy system, including the integration of renewable energy sources and the decarbonization of transport. At the same time, it is influenced by evolving policy objectives, industrial strategies, and changes in consumer preferences. In this sense, the development of ECs cannot be fully understood in isolation, but must be analyzed as part of a complex and interconnected system.

Overall, this chapter aims to provide a clear and structured overview of the current status of ECs, offering both a quantitative and qualitative perspective on their recent evolution. By combining data analysis with contextual interpretation, it seeks to contribute to a deeper understanding of the role of ECs in the ongoing transformation of the transport sector and to highlight the main factors that are shaping their future development.

3.2 Global market evolution

3.2.1 Early development and inflection point

ECs remained a niche market product throughout most of the 2010s, being widely regarded as an emerging technology. During this period, their sales were relatively limited and largely confined to specific market segments. This reduced market presence can be attributed to several key factors, including the high cost of batteries, the limited driving range of available vehicles, and the scarce (or in many cases non-existent) charging infrastructure across most countries [1].

Under these conditions, the growth in EC sales during the 2010s was moderate and geographically concentrated. In particular, adoption was mainly observed in regions such as China, the United States, and certain European countries, where strong institutional support and policy incentives played a decisive role. Among these,

[2] https://about.bnef.com/.

[3] https://www.acea.auto/.

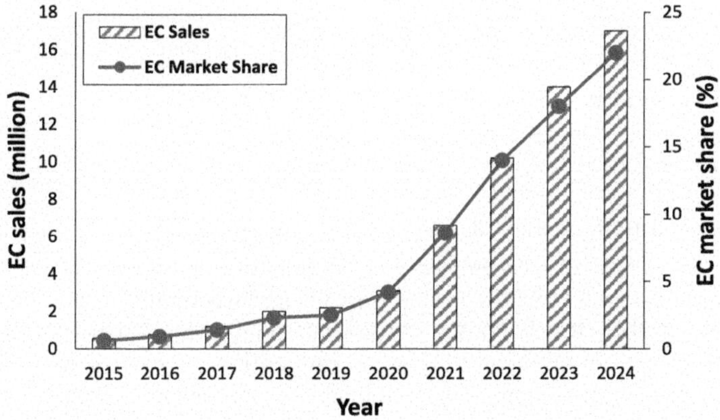

Figure 3.1 Global EC sales and market share over the period 2015–2024. The figure highlights a clear shift in market dynamics from 2020 onwards, with a marked acceleration in both sales volumes and market share, indicating the transition from a gradual growth phase to a period of rapid market expansion.

Norway stands out as a paradigmatic case, achieving significantly higher penetration levels due to a consistent and comprehensive policy framework [2].

However, this moderate growth trajectory changed substantially toward the end of the decade and, more notably, in the early 2020s (see Fig. 3.1). Several reports [3,4] identify this period as a clear inflection point in the market penetration of ECs. At this stage, many of the barriers that had previously constrained adoption were either mitigated or significantly reduced. In parallel, the strengthening of environmental policies, particularly in Europe with initiatives such as the European Green Deal [5], contributed to accelerating the transition.

As a result, the number of available models increased considerably, acting as a catalyst for a rapid expansion in EC sales. This inflection point marks the transition from ECs being perceived as an emerging technology to becoming a consolidated and increasingly dominant solution within the automotive market.

3.2.2 Global sales trends

As shown in Fig. 3.1, the global growth in EC sales over the period 2015–2024 does not follow a linear trajectory. As previously discussed, two clearly differentiated phases can be identified. The first phase extends up to 2020, during which ECs were largely perceived as an emerging technology. This period is characterized by a moderate and approximately linear increase in sales, primarily constrained by the technological and market limitations outlined earlier. From 2020 onwards, however, a pronounced acceleration in sales can be observed, marking the beginning of a new expansion phase.

It is instructive to place the data presented in Fig. 3.1 in the broader context of total vehicle sales, including all propulsion technologies. In this regard, it is noteworthy that total vehicle sales experienced a significant contraction in 2020 due to the impact of the COVID-19 pandemic, with global production declining by approximately 15% [6]. In contrast, EC sales continued to grow during the same period, contributing to a substantial increase in their market share, which reached around 10% in 2021. This divergence highlights the resilience of the electric mobility sector and its progressive consolidation within the automotive market, particularly when compared to conventional ICE vehicles.

The strong shift in market dynamics observed in 2020 has been sustained in subsequent years, with EC market share exceeding 20% by 2024 [7]. This milestone cannot be explained solely by the recovery of the global automotive market following the pandemic. Rather, it reflects the consolidation of several structural drivers that have facilitated the adoption of ECs as a competitive alternative to ICE vehicles. These include the significant increase in the number of available models and the rapid expansion of charging infrastructure [7].

Another relevant aspect illustrated in Fig. 3.1 is the sustained growth in the market share of ECs. This trend reinforces the notion that the observed increase in sales is not merely a consequence of overall market recovery, but rather the result of a genuine shift toward electrified mobility. The continuous rise in market share provides clear evidence of the progressive electrification of the transport sector.

Overall, global sales data indicate that ECs have moved beyond their initial adoption phase and have entered a period of accelerated growth. This evolution suggests that the electrification of road transport is transitioning from an emerging trend to a structural transformation with long term implications for the automotive industry.

3.2.3 Global fleet growth

Beyond the analysis of annual sales, it is also of particular interest to examine the total number of ECs in circulation, as this metric provides a more accurate indication of the actual level of technology deployment within the market. In this regard, Fig. 3.2 presents the global EC stock over the period 2010–2024, disaggregated by technology (BECs, PHECs, and FCECs). As shown in the figure, a marked acceleration in the growth of the global fleet can be observed from 2020 onwards, coinciding with the inflection point previously identified in sales trends (see Fig. 3.1). This rapid expansion leads to a global stock approaching 60 million ECs by 2024 [7].

In the years prior to 2020, the growth of the EC fleet was comparatively moderate, reflecting the sales patterns observed during this period. Nevertheless, due to the cumulative nature of this metric, even relatively low levels of market penetration resulted in a steady increase in the global EC stock. As a consequence, the number of ECs in circulation reached approximately 7 million units by 2019, establishing a foundation for the subsequent phase of accelerated growth.

Another relevant aspect that can be inferred from Fig. 3.2 is the uneven distribution among different electrification technologies. While BECs account for the majority of the global fleet, supposing approximately a 67% of the total of ECs on road, other technologies such as PHECs and FCECs exhibit a significantly lower level of

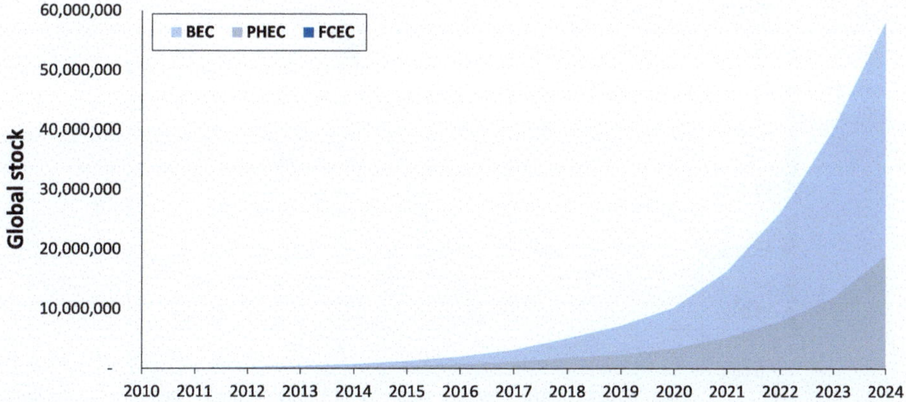

Figure 3.2 Global EC stock over the period 2010–2024, disaggregated by vehicle technology. It should be noted that the number of FCECs remains comparatively low, reaching approximately 70,000 units in 2024, and is therefore not clearly visible in the figure due to scale differences.

penetration (with 33% and 0.1% over the total, respectively). In particular, the number of FCECs remains comparatively small, which limits their contribution to the overall figures.

An additional perspective can be obtained by analyzing the year-on-year increase in sales, which provides insight into the dynamics of market penetration. In this context, Fig. 3.3 illustrates the evolution of the year-on-year growth in EC sales over the period 2015–2024. Although total sales volumes were relatively low between 2015 and 2020, this period was characterized by sustained growth, with increases reaching up to 40% in 2018. This trend was temporarily disrupted in 2019, followed by a strong recovery in subsequent years, particularly in 2021. More recently, a gradual decline in year-on-year growth rates can be observed from 2022 onwards. However, it is important to emphasize that, despite this moderation in growth rates, total sales have continued to increase consistently throughout the decade.

3.2.4 Market share evolution

Market share constitutes one of the most relevant indicators to assess the actual level of penetration of a given technology within a market. It is defined as the proportion of sales of a specific product or technology relative to the total market sales over a given period of time. In the context of electric mobility, it is typically expressed as the percentage of ECs sold over the total number of light-duty vehicles commercialized. In this way, while sales reflect market expansion in absolute terms, market share captures the actual weight of ECs within the global car market, considering all available technologies, and therefore provides a more accurate measure of the ongoing technological transition.

The market share of ECs in recent years is shown in Fig. 3.1. As observed in the figure, the market share of ECs followed a modest and nearly linear growth up to the

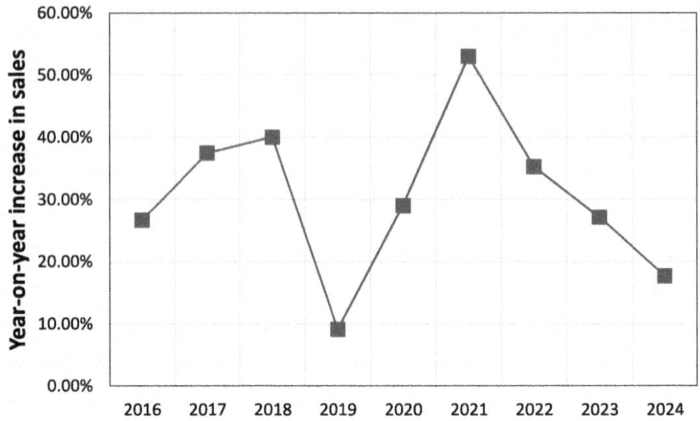

Figure 3.3 Year-on-year increase in global EC sales over the period 2016–2024.

year 2020. During this period, the market share of ECs remained at relatively low levels (between 1% and 3%), thus representing a marginal penetration compared to ICE vehicles, which continued to dominate the market overwhelmingly. From 2020 onwards, a clear change in trend can be identified, coinciding with a sharp increase in sales. From this point, the growth of EC market share has followed a trajectory closer to an exponential trend up to 2024. This shift reflects not only increased consumer adoption, but also a deeper structural transformation of the automotive sector, driven by technological maturity, cost reductions, and regulatory pressure.

At the global level, the market share of ECs has increased from marginal values to a significant fraction of the light-duty vehicle market within just a few years. Specifically, it has grown from approximately 3% in 2019 to more than 20% in 2024. From an economic perspective, this rapid increase signals a transition from an emerging market phase to an early mass adoption stage, in which economies of scale, learning effects, and infrastructure deployment begin to reinforce further growth. However, it is important to note that this transition is highly heterogeneous across regions. Leading markets can be identified where EC penetration has been significantly higher, largely driven by public policies aimed at promoting electric mobility options (see Section 3.3), while in other regions ECs still represent a marginal share. Despite the rapid growth observed in the period (2020–2024), the market share of ECs remains considerably lower than that of ICE vehicles. As shown in Fig. 3.4, ICE vehicles still account for the majority of global sales, highlighting that the substitution process is ongoing rather than complete.

Analyzing market share is particularly useful to assess the relative growth of a technology within the overall market, as an increase in market share does not necessarily imply a proportional reduction in the sales of other technologies such as ICE vehicles. This distinction is especially relevant in a context where the total car market is itself expanding. In recent years, global car sales (including both electric and combustion vehicles) have shown moderate growth, with an increase of 6%

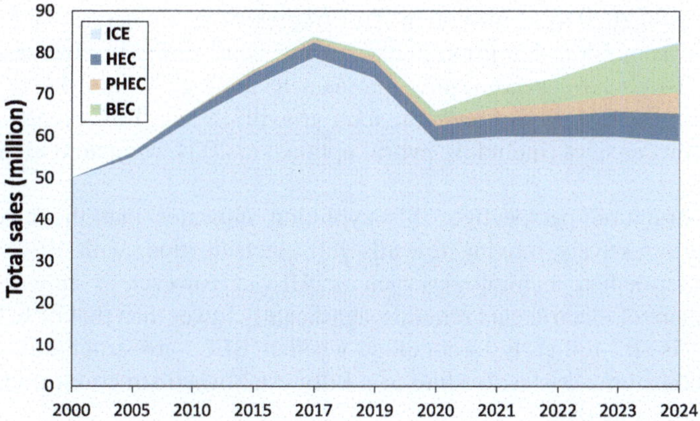

Figure 3.4 Global new car sales by powertrain.

in 2024 [8]. As a result, part of the growth in EC sales reflects market expansion rather than direct substitution. Nevertheless, the sustained increase in EC market share clearly indicates a progressive reallocation of market demand away from ICE technologies and towards electrified alternatives.

Overall, the recent evolution of market share shows that ECs have transitioned from a niche technology to a central element in the transformation of the automotive sector. Although the transition is far from complete, the observed trends clearly point towards a progressive consolidation of ECs as one of the dominant technologies in the global automotive market, with increasing implications for industrial structure, energy systems, and long-term decarbonization pathways.

3.2.5 Segmentation

It is important to note that the evolution of the EC market has not followed a homogeneous trend across the different technologies considered. In particular, it is especially relevant to analyze the sales evolution of two key technologies, namely BECs and PHECs, which have exhibited differentiated growth patterns in recent years.

As shown in Fig. 3.4, during the early stages of EC market development, PHECs represented the technology with the highest sales volumes. Specifically, in 2015, PHEC sales reached 0.5 million units, compared to 0.3 million BECs sold in the same year. These figures indicate an initial preference for hybrid solutions combining thermal and electric propulsion. This consumption pattern reflects a natural user behavior, favoring intermediate solutions between conventional and emerging technologies. In addition, PHECs provided an effective response to some of the main barriers faced by electric mobility at that time, such as the limited availability of charging infrastructure.

However, this trend changed from 2020 onwards, with a significantly faster growth of BECs observed since then. In particular, BEC sales increased from approximately

2 million units in 2020 to 12 million in 2024 (representing an increase of 83%), while PHECs grew from 1.5 to 5 million units over the same period (an increase of 70%). As a result, BECs have become the dominant technology within the EC segment, accounting for the majority of recent market growth. In fact, BECs represented 51% of total electric car sales (including hybrid options) in 2024, compared to only 28% in 2020.

From a structural perspective, this evolution indicates that the electrification process is progressively moving towards full electrification, with BECs gradually displacing intermediate technologies such as PHECs. However, as shown in Fig. 3.4, the market share of electric cars remains significantly lower than that of ICE vehicles. Therefore, it is still too early to conclude whether BECs are structurally displacing internal combustion vehicles, leading to a fully established structural transformation of the automotive market.

3.3 Regional analysis

3.3.1 China

In recent years, China has established itself as the main driving force of the global EC market. This leadership is not only reflected in higher sales volumes compared to other regions, but also in the depth of technological development and the level of market penetration achieved. Unlike other regions, where the adoption of electrified mobility options has been largely driven by regulatory constraints or environmental targets, the Chinese case reflects a comprehensive and coordinated strategy. This strategy integrates public policies with industrial development, expansion of manufacturing capabilities, and direct and indirect market stimulation. As a result, China has not only achieved a leading position in terms of EC sales and market share, but has also played a central role in shaping the global evolution of the sector [9].

As shown in Fig. 3.5, China clearly dominates global EC sales in 2024, reaching volumes exceeding 11 million vehicles and representing approximately two thirds of total global sales. This dominant position is not a recent development. During the second half of the 2010s, China began to distance itself from other major markets such as Europe and the United States. While early growth in these regions was relatively gradual, China experienced a faster and more sustained expansion. This divergence became even more pronounced after 2020, when the global EC market entered a phase of accelerated growth. In this context, China exhibited particularly strong dynamics, with EC sales increasing by around 60% in 2021, compared to 57% in the United States and 44% in Europe. These figures indicate that China is not merely benefiting from global trends, but is actively driving them, acting as the primary engine of market expansion.

Beyond aggregate volumes, it is essential to analyze the internal structure of the Chinese EC market. In this regard, Fig. 3.6 provides a detailed view of the global market share disaggregated by technology (PHECs and BECs). The data show that China not only leads in total sales, but also exhibits a dominant position in specific technological segments. In particular, the PHEC market is highly concentrated in

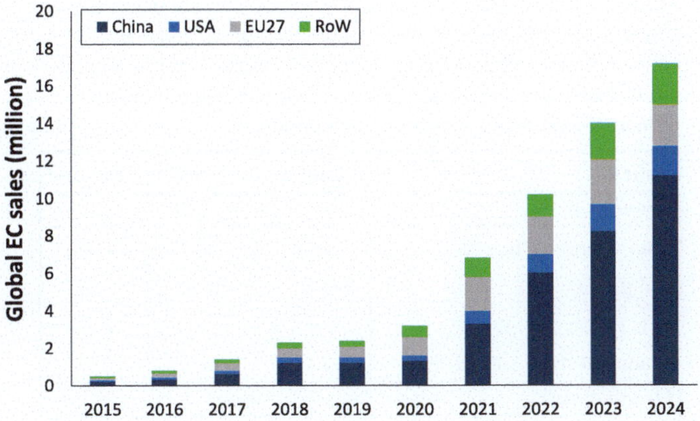

Figure 3.5 Global new car sales by region. EU27: European Union, RoW: rest of the World.

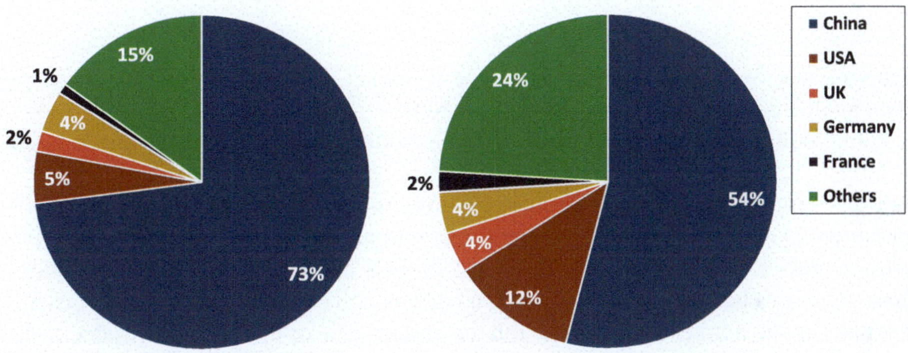

Figure 3.6 Global EC share by region and technology in 2024: PHEC (left) and BEC (right).

China, which accounts for approximately 73% of global sales. In the case of BECs, although the market is more geographically diversified, China still represents around 54% of global demand. This dual dominance suggests that China has developed a diversified technological ecosystem, capable of supporting multiple electrification pathways simultaneously.

From an economic perspective, this diversification reflects a deliberate strategy aimed at managing uncertainty and reducing risks during the early stages of the transition. In particular, the strong presence of PHECs can be interpreted as an intermediate solution that alleviates key constraints such as limited charging infrastructure and concerns related to driving range. At the same time, it allows for the progressive development of electric mobility without requiring an immediate and complete shift towards fully electric vehicles. However, the rapid growth of BECs in recent years indicates a clear transition towards full electrification, in line with global technological trends and cost reductions in battery systems. This evolution

Table 3.1 Comparison of major EV policy instruments across China, the United States, and Europe. Symbol ✓ indicates active policies, while × denote measures that are being phased out or losing relevance. The number of symbols qualitatively reflects the relative strength of each policy.

Policy	**China**	**USA**	**Europe**
Supply side			
Fuel economy targets	✓✓	✓	✓✓✓
EV quotas	✓✓	✓✓	✓
ICE phase out	–	✓	✓✓✓
Manufacturing subsidies	✓✓	✓✓✓	✓
Demand side			
Purchase incentives	× × ×	✓✓✓	× × ×
Company car benefit	–	–	✓✓

highlights the dynamic nature of the Chinese market, which has been able to adapt its technological mix as conditions have evolved.

Public policy has played a central role in enabling this transition. As summarized in Table 3.1, China has implemented a wide range of policy instruments targeting both the demand side and the supply side of the market. During the initial stages of market development, demand side measures, particularly purchase incentives, were instrumental in reducing the effective cost of ECs and stimulating early adoption. These measures helped to overcome initial barriers and create a critical mass of users. However, as the market matured, the relative importance of these incentives declined, and a gradual phase-out has taken place in recent years.

In parallel, there has been a significant strengthening of supply side policies. These include fuel economy targets, EV production quotas, and direct support for manufacturing activities. This shift reflects a transition from short-term market stimulation towards a more structural approach focused on building industrial capacity and technological leadership. By reinforcing the supply side, China has been able to consolidate its position not only as the largest market, but also as a key global producer of ECs and related technologies. This contrasts with other regions, where policy frameworks have often been more fragmented or more heavily focused on demand-side incentives.

The evolution of EC market share in China further reinforces this interpretation. As illustrated in Fig. 3.7, the penetration of ECs in the Chinese market has increased steadily over time, reaching levels close to 40% in 2024. This places China significantly ahead of the United States and broadly in line with Europe. More importantly, projections indicate that this upward trend will continue, with market share levels expected to reach around 60% by 2030. This sustained growth suggests that the transition towards electric mobility in China is not only rapid, but also structurally embedded within the market.

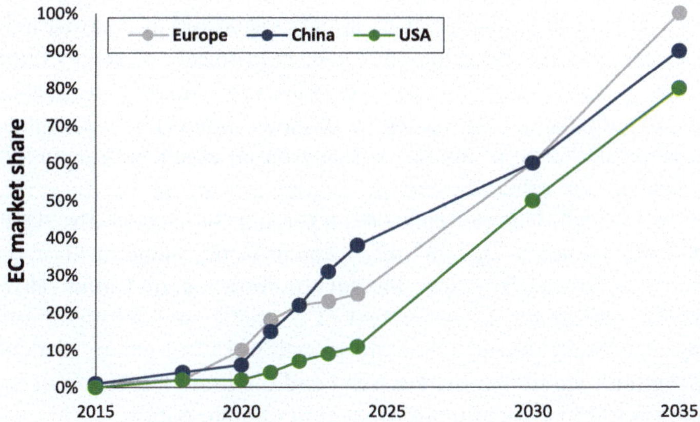

Figure 3.7 Market share of ECs by region. Historical data for the period 2015–2024 are combined with projections consistent with current policy frameworks and market trends up to 2035. This figure contemplates original share projected by Europe for 2035, which has been revised in 2025 [10].

From a comparative perspective, these trends highlight important differences in the dynamics of regional transitions. While Europe is expected to achieve similar levels of market penetration by 2030, this convergence is largely driven by increasingly stringent regulatory frameworks. In contrast, the Chinese trajectory appears to be supported by a combination of industrial policy, market forces, and technological development. The United States, on the other hand, exhibits a more gradual transition, although recent policy developments may accelerate adoption in the coming years.

Overall, the Chinese case illustrates a model of transition towards electric mobility that combines scale, technological diversity, and policy coherence. Unlike other markets, where electrification has been more strongly influenced by regulatory constraints or cost evolution alone, China has developed an integrated approach that encompasses the entire value chain, from production to final adoption. This integrated strategy has enabled not only rapid market growth, but also a strong positioning in key industrial segments, including battery manufacturing and vehicle production.

Therefore, the analysis of China highlights that the transition towards electric mobility should not be understood solely as a technological shift, but rather as a large-scale economic and industrial transformation. In this context, China's leadership is not simply the result of favorable market conditions, but rather the outcome of a long-term structural strategy. This strategy has allowed the country to position itself at the center of the global transformation of the automotive sector, influencing not only current market dynamics, but also the future direction of the industry.

3.3.2 Europe

The European EC market is characterized as a representative case of a leading region in terms of market penetration and regulatory framework. In contrast to the Chinese case, where market development has been largely supported by a strong industrial and technological push, the European market has been primarily driven by climate policy strategies aimed at reducing greenhouse gas emissions [5]. This sustained regulatory effort, particularly intensified during the early 2020s, has enabled Europe to position itself as a leading market in terms of EC penetration, albeit with a smaller absolute market size compared to other regions.

As shown in Fig. 3.5, Europe represents a relevant but secondary player in terms of global EC sales volumes. In 2024, EC sales in Europe amount to approximately 2 million units, significantly below the levels observed in China. However, this figure should be interpreted in the context of a smaller overall automotive market. In fact, Europe has experienced significant growth in EC adoption, particularly since 2020, following both the global market trend and the implementation of increasingly stringent regulatory frameworks in the context of climate policy.

This trend becomes more evident when analyzing market share, as shown in Fig. 3.7. Europe has reached penetration levels close to 25% in 2024, comparable to those observed in China and clearly above those in the United States. These figures highlight that, although Europe does not lead in absolute sales volume, it stands out in terms of adoption intensity. The relatively high market share indicates that electrification has already reached a significant level of maturity within the European automotive market. Even more noteworthy are the projections for the coming decade, which suggest that EC market share could reach around 60% by 2030.

In this context, it is important to note that, although Europe initially proposed a complete phase-out of ICE vehicle sales by 2035, recent revisions of the regulatory framework have introduced greater flexibility. These revisions allow for the partial continuation of combustion-based and hybrid technologies, implying a residual penetration of such vehicles in the long term [10]. This adjustment reflects a more pragmatic approach to the transition, acknowledging both technological and economic constraints.

As previously mentioned, the development of the European EC market is closely linked to the adoption of ambitious climate policies aimed at reducing greenhouse gas emissions. In particular, CO_2 emission standards for light-duty vehicles have acted as a key catalyst for electrification. These regulatory mechanisms have directly influenced manufacturers' strategies, effectively forcing the introduction of EVs in order to comply with fleet emission targets. This represents a clear distinction from the Chinese case, where market development has been driven by a broader combination of industrial, technological, and policy factors.

From a technological perspective, the European market exhibits a relatively balanced distribution between PHECs and BECs, as shown in Fig. 3.6. However, in recent years, a clear shift towards BECs can be observed, driven by the tightening of regulatory requirements and the gradual withdrawal of incentives for hybrid technologies. This transition suggests a convergence towards fully electrified mobility solutions, in line with global trends. In earlier stages of market development, PHECs played a crucial role as a transitional technology, enabling manufacturers and consumers to overcome barriers such as limited driving range and insufficient charging infrastructure. This pattern is consistent with observations in other regions, including China, where hybrid technologies initially facilitated market uptake.

Regarding the regulatory framework, Europe presents a fundamentally different approach compared to China. As reflected in Table 3.1, regulatory instruments play a particularly prominent role in the European context, with strict emission reduction targets acting as the primary driver of market development. This regulatory pressure has created a strong and predictable signal for market participants, accelerating the transition towards electrification. In addition to regulatory measures, demand side incentives have also played a significant role, particularly during the early stages of market development, helping to reduce the cost gap between ECs and conventional ICE vehicles.

However, similar to the Chinese case, a gradual reduction of these incentives can be observed as the market matures. This evolution reflects a transition from policy-driven adoption towards a more self-sustaining market dynamic. Compared to the United States, where demand side incentives continue to play a central role, Europe exhibits a clearer shift towards a regulation-driven model. This difference highlights the diversity of policy approaches across regions and their implications for market development.

Another relevant characteristic of the European market is its pronounced heterogeneity across countries. While some countries, such as Norway, Denmark, and the Netherlands, have achieved very high levels of EC penetration, others, particularly in Eastern Europe, such as Bulgaria or Poland, still exhibit relatively low adoption rates. This fragmentation introduces additional complexity in the analysis of the European market, as aggregate figures may mask significant national disparities. Differences in income levels, infrastructure availability, and policy implementation contribute to this uneven adoption pattern.

From a structural perspective, the European case illustrates a transition model primarily driven by regulation. This approach has proven highly effective in accelerating EC adoption, particularly in contexts where market forces alone would not have been sufficient to drive such rapid change. However, it also raises important challenges related to the economic sustainability of the transition, including the need for continued investment in charging infrastructure and the integration of increasing levels of electrification into the power system.

In summary, Europe represents a key region in the global transition towards electric mobility, distinguished by its high market share and the central role of regulation in shaping the market. Although its contribution in terms of absolute sales volume remains lower than that of China, its ability to drive structural change through policy frameworks makes it a fundamental actor in the evolution of the global EC market.

3.3.3 United States

The EC market in the United States exhibits a lower level of penetration compared to both Europe and China, with a more gradual growth trajectory, although a clear acceleration has been observed in recent years. In contrast to the Chinese market, whose expansion has been closely linked to a comprehensive strategy focused on industrial development and technological innovation, and to the European market, where regulation has played a dominant role, the U.S. market has historically been characterized by relatively low levels of intervention. Instead, its development has been primarily driven by market dynamics and direct economic incentives.

As shown in Fig. 3.5, the United States ranks as the third largest EC market globally in terms of sales volume, behind China and Europe. In 2024, EC sales in the United States amount to approximately 1.5 million units, remaining below the levels observed in the two leading regions. This outcome reflects not only a lower degree of market penetration, but also structural characteristics of the U.S. automotive market, including a strong consumer preference for larger vehicles, where electrified alternatives are still comparatively limited, as well as differences in cost structures.

This lower level of market penetration is clearly illustrated in Fig. 3.7, where the EC market share in the United States is estimated at around 10–15% in 2024, significantly below the levels reached in China and Europe. However, projections indicate a substantial expansion in the coming years, with penetration levels potentially approaching 50% by 2030. This trend highlights the significant growth potential of the U.S. market in the medium term, particularly as technological and economic conditions continue to evolve.

From a technological perspective, the U.S. market shows a clear preference for fully electrified solutions, as evidenced in Fig. 3.6. In contrast to other regions, where hybrid technologies such as PHECs played a significant role during the early stages of market development, the adoption of PHECs in the United States has been relatively limited. This difference can be explained by factors such as the development of charging infrastructure in certain states, as well as consumer preferences that favor fully electric vehicles. As a result, the electrification pathway in the U.S. has been more directly oriented towards BECs, bypassing, to some extent, intermediate technological solutions.

Regarding public policy, the U.S. market has been predominantly influenced by economic incentive mechanisms, in contrast to the more regulation-driven approaches observed in China and Europe. As shown in Table 3.1, the U.S. policy framework has largely relied on tax credits and financial incentives to stimulate demand. While these measures have contributed to market growth, they have resulted in a less structured and less predictable regulatory environment compared to other regions. Although the role of policy has increased in recent years, its historical impact has been more limited in shaping the overall trajectory of the market.

The introduction of the Inflation Reduction Act represents a significant shift in this context, as it includes a comprehensive set of measures aimed at promoting both demand and domestic production of ECs. These measures include purchase incentives linked to local manufacturing requirements, as well as support for battery production and supply chains. However, recent developments in the regulatory and political landscape have introduced a degree of uncertainty regarding the long-term continuity of these policies. From 2025 onwards, discussions have emerged around revising this approach, placing greater emphasis on technological neutrality and increased support for traditional automotive industries.

From a structural standpoint, the U.S. market can be interpreted as a transition model primarily driven by market forces rather than by strong and stable regulatory frameworks. This has resulted in a slower and more progressive transition compared to China and Europe. However, this market-driven approach may also provide the foundation for a more flexible and adaptive system in the long term. In this context, the U.S. represents a market with substantial growth potential, whose future evolution will depend on the interaction between policy signals, technological progress, and consumer behavior.

3.3.4 Other regions

Focusing on markets beyond China, Europe, and the United States, a significantly lower level of EC penetration and market share can be observed. These regions mainly include emerging economies in Asia, Latin America, and the Middle East. Although their current contribution to the global EC market remains limited, their growth potential is considerable and they are expected to play an increasingly relevant role in the future.

As shown in Fig. 3.5, the volume of EC sales in these regions is still modest compared to the leading markets, representing a relatively small share of global sales in 2024. This reflects a lower level of market maturity, as well as the presence of structural barriers that constrain the development of electric mobility. Despite this, recent trends indicate that some emerging markets are beginning to experience accelerated growth, driven in part by the increasing availability of more affordable EVs and international market expansion [7].

Several key barriers explain the limited penetration of ECs in these regions. First, the high upfront cost of EVs remains a major constraint, particularly in markets with lower purchasing power. In addition, charging infrastructure is often scarce or underdeveloped, which limits usability and consumer confidence. In some regions, further challenges such as limited access to reliable electricity supply also play a critical role in slowing adoption [7]. As a result, ICE vehicles continue to dominate the automotive market.

From a technological perspective, no uniform adoption pattern can be identified. In some markets, fully electric solutions dominate, while in others hybrid technologies represent a larger share of the electrified segment, potentially acting as transitional solutions. Furthermore, electrification is often concentrated in specific segments, particularly in public transport or commercial fleets, where total cost of ownership can be more favorable and policy support is more targeted.

In terms of public policy, although some initiatives have been implemented, regulatory frameworks are generally less developed and less consistent than in leading markets. Limited fiscal capacity often restricts the deployment of large-scale incentive schemes, reducing the effectiveness of policy-driven market expansion. Consequently, the transition in these regions tends to be slower and more dependent on external factors such as technology transfer, international financing, and the availability of low-cost vehicle imports.

In conclusion, regions beyond the main global leaders represent the largest untapped growth segment for the EC market. While their current contribution remains limited, their future development will be critical for achieving a fully electrified and low-emission transport sector at a global scale. This transition will depend not only on the implementation of more robust policy frameworks, but also on continued cost reductions, expansion of charging infrastructure, and the introduction of vehicle models adapted to local market conditions.

3.4 Technology and market trends

The development of the EC market cannot be understood without considering the central role played by technological progress. Beyond the sales data and regional market shares analyzed in previous sections, it is essential to examine how technological advancements have effectively acted as a catalyst for market growth. These developments have progressively contributed to the maturation of electric mobility solutions, making them an increasingly competitive alternative to traditional ICE vehicles.

The objective of Section 3.4 is therefore not to provide an in-depth analysis of the technical foundations of these technologies, which are addressed in detail in subsequent chapters, but rather to assess how their evolution has shaped the dynamics of the EC market. In this context, technology is not treated as an end in itself, but as a key enabler of adoption.

From this perspective, this section adopts an integrated techno-economic approach, aimed at understanding how cost reductions, the emergence of economies of scale, performance improvements, and industrial maturation have jointly influenced both consumer perception of ECs and the strategic decisions of manufacturers. This approach provides a coherent framework to interpret not only the recent expansion of the market, but also its expected evolution in the medium term.

3.4.1 Cost reduction as primary driver

When analyzing the recent evolution of the EC market, particular attention must be paid to the progressive reduction in manufacturing costs associated with EVs. Among the different components that shape the overall cost structure of an EC, energy storage systems have played a fundamental role, as they account for approximately 30–40% of the total vehicle cost [11]. As discussed in Chapter 1, battery costs experienced a substantial decline during the 2010s, with cumulative reductions close to 90%. More recent estimates indicate that battery pack costs have approached levels below 100 $/kWh, reflecting the continued maturation of the technology [12].

This cost reduction has progressively narrowed the price gap between ECs and ICE cars, making electric alternatives increasingly competitive from a purchase cost perspective. This aspect is particularly relevant for market adoption, as the upfront cost remains one of the main barriers to the widespread adoption of ECs compared to ICE vehicles [7]. As this gap continues to decrease, ECs become accessible to a broader segment of consumers, which directly contributes to the acceleration of market penetration.

However, the rapid increase in EC market share cannot be attributed solely to technological advancements and cost reductions. Industrial development has also played a key role, enabling large-scale manufacturing, process optimization, and the consolidation of supply chains. These factors have significantly contributed to lowering unit costs, following well-established industrial learning dynamics observed in other emerging technologies. This evolution indicates that ECs are progressively transitioning from an emerging technology to a mature industrial product.

This trend is clearly reflected in Fig. 3.5, where a sharp acceleration in global EC sales can be observed from 2020 onwards. This inflection point coincides with a period of rapid expansion in manufacturing capacity and continued cost reductions, suggesting a strong interaction between industrial scale and market growth.

From a structural perspective, the combination of cost reductions and industrial scaling represents a fundamental shift in the automotive sector. If the trends observed during the 2010s and 2020s persist, the EC market may become less dependent on policy-driven incentives and increasingly governed by intrinsic market dynamics, similar to those observed in other mature technologies.

3.4.2 Performance improvements and user acceptance

Not only cost reductions have contributed to the rapid development of the EC market. In this context, the progressive improvement in vehicle performance has played a fundamental role in the acceptance of this technology by users. Beyond purely economic considerations, the adoption of electric mobility is strongly influenced by how users perceive the functionality, reliability, and usability of ECs in real-world conditions.

Among the different performance attributes, driving range has traditionally been one of the most critical factors from the consumer perspective [13]. In the early stages of market development, when the technology was still immature, limited driving range generated a strong perception of risk among users, commonly referred to as "range anxiety". This phenomenon acted as a significant psychological barrier, discouraging potential buyers due to concerns about the ability to complete daily trips or longer journeys without running out of energy.

However, as shown in Fig. 3.8, the average driving range of ECs has experienced a sustained increase over the last two decades. This improvement has been primarily driven by technological advances, particularly in battery systems, as well as by cost reductions that have enabled the deployment of larger battery capacities. As a result, the average range of ECs has increased from approximately 100 km in 2010 to more than 400 km in 2024. This level of performance allows users to undertake long-distance trips with limited need for intermediate charging, effectively aligning EC capabilities with conventional mobility patterns.

From a behavioral standpoint, this evolution has had a profound impact on user perception. Driving range has gradually shifted from being a perceived critical

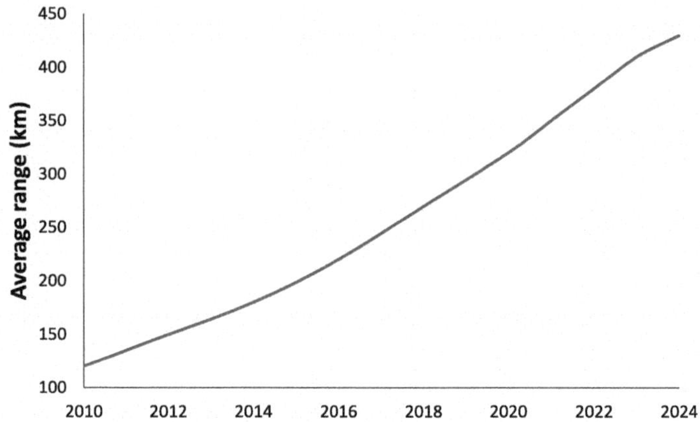

Figure 3.8 Evolution of average electric range of ECs in the period 2010–2024.

limitation to a secondary consideration in the purchasing decision process. As a result, the importance of range as a barrier to adoption has diminished over time, enabling ECs to compete more directly with ICE vehicles in terms of perceived usability.

In parallel with range improvements, reductions in charging time have also contributed significantly to enhancing user acceptance. As discussed in Chapter 6, current fast-charging technologies enable ECs to recover a substantial portion of their battery capacity in relatively short periods, typically allowing up to 80% charge in approximately 15–20 minutes under optimal conditions [7]. Although these charging times remain longer than refueling times for ICE vehicles, they no longer represent a prohibitive constraint for most use cases.

This shift is particularly relevant when considering actual usage patterns. Most daily trips involve relatively short distances, meaning that fast charging is primarily required for occasional long-distance travel. As a result, charging time has transitioned from being a major barrier to a manageable aspect of vehicle operation. This transformation reflects a broader adaptation process in which user expectations evolve alongside technological improvements.

In addition to vehicle-related factors, the expansion of charging infrastructure has played a complementary role in shaping user perception. The increasing availability of public charging points has improved the feasibility of long-distance travel and reduced concerns regarding energy availability. In recent years, the number of charging stations has grown significantly at the global level, although with notable regional disparities [14]. In particular, regions such as Europe have experienced rapid infrastructure deployment driven by targeted public policies, while other markets remain at earlier stages of development.

Importantly, the impact of infrastructure expansion extends beyond its physical availability. The mere visibility and perceived accessibility of charging points contribute to reducing uncertainty and reinforcing user confidence. This highlights the role of infrastructure not only as a technical enabler, but also as a psychological factor influencing adoption.

Beyond objective improvements in performance and infrastructure, the perception of these attributes by users has become a key determinant of market adoption. In this regard, communication and marketing strategies implemented by manufacturers and distributors have played an increasingly important role. Several studies indicate that consumer perceptions of key attributes such as vehicle performance, environmental benefits, and ease of use significantly influence purchase intentions [15].

Manufacturers have progressively adapted their communication strategies to address the main concerns associated with electric mobility. Rather than focusing exclusively on technical specifications, marketing campaigns have increasingly emphasized real-world usability, highlighting aspects such as driving range under typical conditions, charging convenience, and overall driving experience. This shift reflects a transition from a technology-centered narrative to a user-centered value proposition.

Furthermore, the growing visibility of ECs in the market has contributed to their normalization. The increasing presence of ECs in urban environments, combined with word-of-mouth effects and indirect user experience, has played a significant role in reducing uncertainty. Exposure to the technology, even without direct ownership, enhances familiarity and lowers perceived risk, which in turn supports adoption.

From a structural perspective, these dynamics indicate that the diffusion of ECs is not driven solely by technological or economic factors, but also by social and behavioral processes. The interaction between performance improvements, infrastructure availability, and user perception creates a reinforcing cycle that accelerates market adoption.

Overall, the combined effect of improved vehicle performance, reduced charging constraints, expanded infrastructure, and evolving consumer perception has contributed to the removal of key non-economic barriers. As a result, ECs have transitioned from being perceived as a niche or experimental technology to a mature and viable alternative within the automotive market. This transformation has enabled the expansion of EC adoption across multiple market segments, including private users, commercial fleets, and public transport applications, further reinforcing their role in the ongoing transition towards sustainable mobility.

3.4.3 Evolution of vehicle offer and segments

The evolution of the EC market has been accompanied by a significant transformation in the range of available vehicle models. In the early stages of market development, the offer of ECs was largely limited to small-sized vehicles primarily designed for urban use. As shown in Fig. 3.9, until around 2020, the vast majority of ECs were concentrated in sedan or compact segments, typically characterized by limited performance in terms of driving range and power. As a result, the available offer was targeted at a niche group of users with specific mobility needs, mainly associated with short-distance or urban travel. This initial market configuration was not only driven by commercial strategies but also by technical constraints, as the electrification of larger vehicles posed significant challenges from both an energy and cost perspective.

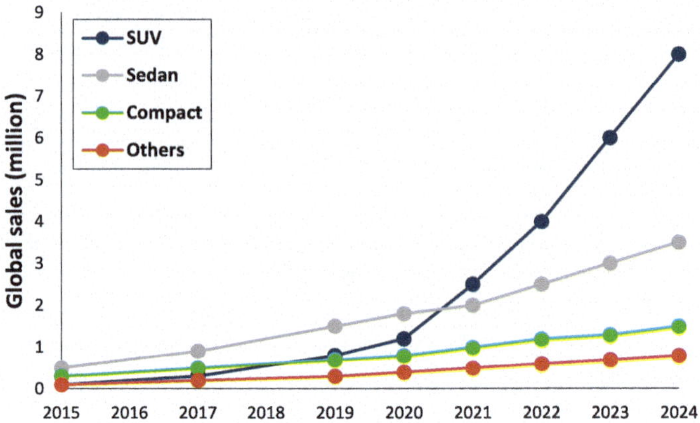

Figure 3.9 Evolution of global electric vehicle sales by vehicle segment. The rapid growth of electric SUVs reflects the expansion of electrification into high-demand market segments and the increasing maturity of EV technology.

However, as the technology has matured and production costs have decreased, manufacturers have progressively diversified their EC portfolios, expanding into segments that were previously dominated by ICE vehicles. This process has enabled the electrification of larger and higher-performance vehicles, whose technical and economic feasibility was limited in earlier stages. In this sense, the expansion of the product offering should not be interpreted solely as a response to increasing demand, but also as a clear indicator of the technological maturity achieved by the sector.

In particular, the sustained growth of the electric SUV segment since 2020 stands out as one of the most relevant trends, as illustrated in Fig. 3.9. SUVs already dominated the global automotive market in their conventional form, but their electrification was not viable until significant advances were achieved in battery capacity, energy efficiency, and cost reduction. As a result, manufacturers have increasingly focused their strategies on this segment, significantly expanding the number of available models and sales volumes. This trend reflects a clear convergence between the evolution of the EC market and the structural dynamics of the traditional automotive sector, where SUVs represent a substantial share of total demand.

From an economic perspective, this transition is particularly relevant, as it indicates that electrification has moved beyond niche applications to become a cross-cutting strategy across the industry. The ability to electrify high-demand and high-value segments such as SUVs represents a critical step towards the consolidation of ECs as a competitive alternative to internal combustion engine vehicles.

In parallel, a growing diversification in price ranges has been observed, encompassing both high-end premium models and more affordable options targeted at mass-market segments. In this context, recent years have seen the introduction of ECs in the €25,000–30,000 price range [16], while high-end models with prices exceeding €60,000 remain widely available, reflecting a substantial expansion of the market spectrum. This diversification is not only the result of commercial strategies, but also of ongoing cost reductions and increasing competition among manufacturers, which have contributed to improving affordability. As a result, one of the main limitations of the early market stages—the lack of variety—has been progressively overcome.

Additionally, the expansion of the EC offer has been reinforced by the development of the second-hand vehicle market. This dynamic follows a clear structural logic: as the number of ECs in circulation increases, the availability of used electric vehicles also grows. This segment, which was virtually non-existent at the beginning of the 2020s, has experienced significant growth in recent years, with strong year-on-year increases and a growing presence within the overall used car market [17]. From an economic standpoint, this phenomenon is particularly relevant, as it reduces the entry barrier associated with the high upfront cost of ECs.

Overall, the evolution of the EC offering reflects a clear transition from a niche market to a mass-market paradigm. The diversification across segments, price ranges, and vehicle types highlights not only technological progress but also the industrial consolidation of the sector. The ability of manufacturers to align their product portfolios with consumer preferences and broader market dynamics constitutes a key factor in explaining the recent growth of ECs and their expected expansion in the medium and long term.

3.4.4 Cost competitiveness

ECs have historically faced a significant barrier in their economic competitiveness with respect to ICE vehicles, their main competitors. In the early stages of market development, this technology was considerably less economically attractive than ICE vehicles, a much more mature technology. This gap effectively confined ECs to a niche market, as previously discussed. However, this economic disparity between technologies has been substantially reduced during the 2020s.

To properly assess the cost competitiveness of different technologies, it is useful to compare their Total Cost of Ownership (TCO), which accounts for both the vehicle acquisition cost and the expenses associated with energy consumption and maintenance over its lifetime. From this perspective, ECs still exhibit a higher upfront cost than ICE vehicles, mainly due to the cost of their battery systems. Nevertheless, the reduction in lithium-ion battery costs (see Section 3.4.1) has contributed to narrowing this gap, bringing the purchase cost of ECs closer to that of ICE vehicles.

On the other hand, ECs benefit from lower operating costs. In particular, the energy cost per kilometre for an EC typically lies in the range of 0.03–0.06 €/km, depending on whether the vehicle is primarily charged at home, whereas this cost is in the range of 0.08–0.12 €/km for an ICE vehicle [18]. In addition, maintenance costs for ECs can be up to 20–40% lower than those of ICE vehicles due to their simpler architecture [19]. As a result, the higher initial cost can be offset over time, leading to a break-even point beyond which ECs become more cost-effective than ICE vehicles. However, this break-even point varies significantly depending on factors such as energy prices. In this context, ECs are particularly attractive in high-utilization applications, which explains their early adoption in commercial fleets.

Overall, this analysis points towards a TCO parity between ECs and ICE vehicles. The cost gap between both technologies has been steadily narrowing, with parity already achieved in some segments, particularly in compact passenger cars. This trend is expected to extend to other market segments, although reaching full parity will strongly depend on boundary conditions such as energy costs and regional regulatory frameworks.

3.5 Current challenges and limitations

3.5.1 Technological and performance limitations

ECs still exhibit a number of technical limitations with respect to ICE vehicles, which can restrict or complicate their use in certain applications. Although the technological progress achieved during the 2020s has significantly mitigated many of these constraints and enabled the deployment of ECs across a broader range of use cases, these limitations remain an important barrier to their widespread adoption.

One of the most significant technical differences lies in the energy density of the fuels employed. Energy density can be defined on a gravimetric or volumetric basis, and provides an indication of the storage mass or volume required per unit of energy. In this regard, fossil fuels such as gasoline or diesel exhibit energy densities on the order of 10–20 times higher than those of current lithium-ion batteries used in ECs [20]. As a result, an ICE vehicle requires substantially less onboard energy storage than an equivalent EC to achieve a similar driving range. This limitation translates into larger and heavier energy storage systems in ECs, which directly impacts vehicle design, packaging constraints, and overall efficiency.

Another relevant aspect is charging time. While fast-charging technologies have advanced considerably and can reduce charging times to the order of minutes under optimal conditions, refuelling times for ICE vehicles remain significantly shorter. This difference becomes particularly critical in long-distance travel or in applications requiring high vehicle availability, where charging requirements may affect route planning and increase total travel time.

In addition, the performance of ECs is inherently more variable than that of ICE vehicles. This is mainly due to the behavior of battery systems, whose available capacity and power capability depend on operating conditions such as temperature, state of charge, and ageing. Battery degradation over time further contributes to this variability, potentially reducing the effective driving range and performance of the vehicle. Consequently, ECs may exhibit less predictable behavior under non-ideal conditions, which can negatively impact user experience and perceived reliability.

3.5.2 Infrastructure and system-level challenges

The large-scale adoption of ECs poses significant challenges at both infrastructure and energy system levels, which may hinder their deployment beyond the intrinsic limitations of the vehicle itself (see Section 3.5.1).

ECs rely on the availability of an adequate charging infrastructure, whose deployment remains uneven in terms of both accessibility and geographic distribution. In contrast, ICE vehicles benefit from a mature and extensive refuelling network, developed over several decades. Although charging infrastructure has expanded rapidly in recent years, this growth has been markedly uneven, with significant disparities across regions. In particular, areas with lower population density often exhibit a limited availability of charging points. This structural limitation negatively affects user perception, as ECs are still often regarded as less reliable in terms of driving range and charging availability, especially for long-distance travel.

In addition to the limited availability of public charging infrastructure, access to private charging represents another important barrier. The installation of residential chargers typically requires access to dedicated parking spaces, such as private garages, which are not universally available. As a result, a significant portion of potential users (particularly those living in multi-unit dwellings) face technical, economic, and regulatory challenges when attempting to install charging infrastructure. This constraint increases reliance on public charging networks, which are generally less convenient and often more expensive.

From the perspective of the electrical system, the integration of ECs introduces substantial challenges, primarily due to the increase in electricity demand. More importantly, charging demand is typically not uniformly distributed over time, but tends to be concentrated during specific periods, particularly in the evening hours. This coincides with existing demand peaks in many power systems, leading to steeper load gradients and increased stress on distribution networks. The superposition of charging demand with conventional demand peaks can result in local congestion issues and may require significant grid reinforcements.

Furthermore, widespread nighttime charging has the potential to alter traditional demand profiles, affecting generation scheduling and system operation. These changes increase the need for system flexibility, demand-side management strategies, and more advanced coordination mechanisms between generation and consumption. Consequently, the integration of ECs requires not only infrastructure expansion but also a transformation in the way electricity systems are planned and operated.

3.5.3 Economic and market barriers

Although the economic competitiveness of ECs has improved during the 2020s (see Section 3.4.4), a significant gap with respect to ICE vehicles still remains, hindering their widespread adoption. This limitation is particularly evident in certain regions, reflecting the heterogeneous nature of the EC market.

As discussed in Section 3.4.4, ECs continue to exhibit higher upfront costs than their ICE counterparts, especially in compact and lower-cost vehicle segments. This barrier is particularly relevant for cost-sensitive users, who tend to prioritize the initial purchase price and may not fully account for lifecycle cost savings. As a result, the higher acquisition cost remains a key obstacle to broader adoption.

This disparity in upfront cost makes the economic competitiveness of ECs highly dependent on external conditions that influence their operating costs, such as

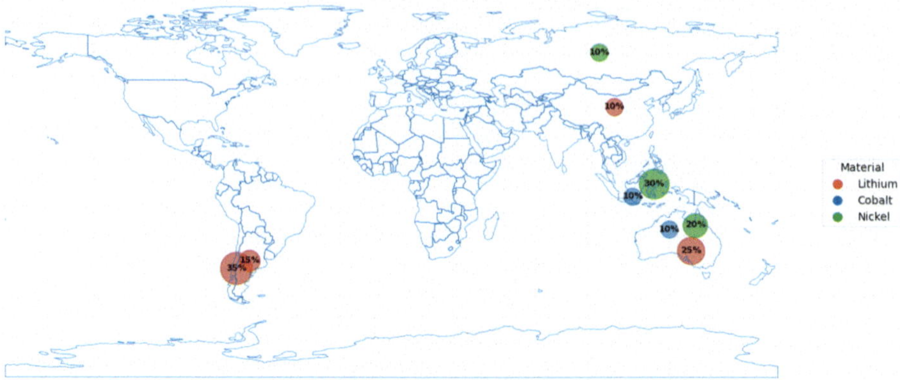

Figure 3.10 Global distribution of key battery materials (lithium, cobalt, and nickel) using proportional markers. The figure highlights some of the main producing regions to show the strong geographical concentration of critical resources, although additional countries also contribute to global reserves. This concentration introduces potential supply chain risks and geopolitical dependencies associated with large-scale EC deployment.

electricity prices, fuel costs, and maintenance expenses. In this context, TCO parity between ECs and ICE vehicles is only achieved under certain conditions, introducing economic uncertainty and making EC adoption dependent on boundary factors such as energy prices and policy incentives.

Another important aspect is the residual value of ECs. This factor is inherently more difficult to estimate than for ICE vehicles, not only due to the lack of long-term historical data, but also because of influencing factors such as battery degradation and rapid technological evolution. This uncertainty introduces additional complexity and perceived risk for both private users and market actors, including manufacturers and fleet operators.

Finally, significant regional differences further contribute to the heterogeneity of the market. Variations in energy prices, market maturity, and regulatory frameworks result in different levels of economic attractiveness for ECs across regions. Consequently, while ECs may already be competitive in some contexts, they remain less attractive in others, ultimately limiting the overall pace of market penetration.

3.5.4 Sustainability and supply chain constraints

In contrast to conventional vehicles, the manufacturing of ECs is strongly dependent on a set of critical materials, namely lithium, nickel, and cobalt, among others. These elements are fundamental for current battery chemistries and largely determine the performance, cost, and scalability of EC technologies. Their availability is not only constrained, but also geographically concentrated in a limited number of countries, as illustrated in Fig. 3.10. This concentration introduces structural vulnerabilities in the supply chain, increasing exposure to geopolitical risks, trade restrictions, and price volatility, all of which may hinder large-scale deployment.

In addition to this structural dependency, a significant carbon footprint is associated with the extraction and processing of these materials. Mining and metallurgical activities are inherently energy-intensive and require substantial water resources, often leading to notable environmental impacts at the local level. Consequently, the manufacturing phase of ECs is considerably more carbon-intensive than that of ICE vehicles. In quantitative terms, the carbon footprint of EC manufacturing can be approximately twice that of an ICE vehicle, with battery production accounting for around 40–60% of total manufacturing emissions [21]. As a result, although ECs offer substantial emission reductions during the use phase, these benefits must be evaluated against the higher initial carbon footprint, which is largely concentrated in the production stage.

Within this context, the development of efficient recycling processes for traction batteries becomes a key enabler for improving overall sustainability. Recycling allows the recovery of valuable materials, reducing both the reliance on primary extraction and the environmental burden associated with raw material processing. However, battery recycling remains a technically complex and economically challenging process, requiring advances in collection, sorting, and material recovery technologies. Therefore, its large-scale implementation is still evolving and constitutes a critical area for future development in the EC value chain.

References

[1] O. Egbue, S. Long, Barriers to widespread adoption of electric vehicles: an analysis of consumer attitudes and perceptions, Energy Policy 48 (2012) 717–729, https://doi.org/10.1016/j.enpol.2012.06.009, special section: Frontiers of Sustainability.

[2] A. Yang, C. Liu, D. Yang, C. Lu, Electric vehicle adoption in a mature market: a case study of Norway, Journal of Transport Geography 106 (2023) 103489, https://doi.org/10.1016/j.jtrangeo.2022.103489.

[3] I.E. Agency, Global ev outlook 2023, Tech. Rep., IEA, Paris, 2023, https://www.iea.org/reports/global-ev-outlook-2023. (Accessed 27 March 2026).

[4] BloombergNEF, Electric vehicle outlook 2023, Tech. Rep., BloombergNEF, 2023, https://about.bnef.com/electric-vehicle-outlook/. (Accessed 27 March 2026).

[5] E. Commission, The European green deal, Tech. Rep. COM(2019) 640 final, European Commission, 2019, https://eur-lex.europa.eu/legal-content/EN/TXT/?uri=CELEX:52019DC0640. (Accessed 27 March 2026).

[6] E.A.M. Association, Economic and market report: full-year 2020, https://www.acea.auto/files/Economic_and_Market_Report_full-year_2020.pdf, 2021. (Accessed 27 March 2026).

[7] I.E. Agency, Global ev outlook 2025, Tech. Rep., IEA, Paris, 2023, https://www.iea.org/reports/global-ev-outlook-2025. (Accessed 27 March 2026).

[8] I.E. Agency, What next for the global car industry, https://www.iea.org/reports/what-next-for-the-global-car-industry, 2025. (Accessed 27 March 2026).

[9] P. Zhu, Z. Wang, R. Singh, X. Tan, China's model of technology leapfrog: a case study of electric vehicle policies and the development of green technology, Renewable and Sustainable Energy Reviews 226 (2026) 116414, https://doi.org/10.1016/j.rser.2025.116414.

[10] argusmedia, Eu eases 2035 co2 targets for cars, allowing continued role for ice and hybrid vehicles, https://www.argusmedia.com/en/news-and-insights/latest-market-news/2766335-eu-eases-ice-phase-out-with-new-2035-co2-car-target, 2025. (Accessed 27 March 2026).
[11] Wikipedia contributors, Electric vehicle battery, https://en.wikipedia.org/wiki/Electric_vehicle_battery, 2026. (Accessed 6 April 2026).
[12] Our World in Data, Average battery cell price, https://ourworldindata.org/grapher/average-battery-cell-price, 2026. (Accessed 6 April 2026).
[13] K.T. Gillingham, A.A. van Benthem, S. Weber, M.A. Saafi, X. He, Has consumer acceptance of electric vehicles been increasing? Evidence from microdata on every new vehicle sale in the United States, AEA Papers and Proceedings 113 (2023) 329–335, https://doi.org/10.1257/pandp.20231065.
[14] International Energy Agency, Global ev data explorer, Tech. Rep., International Energy Agency, 2024, https://www.iea.org/data-and-statistics/data-tools/global-ev-data-explorer.
[15] W. Jia, Z. Ma, Y. Zhao, M. Luo, Understanding the role of electric vehicle perception in shaping ev adoption intention, Transport Policy 164 (2025) 255–264, https://doi.org/10.1016/j.tranpol.2025.02.003.
[16] G. Herbert, Affordable evs around €25,000 in Europe, Euronews (2025), https://www.euronews.com/next/2025/04/26/which-are-the-cheapest-evs-6-of-the-best-budget-friendly-options-for-around-25k.
[17] G. Guillaume, M. Mannes, Used ev sales jump in Europe as fuel prices rise, Reuters (2026), https://www.reuters.com/business/energy/used-ev-sales-jump-europe-iran-war-drives-up-petrol-prices-2026-03-26/.
[18] The International Council on Clean Transportation (ICCT), Realizing the benefits of electric vehicles: a 2035 perspective on costs and benefits, White paper, International Council on Clean Transportation, Washington, DC, https://theicct.org/wp-content/uploads/2022/10/ev-cost-benefits-2035-oct22.pdf, October 2022.
[19] J.T. Sánchez, J.A. del Río, A. Sánchez, Economic feasibility analysis for an electric public transportation system: two cases of study in medium sized cities in Mexico, PLoS ONE 17 (8) (2022) 1–23, https://doi.org/10.1371/journal.pone.0272363.
[20] U.S. Department of Energy, Fuel properties comparison, https://afdc.energy.gov/fuels/properties, 2023. (Accessed 7 April 2026).
[21] M. Hasan, Ev manufacturing vs ice manufacturing: carbon footprint comparison, https://evenergyhub.com/ev-manufacturing-vs-ice-manufacturing/, 2025. (Accessed 7 April 2026).

Unmanned aerial vehicles (UAVs)

4

4.1 Definition

An UAV, commonly referred to as a drone, is an aircraft that does not require a human crew on board and can therefore be operated either remotely by a human operator or through autonomous onboard control systems. This absence of onboard crew fundamentally differentiates UAVs from conventional manned aircraft, enabling alternative operational paradigms in which control, supervision, and decision-making processes may be partially or fully decoupled from the physical platform.

It is important to distinguish between the concept of a UAV and that of an Unmanned Aircraft System (UAS). In this context, the UAV should be understood as a constituent element within the broader UAS framework. Specifically, the term UAS refers to a more comprehensive system that encompasses not only the aerial vehicle itself, but also all the necessary supporting elements required for its operation. These elements include, among others, the ground control station from which the vehicle is monitored and commanded, the communication links that enable the exchange of data and control signals between the operator and the aircraft, and the auxiliary equipment that ensures proper system functionality.

This distinction is conceptually relevant, as it emphasizes that the operation of a UAV cannot be fully understood by considering the aerial platform in isolation. Instead, its functionality and effective use depend on the coordinated interaction of all components within the UAS, which together enable the execution of missions in a controlled and reliable manner.

4.2 Evolution and electrification of UAVs

4.2.1 Historical evolution of UAVs

Although conceptual precedents of unmanned aerial devices can be traced back to ancient times, these artifacts were highly rudimentary and far removed from the modern concept of UAVs. The contemporary UAV has its origins in the early twentieth century, when the first systems began to be developed primarily for military and experimental purposes. In these initial stages, UAVs exhibited very limited operational capabilities, facing significant technological barriers, particularly due to the immaturity of control and communication systems. As a result, their functionality was largely restricted to simple missions and basic operational tasks.

These early UAVs were predominantly based on ICEs. This was mainly due to the limited development of electrical energy storage systems at the time. The batteries available in the early twentieth century provided very low energy density, resulting in extremely limited endurance, and in many cases offered limited rechargeability or required long and impractical recharging processes. Under these conditions, battery-based solutions were not suitable for most practical UAV applications. Consequently,

Electric Mobility Systems. https://doi.org/10.1016/B978-0-44-349397-3.00012-4

UAV architectures during this period were largely based on conventional manned aircraft designs, incorporating the necessary modifications to enable remote or autonomous operation.

Throughout the twentieth century, UAVs experienced a gradual technological evolution, largely driven by military applications. However, most systems, particularly those designed for larger scale and longer range missions, continued to rely on ICE-based propulsion, while electric UAVs remained largely confined to experimental or demonstrative platforms.

4.2.2 Emergence of electric UAVs

The convergence of several technological advances in the early twenty-first century enabled the emergence of electric UAVs. In particular, progress in power electronics, the miniaturization of sensors, and improvements in computational capabilities have supported the development of compact, cost-effective, and high-efficiency flight control systems. These advances have acted as key enablers for enhancing the capabilities of small-scale UAV platforms, allowing improved performance, stability, and operational reliability.

Another significant development that has facilitated the emergence of electric UAVs is the advancement of lithium-based energy storage systems (see Chapter 1). The reduction in battery costs, together with technological improvements that have increased energy density, has enabled the deployment of electric UAVs with progressively higher levels of endurance. This evolution has been essential in expanding the range of feasible applications beyond purely experimental use.

In addition, the development of high-efficiency electric motors, combined with advanced electronic controllers, has contributed to improving the overall efficiency of electric UAV systems over recent decades. Collectively, these technological advances have supported the transition of electric UAVs toward increasingly practical and capable operational platforms.

4.2.3 ICE vs electric UAVs: technological differentiation

The development of UAVs throughout the twentieth century and the early twenty-first century has led to a clear differentiation between two main categories according to their propulsion system: ICE-based UAVs and electric UAVs. Each paradigm exhibits distinct characteristics, with specific advantages, limitations, and preferred application domains, as summarized in Table 4.1.

ICE-based UAVs are characterized by their high endurance, extended range, and the relative simplicity and speed of refueling. These features make them particularly suitable for long-duration missions requiring high persistence and the ability to operate over large distances. Consequently, they are commonly employed in applications where sustained operation is a critical requirement. However, these systems also involve greater mechanical complexity, as well as higher levels of noise and emissions, which may limit their suitability in certain operational environments.

Table 4.1 ICE vs Electric UAVs: Conceptual Comparison.

Category	ICE-based UAVs	Electric UAVs
Advantages	• High endurance and range • Superior payload capability • Suitability for persistent operations	• Reduced system complexity • Low acoustic signature • High efficiency at small scale
Disadvantages	• Increased mechanical complexity • Higher maintenance requirements • Noise and emissions impact	• Limited endurance • Restricted payload capacity • Dependence on battery performance
Applications	• Long-duration missions • Wide-area operations • Payload-intensive tasks	• Short- to medium-duration missions • Precision and proximity operations • Noise-sensitive environments

In contrast, electric UAVs are characterized by simpler system architectures, high efficiency at small scale, and significantly reduced noise and emissions. These properties make them especially suitable for short- to medium-duration missions that require high precision or operation in noise-sensitive environments, particularly in the presence of people. Nevertheless, the relatively lower energy density of lithium-ion batteries compared to fossil fuels (see Chapter 3) imposes limitations in terms of endurance and range, which constrains their applicability in long-duration missions.

4.2.4 Electric UAVs within the EV ecosystem

In the context of this book, the analysis focuses on electric UAVs, which currently account for the majority of civil applications. As discussed in Table 4.1, these platforms are particularly well suited for short- to medium-duration missions where high efficiency and low noise and emissions are required.

Electric UAVs have experienced significant growth and development throughout the twenty-first century. It is estimated that tens of millions of UAVs are currently in operation worldwide [1], representing an increase of several orders of magnitude compared to the figures observed at the beginning of the century. This rapid expansion reflects both technological progress and increasing market adoption.

From a technological perspective, advances in key components have played a central role in this evolution. Improvements in efficiency and reductions in cost have enabled the development of more capable and accessible commercial platforms, particularly in terms of endurance and operational performance. While at the beginning of the century UAVs were largely restricted to specialized niche

Figure 4.1 DJI Phantom 4 Pro+ in flight. (Image courtesy of Wikimedia Commons.)

applications, the 2010s marked a turning point with the emergence of consumer-oriented platforms. Representative examples such as the DJI Phantom 4, released in 2016 (see Fig. 4.1), with a retail price in the range of 1000–1500 $, contributed significantly to the popularization of UAVs and their penetration across multiple market segments.

This widespread adoption has enabled UAVs to expand beyond their traditional association with defense applications. Electric UAVs are now widely used in areas such as infrastructure inspection, environmental monitoring, precision agriculture, mapping, emergency response, and last-mile logistics.

From a technological standpoint, electric UAVs share important similarities with other EVs, such as BECs, particularly in terms of subsystem integration, including electric propulsion, energy storage, and energy management. However, the aerial environment imposes stringent constraints in terms of weight, endurance, and system redundancy, which strongly influence their design and operation.

4.3 UAV classification and typologies

4.3.1 Classification framework and criteria

Modern UAVs span a wide range of applications, configurations, and sizes, making it difficult to establish rigid and universally accepted classification criteria. Instead, UAVs are typically categorized according to the operational context and the level of analysis required, following a multidimensional approach based on complementary criteria.

In this book, three primary classification axes are adopted. First, UAVs are classified according to their airframe configuration, distinguishing between fixed-wing, rotary-wing, and hybrid platforms. Second, a classification based on size or

Figure 4.2 Example of a fixed-wing UAV. (Image courtesy of Wikimedia Commons.)

weight is introduced, using the maximum take-off weight (MTOW) as a unified metric to consistently categorize UAVs under this criterion. Finally, a classification based on endurance and range is considered, as these are critical parameters that largely determine the performance and primary application domains of the aircraft.

4.3.2 Classification by airframe configuration

4.3.2.1 Fixed-wing UAVs

In fixed-wing UAVs, the wing remains rigidly attached to the fuselage (see Fig. 4.2), generating lift through the relative airflow over the airfoil. As such, fixed-wing UAVs follow the same aerodynamic principles as conventional manned aviation. Lift is achieved as a function of forward velocity, which implies that these platforms require continuous forward motion, as well as dedicated take-off and landing systems such as runways or launch and recovery mechanisms.

This configuration enables high levels of aerodynamic efficiency, allowing extended endurance even with limited onboard energy sources. As a result, fixed-wing UAVs are particularly well suited for long-duration missions such as surveillance and mapping, where efficient coverage of large areas is essential. However, their inability to hover makes them less suitable for operations in confined or complex environments.

4.3.2.2 Rotary-wing UAVs

Unlike fixed-wing UAVs, rotary-wing UAVs incorporate one or more rotors, generating both thrust and lift simultaneously through rotor motion, thus eliminating the need for a fixed wing. This configuration enables a high degree of control over

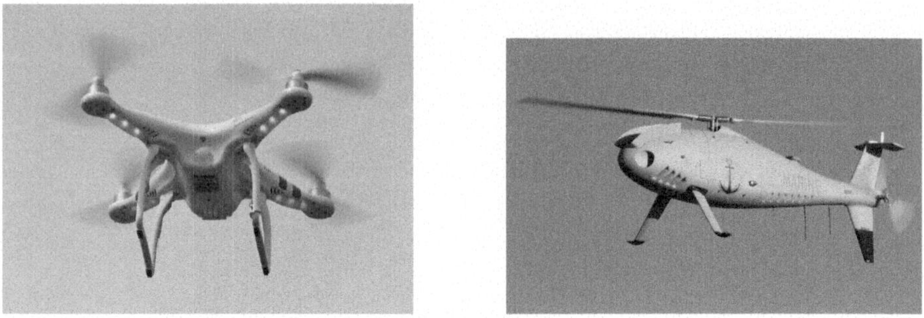

Figure 4.3 Example of (a) multirotor and (b) single-rotor rotary-wing UAVs. (Images courtesy of Wikimedia Commons.)

the aircraft's motion, particularly at low speeds or during stationary flight. Within this category, two main configurations can be distinguished:

- **Multirotor (Fig. 4.3a)**: these UAVs employ multiple propellers, leading to configurations with four (quadcopters), six (hexacopters), or eight (octocopters) rotors. They are characterized by high maneuverability, mechanical simplicity, and precise control, and are widely used in civil and commercial applications such as inspection, photography, and short-range monitoring.
- **Single-rotor (Fig. 4.3b)**: these UAVs feature a single main rotor that provides lift, typically complemented by a secondary rotor for yaw control. They offer higher aerodynamic efficiency and payload capacity than multirotor systems, although at the cost of increased mechanical complexity and maintenance requirements.

The main advantages of rotary-wing UAVs over fixed-wing platforms are their vertical take-off and landing (VTOL) capability and their ability to hover. These features make them particularly suitable for operations in confined or constrained environments. However, this operational flexibility comes at the expense of lower energy efficiency, as continuous power is required to sustain lift, limiting endurance and range compared to fixed-wing UAVs.

4.3.2.3 Hybrid configuration

Hybrid UAVs (also referred to as VTOL UAVs) are aircraft designed to combine the advantages of both fixed-wing and rotary-wing configurations. In this sense, their operation relies on the ability to alternate between two distinct flight modes. On the one hand, they exploit vertical flight capabilities for take-off and landing, eliminating the need for complex or dedicated infrastructure. Once this phase is completed, the aircraft transitions to a horizontal cruise flight mode, typical of fixed-wing UAVs, in order to benefit from their higher aerodynamic efficiency during forward flight.

Several architectures can be identified within hybrid UAVs. On the one hand, some platforms achieve the transition by reorienting key components, such as in tilt-rotor and tilt-wing configurations, where rotors or wings change their orientation

Table 4.2 Typical MTOW ranges for UAV classification by size and weight. Values are indicative and may vary depending on regulatory frameworks and application domains (e.g., see [2,3]).

Category	**MTOW Range**
Nano UAVs	< 0.25 kg
Micro UAVs	0.25–2 kg
Small UAVs	2–25 kg
Tactical UAVs	25–150/200 kg
MALE UAVs	$\sim$ 1000–5000 kg
HALE UAVs	$>$ 5000 kg

during flight, respectively. On the other hand, lift+cruise configurations incorporate independent propulsion systems for vertical take-off and landing and for horizontal flight.

By combining the flight capabilities of fixed-wing and rotary-wing UAVs, hybrid configurations provide both improved aerodynamic efficiency and operational flexibility. However, these advantages are achieved at the expense of increased mechanical and control system complexity, which may result in higher costs, greater integration challenges, and potentially reduced system reliability.

4.3.3 Classification by size and weight

One of the most widely adopted classification approaches for UAVs is based on their weight. This criterion is particularly relevant from both a design and operational perspective, as the MTOW largely determines the structural complexity of the aircraft, its payload capacity, and the associated operational requirements.

MTOW is typically used as a unified metric for classification, as it provides an objective measure of system capabilities. Based on this parameter, UAVs can be grouped into several categories, whose typical MTOW ranges are summarized in Table 4.2:

- **Nano / Micro UAVs**: these platforms exhibit the lowest MTOW values and are generally limited to short-range operations in indoor or urban environments, with very constrained payload and endurance capabilities.
- **Small UAVs**: lightweight systems that offer a balance between portability and performance, and are widely used in civil and commercial applications such as inspection, mapping, and monitoring.
- **Tactical UAVs**: larger platforms typically employed in professional or military contexts, where increased endurance and payload capacity are required.

Table 4.3 Indicative classification of UAVs based on endurance and range. Values are approximate and not standardized, and may vary depending on the platform and mission profile (e.g., see [2]).

Category	**Endurance**	**Range**
Short	< 2 h	< 20 km
Medium	2–10 h	20–200 km
Long	10–24 h	> 200 km
Ultra-long	> 24 h	> 500 km

- **MALE / HALE UAVs**: large-scale systems designed for medium- and high-altitude operations, characterized by long endurance and the ability to carry complex and often mission-specific payloads.

Beyond classification, the size and weight of a UAV directly constrain its payload, understood as the mass available for mission-specific equipment within the MTOW limit. Furthermore, MTOW influences endurance through a trade-off between increased onboard energy capacity and higher power consumption. Finally, larger UAVs are generally subject to more stringent regulatory requirements, reflecting their higher operational risk and system complexity.

4.3.4 Classification by endurance and range

Another key aspect for classifying UAVs is their endurance and range. Although these two concepts are closely related, it is important to highlight their fundamental differences. Specifically, endurance is defined as the maximum time a UAV can remain in flight, whereas range refers to the maximum distance the aircraft can travel or, more generally, its operational radius. While these differences may appear subtle, they have important practical implications. In particular, although both endurance and range are strongly influenced by the onboard energy storage system, other factors such as cruise speed, aerodynamic efficiency, and mission profile can affect them in different ways, leading to distinct operational trade-offs.

Although there is no universally accepted or standardized classification of UAVs based on endurance and range, they are often informally categorized as short-, medium-, and long-endurance/range systems, as summarized in Table 4.3. These categories should be understood as indicative rather than strictly defined, as actual performance depends on both the platform design and the specific mission conditions.

The relevance of these metrics depends strongly on the intended mission or application. For instance, surveillance missions typically prioritize long endurance over extended range, as persistent observation over a given area is required. In contrast, delivery missions tend to prioritize range and payload capacity, as they involve transporting goods over a defined distance. Other applications, such as inspection

Table 4.4 Comparative summary of operational capabilities across UAV typologies.

Aspect	Fixed-wing	Rotary-wing	Hybrid
Take-off / landing	Runway / launch system	VTOL	VTOL
Flight envelope	Efficient cruise	Hover / low speed	Hover + cruise
Control complexity	Low–medium	Medium	High
Wind robustness	High (cruise)	Low	Medium
Mission profile	Long-range	Local / inspection	Versatile
Infrastructure	Medium–high	Low	Low–medium

tasks, usually require shorter endurance but demand high maneuverability and the ability to operate in constrained or complex environments.

This diversity of operational requirements highlights the need to carefully match UAV characteristics to the mission profile, for which the classifications and typologies discussed in Section 4.3 provide a useful framework.

4.3.5 Operational differences across UAV typologies

The classification of UAVs according to different criteria provides a concise and effective way to identify their operational characteristics. Among the various classification approaches, the one based on airframe configuration enables a clearer differentiation in terms of operational capabilities, which are summarized in Table 4.4.

One of the most relevant operational aspects relates to take-off and landing requirements. Rotary-wing UAVs inherently provide VTOL capability, allowing operations in confined or infrastructure-limited environments. In contrast, fixed-wing UAVs typically require launch and recovery systems, such as runways, catapults, or equivalent solutions, which restrict their deployment in constrained areas.

The flight envelope, understood here as the range of flight modes and operating conditions in which the UAV can function effectively (e.g., speed, altitude, and general operating conditions), also varies significantly depending on the aircraft configuration. Fixed-wing UAVs operate efficiently in steady cruise flight at relatively high speeds, making them well-suited for long-range missions. Conversely, multirotor UAVs perform optimally in hover and low-speed conditions, offering high maneuverability and precise positioning capabilities. Hybrid platforms aim to bridge these two regimes by combining hover and cruise performance, albeit at the expense of increased structural and system complexity.

This increased structural complexity implies more demanding flight control systems, particularly in hybrid configurations that involve multiple propulsion subsystems and transition phases between flight modes. Another critical aspect is the robustness of the platform to external disturbances. Fixed-wing UAVs generally exhibit higher stability during sustained flight, whereas rotary-wing UAVs tend to be more sensitive to wind gusts and turbulent conditions.

Finally, mission planning and infrastructure requirements must also be considered. Fixed-wing UAVs typically require larger operational areas for take-off and landing, increasing mission planning complexity and logistical constraints. In contrast, rotary-wing and hybrid UAVs enable more flexible deployment with reduced infrastructure requirements, simplifying mission execution across a wide range of environments.

4.3.6 Advantages and limitations of each typology

As a concluding remark, a clear comparison can be established by synthesizing the main advantages and limitations of each UAV configuration. Fixed-wing UAVs are particularly well suited for missions requiring high endurance and long-range operation. Their superior aerodynamic efficiency reduces energy consumption, making them especially appropriate for long-duration missions such as mapping or large-scale surveillance. However, their reliance on launch and recovery systems limits their use in confined or complex environments and increases logistical requirements.

In contrast, rotary-wing UAVs overcome many of these limitations by offering high maneuverability, particularly at low speeds, as well as inherent VTOL capability without the need for dedicated infrastructure. These features make them ideal for precision tasks and operations in constrained environments. Nevertheless, this operational flexibility comes at the cost of lower aerodynamic efficiency, which significantly limits their endurance and range compared to fixed-wing platforms.

Finally, hybrid configurations aim to combine the advantages of both fixed-wing and rotary-wing UAVs, enabling both hover and efficient cruise flight. This compromise, however, results in increased mechanical and control complexity, which may lead to higher system costs, greater integration challenges, and potentially reduced reliability.

4.4 Core components and system architecture

4.4.1 System-level architecture overview

From a systems engineering perspective, an UAV can be understood as a set of interconnected functional blocks, each fulfilling a specific role within the system while remaining strongly coupled to the others. The overall objective of this integrated system is to ensure controlled flight and the successful execution of the assigned mission.

As shown in Fig. 4.4, an UAV can be decomposed into five main subsystems:

- **Propulsion system:** Responsible for generating the forces required for lift and motion, enabling the vehicle to maintain flight and perform maneuvers.
- **Energy storage and management system:** Comprises the onboard energy sources (typically batteries) and is responsible for supplying, regulating, and distributing electrical power across the system.
- **Flight control system (FCU):** Ensures stable flight conditions and trajectory tracking by regulating variables such as attitude, altitude, and velocity.

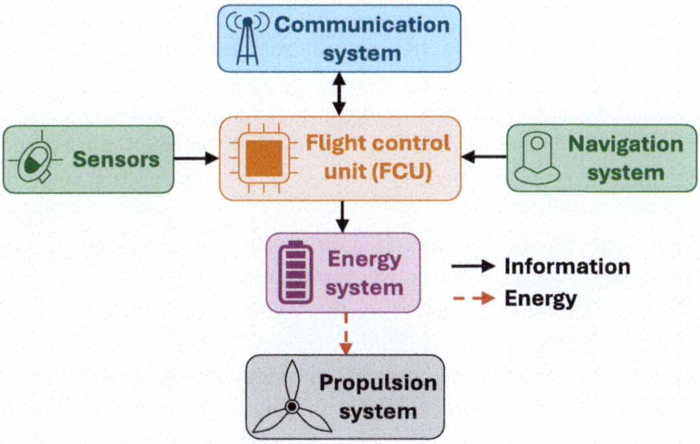

Figure 4.4 Simplified system-level architecture of an UAV.

- **Sensing and navigation subsystem:** Provides information about the UAV state and its environment, including onboard sensor measurements and positioning data required for state estimation and control.
- **Communication system:** Enables interaction with external agents, such as ground control stations or other vehicles, allowing the exchange of commands, telemetry, and mission data.

The interactions between these subsystems, including both information and energy flows, are illustrated in Fig. 4.4. Within this architecture, the FCU acts as the central element, integrating inputs from sensors, navigation systems, and external communications. Based on this information, it generates the control signals required to command the propulsion system and maintain the desired flight conditions.

In parallel, the energy flow supports the operation of all subsystems. Electrical power is delivered from the energy storage system and distributed to the propulsion system and onboard electronics. Unlike information flows, energy flow is not inherently closed-loop; however, it is monitored and managed through feedback mechanisms within the control and sensing systems to ensure safe and efficient operation under varying conditions.

It is worth noting that this functional architecture presents clear similarities with the one described for BECs in Chapter 2. However, in UAVs, constraints related to mass, volume, and energy availability are significantly more stringent, playing a critical role in system design, integration, and overall performance.

4.4.2 Propulsion system

The propulsion system constitutes one of the primary subsystems within the UAV architecture. Its main function is to convert the energy supplied by the onboard energy storage system into mechanical power, which in turn generates lift in rotary-wing UAVs or provides longitudinal thrust in fixed-wing platforms.

In most modern UAVs, the propulsion system relies on electric drives based on brushless DC (BLDC) motors [4]. In contrast to brushed DC motors, where commutation is achieved mechanically through physical contact between brushes and a commutator, leading to friction losses, wear, and reduced efficiency, BLDC motors perform commutation electronically. This eliminates the need for brushes, thereby reducing mechanical losses, improving efficiency, and increasing system reliability and lifetime.

BLDC motors are inherently coupled with electronic subsystems responsible for both commutation and operational control, commonly referred to as Electronic Speed Controllers (ESCs). These devices act as the interface between the energy source (typically batteries) and the motor. Their primary function is to convert the DC electrical power into a three-phase modulated signal using pulse-width modulation (PWM) techniques. This signal provides the required electronic commutation while enabling precise control of the motor rotational speed. In addition, modern ESCs may incorporate advanced control algorithms, such as sensorless rotor position estimation based on back electromotive force, reducing system complexity and weight. The ESC also plays a key role in overall system efficiency, as switching losses and thermal management within the power electronics directly impact performance and reliability.

The configuration of the propulsion system varies depending on the UAV type. Multirotor platforms typically employ multiple motor–propeller units to achieve direct control of attitude through differential thrust. In such configurations, ESC parameters such as switching frequency, control resolution, and response latency are critical for ensuring flight stability and control performance. In contrast, fixed-wing UAVs usually incorporate one or, at most, two propulsion units, generally optimized for cruise efficiency rather than rapid dynamic response.

4.4.3 Energy storage and power management

All onboard systems, such as sensors and the propulsion system, require electrical energy for their operation. Within the UAV architecture shown in Fig. 4.4, the subsystem responsible for storing energy and distributing it appropriately among the different onboard systems is referred to as the energy storage and power management subsystem. This subsystem therefore acts as the central energy node within the vehicle architecture. Its design has a significant impact on UAV endurance, overall performance, and operational feasibility.

Within this subsystem, energy storage in electric UAVs is typically achieved through onboard battery systems, most commonly based on lithium technologies. From a system design perspective, the specific battery chemistry is less critical than the performance metrics it enables. In this context, two key parameters define the behavior of the storage subsystem: energy density, which determines the total amount of energy available onboard, and power density, which characterizes the capability to deliver power over short time intervals. While energy density is directly related to vehicle endurance, power density is more closely associated with dynamic performance and transient response.

The energy storage system requires continuous monitoring and control. Similarly to ECs (see Chapter 2), the BMS acts as the "brain" of the energy storage subsystem in UAVs. It monitors key variables such as cell and pack voltage, current, and temperature, ensuring operation within safe limits. The BMS also implements protection mechanisms against overcharge, overdischarge, overcurrent, and adverse thermal conditions. In addition, it estimates the SoC and, in more advanced implementations, the State of Health (SoH), which reflects the relative available capacity of the battery over its lifetime (see Chapter 7). It should be noted that, particularly in small UAVs, BMS functionalities may be partially simplified or distributed due to constraints in weight, cost, and system integration.

Another primary function of the energy storage and power management subsystem is to regulate and distribute electrical power among the different subsystems defined in Fig. 4.4. In typical UAV architectures, a main DC power bus distributes energy throughout the vehicle, supplying different loads through dedicated DC-DC converters that adapt voltage and current levels to the specific requirements of each subsystem (e.g., propulsion, avionics, and payload). From an engineering standpoint, aspects such as voltage regulation, conversion efficiency, electromagnetic compatibility, and stability under dynamic load conditions are critical design considerations. Furthermore, peak power demand (particularly from the propulsion system) must be properly managed to avoid voltage drops and performance degradation. In more advanced UAVs, energy management strategies can be implemented to optimize power allocation, prioritizing critical loads and improving overall system efficiency.

4.4.4 Fligth control unit (FCU)

The FCU constitutes the functional core of the UAV. Its primary mission is to generate the control signals required to ensure stable flight and compliance with the mission objectives. To achieve this, the FCU must process a large amount of information originating both from the vehicle itself and from its surrounding environment. This information is provided by several onboard sensing systems such as Inertial Measurement Units (IMUs) or Global Navigation Satellite Systems (GNSSs) (see Section 4.4.5).

UAV control is typically implemented using a hierarchical structure with two main control loops, as shown in Fig. 4.5. The outer loop is responsible for position control, generating the required position or trajectory tracking behavior according to mission-level references. Based on the position error, this loop computes the desired attitude and thrust commands, typically using proportional–integral–derivative (PID) controllers or equivalent strategies. These references are then passed to the inner loop, which is responsible for attitude (roll, pitch, and yaw) control. The inner loop generates actuator commands by minimizing the error between the desired and measured attitude. This cascaded control architecture enables partial decoupling of the system dynamics and simplifies controller design. Typically, the inner attitude loop operates at a significantly higher bandwidth than the outer position loop, allowing for fast stabilization of the vehicle dynamics, while the slower outer loop focuses on trajectory tracking and disturbance rejection at a larger time scale.

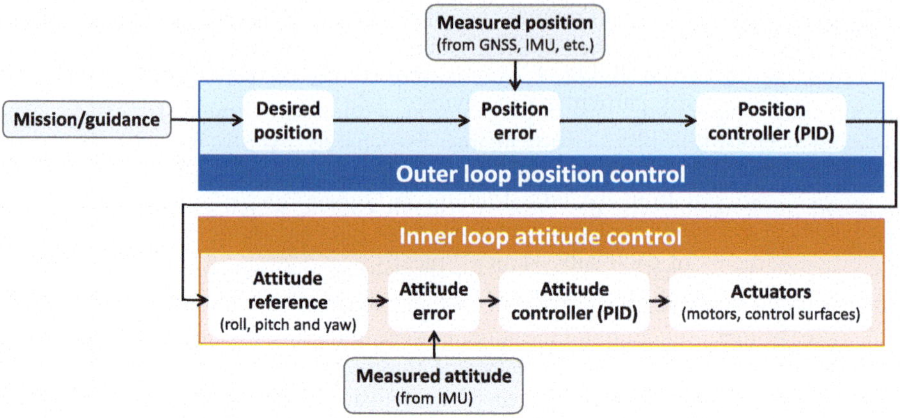

Figure 4.5 Hierarchical flight control architecture in UAVs based on cascaded control loops, where an outer position loop generates attitude and thrust references, and an inner attitude loop ensures fast stabilization and tracking.

The control architecture shown in Fig. 4.5 is generally applicable to different UAV configurations. However, there are important differences in control strategies depending on the vehicle type. In multirotor UAVs, control is achieved through differential thrust generated by multiple motor–propeller units, allowing for direct and highly responsive attitude control, albeit with higher sensitivity to disturbances and increased energy consumption. In contrast, fixed-wing UAVs rely on aerodynamic control surfaces (ailerons, elevator, and rudder), which introduce more coupled dynamics and require forward velocity to be effective, but are generally more efficient, particularly during cruise flight.

4.4.5 Sensing and navigation systems

The FCU relies on information obtained both from the environment and from the vehicle itself. This information is provided by the sensing and navigation subsystem, which is responsible for perceiving the UAV state and its surrounding environment, as well as transmitting this information to the FCU. Based on these inputs, the FCU generates the control commands required to successfully execute the assigned mission (see Section 4.4.4).

To perceive its state and environment, the sensing and navigation subsystem incorporates a variety of sensors, among which the most relevant are:

- **IMU:** integrate accelerometers and gyroscopes to measure linear accelerations and angular rates, providing essential information for estimating the UAV attitude (roll, pitch, and yaw).
- **GNSS:** provide position and velocity information based on satellite signals.

Although these sensors constitute the basis for state estimation, many UAVs incorporate additional sensing systems to enhance perception capabilities. Among these, environmental sensors such as barometers are commonly used to estimate altitude from atmospheric pressure measurements. Other sensing systems, such as cameras or LiDAR, provide detailed information about the surrounding environment, enabling tasks such as obstacle detection and environment mapping.

One of the main challenges of the sensing and navigation subsystem is the management and integration of information from multiple sensors with different underlying technologies. This heterogeneity results in measurements with varying levels of accuracy, noise characteristics, and update rates. In order for this information to be effectively processed by the FCU, it must be properly combined and made consistent. This is typically achieved through sensor fusion techniques, for which a detailed treatment can be found in [5].

It is important to highlight that the sensing and navigation subsystem plays a fundamental role in determining the level of autonomy of the UAV. Remotely operated UAVs may require less sophisticated sensing systems or a reduced number of sensing modalities, as part of the decision-making process is handled by a human operator. In contrast, higher levels of autonomy demand more advanced and reliable sensing capabilities, increasing both the complexity and the importance of this subsystem.

4.4.6 Communication systems

Communication systems are responsible for enabling information exchange between the UAV and its external environment, acting as the interface between the vehicle and external systems. Typically, these systems support communication between the UAV and the control station, understood as a dedicated platform from which the UAV is monitored and, when required, controlled.

Within communication systems, it is useful to distinguish different communication channels:

- **Control link:** enables the transmission of commands from the control station to the UAV, being essential for non-autonomous operation and for handling emergency situations.
- **Telemetry links:** enable the transmission of data from the UAV to the control station, including variables such as velocity, position, and battery system status, among others.
- **Data links:** provide communication between the payload and the control station, enabling the transmission of images, video streams, or other sensor-derived data.

The performance and capabilities of communication systems are constrained by several key factors:

- **Latency:** defined as the time delay between the transmission of a signal or data and its reception or effect on the system. High latency may degrade real-time operability and control performance.
- **Range:** determines the maximum distance between the UAV and the control station beyond which reliable communication cannot be maintained.
- **Susceptibility to interference:** communication links may be affected by electromagnetic interference, particularly in environments with high signal density, potentially degrading reliability and data integrity.

4.5 Energy and performance considerations

4.5.1 Energy constraints and their impact on UAV operation

As in other EVs, the performance of an electric UAV is strongly linked to its energy storage system. However, unlike ground vehicles such as ECs, the energy storage system in a UAV not only determines its endurance, but also directly constrains the payload that can be carried within the MTOW limit, ultimately affecting the feasibility of the assigned mission.

Focusing on mass, beyond its influence under conditions similar to those of terrestrial vehicles, in UAVs mass plays a direct role in lift generation and, consequently, in the power required to sustain flight. This results in a strong coupling between mass and energy demand, such that any increase in UAV mass (whether due to a larger battery system or additional payload) leads to higher power requirements and increased energy consumption. This coupling is particularly critical in hovering flight, where the vehicle must continuously generate sufficient thrust to counteract its weight.

In UAVs, it is especially relevant to distinguish between energy capacity and power delivery capability. The former directly determines endurance, understood as the maximum achievable flight time. In contrast, power delivery capability defines the ability of the UAV to respond to instantaneous demands, which is particularly important in rotary-wing UAVs operating in hover or during aggressive maneuvers. In fixed-wing UAVs, although power requirements are lower during steady cruise, sufficient power availability remains essential during phases such as takeoff and climb.

As a result, the design of electric UAVs is typically approached from an integrated perspective, where decisions related to energy storage, mass distribution, and performance are tightly coupled from the early design stages. This system-level viewpoint is essential to ensure that mission requirements can be met within the inherent energy constraints of the vehicle.

4.5.2 Design trade-offs in UAV systems

UAV design is strongly conditioned by the need to balance a number of key variables through the adoption of design trade-offs. These trade-offs are primarily driven by the intrinsic energy constraints of the vehicle (see Section 4.5.1). The interaction between these variables defines the design space of the UAV and significantly influences both its achievable performance and its range of applications. As a result, UAV design cannot be approached by optimizing individual subsystems in isolation, but rather requires a system-level perspective.

The most critical trade-off in an electric UAV is the one relating total vehicle mass and the available onboard energy. Increasing the available energy typically requires a higher-capacity energy storage system, which in turn increases the overall mass of the vehicle. As a consequence, the power required to generate sufficient lift also increases, leading to higher energy consumption during operation. This feedback effect limits the net gain in endurance and necessitates the identification of an optimal balance between energy capacity and total mass within MTOW constraints.

Similarly, there exists a strong trade-off between payload and maximum flight time, which becomes particularly critical in applications such as logistics, where payload represents the primary value of the UAV. Increasing payload reduces the mass available for the energy storage system in order to comply with MTOW limitations. As a result, higher payload capacity typically leads to reduced onboard energy and, consequently, shorter endurance, requiring a compromise between mission duration and transported load. This trade-off directly impacts mission planning and operational flexibility.

Another important trade-off arises between power capability and efficiency. High power requirements imply larger energy flows within the system, which tend to increase electrical and conversion losses, thereby reducing overall efficiency. However, sufficient power delivery capability is essential to ensure adequate dynamic response, particularly in demanding flight conditions such as takeoff, climb, or disturbance rejection.

Considering these interdependencies, UAV design is often formulated as a constrained optimization problem, where performance metrics such as endurance or energy consumption are optimized subject to constraints including MTOW, mission requirements, and power delivery capabilities. The presence of competing objectives naturally leads to multi-objective optimization approaches, in which a set of trade-off solutions is explored rather than a single optimal design [6].

4.5.3 Energy efficiency across UAV configurations

The aerodynamic configuration and propulsion principle strongly determine the energy efficiency of UAVs. This close coupling explains the coexistence of multiple UAV configurations, as discussed in Section 4.3.2. As previously introduced, each configuration exhibits distinct characteristics in terms of energy consumption and lift generation principles.

Fixed-wing UAVs generate lift through forward motion, relying on the relative airflow over their wings. This aerodynamic lift mechanism significantly reduces the required power compared to configurations that must actively generate lift. As a result, fixed-wing UAVs are generally characterized by high energy efficiency, particularly during steady cruise flight. However, their operation is constrained by the need to maintain a minimum airspeed to sustain lift, which limits their ability to hover and reduces their operational flexibility in confined environments.

In contrast, rotary-wing UAVs generate lift directly through rotor thrust. This requires continuous power delivery even in hover, leading to inherently higher energy consumption. Consequently, rotary-wing configurations typically exhibit lower overall energy efficiency than fixed-wing UAVs. Nevertheless, they provide superior maneuverability, including hover capability and VTOL operation, making them well-suited for missions requiring precise positioning and high agility. This enhanced

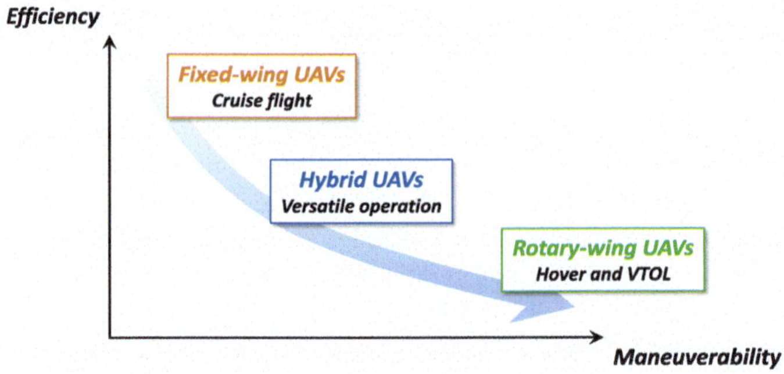

Figure 4.6 Conceptual relationship between energy efficiency and maneuverability across different UAV configurations. Fixed-wing UAVs achieve higher energy efficiency due to aerodynamic lift in cruise conditions while rotary-wing UAVs provide high maneuverability. Hybrid configurations lie between both extremes.

maneuverability comes at the cost of increased power demand, as discussed in Section 4.5.2.

Hybrid UAV configurations aim to combine the advantages of both fixed-wing and rotary-wing systems. These platforms enable dual-mode operation, typically involving vertical take-off and landing followed by efficient forward flight. While this versatility expands their operational envelope, it introduces additional mechanical and control complexity. Moreover, the integration of multiple propulsion modes often results in increased weight and suboptimal efficiency in both hover and cruise when compared to specialized configurations.

A conceptual relationship can be established between energy efficiency and maneuverability, as illustrated in Fig. 4.6. In general, configurations that prioritize maneuverability tend to exhibit lower energy efficiency, while those optimized for efficiency sacrifice operational flexibility. Although this relationship is not strictly linear, it provides a useful framework for understanding the coexistence of different UAV architectures and their alignment with specific mission requirements.

4.5.4 Comparison with other EVs

Although UAVs share a common technological foundation with other EVs, their operational regime is significantly more demanding than that of ground vehicles. This is primarily due to the requirement to continuously generate lift. As a result, the energy consumption of UAVs is not only devoted to propulsion, as in other EVs, but also to directly counteracting the weight of the vehicle.

Consequently, UAVs operate under more severe energy constraints and with a reduced operational margin compared to other EVs. Their performance is highly sensitive to variations in operating conditions, such as changes in payload, environmental factors, or flight profile, which can lead to substantial reductions in mission endurance. In this context, system mass becomes a critical design parameter. Trade-offs between available energy and weight must be carefully balanced (see

Section 4.5.2), as any increase in mass directly translates into higher power and energy requirements.

By contrast, ground EVs are generally more tolerant to increases in mass and exhibit less restrictive operating profiles. In these vehicles, energy is primarily used to overcome resistive forces such as rolling resistance and aerodynamic drag, without the need to sustain lift. This fundamental difference results in less stringent energy constraints and greater flexibility in system design.

Despite these differences, UAVs and other EVs share key technological similarities that shape their system architecture. Both rely on electrochemical energy storage systems and involve the use of critical materials, as discussed in Chapter 3. Moreover, their operation is based on the interaction of multiple subsystems, requiring coordinated power distribution strategies. These commonalities enable the transfer of design principles and optimization approaches from other EV domains, although they must be adapted to meet the stricter constraints imposed by aerial operation.

4.6 Regulatory framework and operational constraints

4.6.1 Regulatory objectives and framework

Due to their integration into the airspace and their direct interaction with third parties, the development and deployment of UAVs are subject to a more stringent regulatory framework than other EVs. UAVs operate within a highly structured three-dimensional environment, in which safety is addressed from a system-level perspective, depending not only on the intrinsic reliability of the vehicle, but also on its ability to safely coexist and interact with other airspace users.

As a result, the regulatory framework governing UAVs imposes a set of technical and operational constraints that must be considered from the early design stages through to actual flight operations. In this context, regulation should not be interpreted merely as a legal boundary condition, but rather as a set of engineering-driven requirements with direct implications on system architecture and performance. From a general standpoint, the objectives of UAV regulatory frameworks can be structured around three main axes:

- Operational safety.
- Protection of people and property on the ground.
- Safe and efficient integration into shared airspace.

These regulatory objectives translate into specific system-level requirements and constraints, affecting aspects such as the permissible flight envelope, stringent reliability targets, precise navigation and control capabilities, and robust communication links to ensure correct operation under nominal and degraded conditions. Furthermore, similarly to other aerial platforms, UAVs are typically required to incorporate redundancy at both system and subsystem levels in order to guarantee predictable and safe behavior, even in the presence of faults or unexpected events.

Therefore, in the context of UAVs, regulation acts as a key design driver, directly constraining the feasible solution space, which is not solely determined by conventional technical considerations but also by compliance with operational and safety requirements.

4.6.2 Civil UAV regulations and operational categories

In civil environments, the regulatory framework governing UAV operations is not primarily structured around the type of aircraft, but rather around the level of risk that a given operation poses to the environment and to third parties. This paradigm establishes a direct relationship between the operational context, its intrinsic risk, and the technical requirements imposed on the system.

Under this approach, certification is not limited to the vehicle itself, but extends to the entire mission. This includes factors closely associated with the operation, such as proximity to people, environmental density (urban versus rural), airspace complexity, and the expected level of autonomy. Based on these criteria, operations are typically classified into distinct categories according to their risk level (e.g., low, medium, or high), each of which determines a corresponding set of regulatory requirements and operational constraints.

As a consequence, a number of limitations are imposed that directly condition the use of UAVs in civil applications. These constraints typically affect key mission parameters, including maximum allowable flight altitude, the requirement to maintain the aircraft within the visual line of sight (VLOS) of the operator in lower-risk scenarios, and the need for additional authorizations and technical capabilities when conducting beyond visual line of sight (BVLOS) operations. Furthermore, restrictions are placed on minimum separation distances from uninvolved persons, as well as on operations in sensitive geographical areas, such as controlled airspace, airport surroundings, or densely populated urban environments.

Although the regulatory landscape for civil UAV operations remains heterogeneous across different regions, and is still evolving in many countries, a clear convergence can be observed in the fundamental criteria used to define operational limitations. As an illustrative example, Table 4.5 presents a simplified comparison between key aspects of the European framework and that of the United States.

From an engineering standpoint, civil UAV regulations impose significant constraints on both system design and operational planning. They define minimum requirements in terms of sensing, navigation, and communication capabilities, particularly for higher-risk operations, and may necessitate features such as enhanced situational awareness, detect-and-avoid functionalities, and robust command and control links. In addition, regulatory constraints directly limit operational parameters such as altitude, range, and level of autonomy, thereby shaping the feasible design space and influencing key architectural decisions.

4.6.3 Military UAV operations and regulatory differences

In the military domain, the regulatory framework governing UAV operations differs significantly from that of civil aviation. While civil regulations primarily emphasize

Table 4.5 Simplified comparison of civil UAV operational constraints: Europe (Spain) vs United States.

Operational Aspect	Europe	United States
Maximum altitude	~120 m AGL	~400 ft AGL (~120 m)
VLOS requirement	Mandatory in low-risk operations; BVLOS subject to authorization	Mandatory by default; BVLOS allowed via waivers
Operations over people	Strongly restricted depending on operational category	Allowed under specific conditions and aircraft classification
Geographical restrictions	Strict limitations in urban areas and controlled airspace	Similar restrictions with defined controlled airspace grids
Communication requirements	Reliable control link required; additional capabilities for advanced operations	Reliable control link required; extended requirements for advanced operations

safety, particularly the protection of third parties and safe integration into shared airspace, military frameworks prioritize mission accomplishment, even under adverse conditions or high levels of uncertainty. This fundamental difference results in reduced regulatory constraints and greater operational flexibility compared to the civil domain.

Consequently, military UAV operations are conducted within a substantially different operational context. These operations typically take place in segregated airspace with limited or no civil traffic, as well as in non-cooperative environments where conventional air traffic management infrastructures may be unavailable. Under such conditions, many of the constraints imposed on civil UAV operations can be relaxed, particularly those related to proximity to people, geographical restrictions in sensitive areas, and requirements associated with BVLOS operations.

From an engineering perspective, this regulatory flexibility significantly expands the design space. Military UAVs can adopt architectures and technological solutions that would be difficult to certify in civil applications, such as higher levels of autonomy, mission-specific system optimization, or more aggressive flight profiles. In addition, the reduced dependence on formal certification processes enables faster development and deployment cycles, facilitating the rapid integration of emerging technologies.

However, reduced regulatory constraints do not imply the absence of operational requirements. On the contrary, military UAV frameworks often impose stringent constraints related to system robustness, resilience against interference, and the ability to operate reliably in degraded or contested environments. These requirements shift the focus from compliance with standardized civil regulations toward ensuring mission effectiveness and system survivability under demanding operational conditions.

4.6.4 Operational constraints beyond regulation

UAV operations are subject to additional constraints that arise primarily from the physical environment, technological limitations, and interaction with existing infrastructures. Although these aspects are partially reflected in regulatory frameworks, many of the limitations derived from them are not formally standardized and must be explicitly considered to ensure mission feasibility and operational safety.

The main sources of non-regulatory constraints include:

- **Meteorological conditions:** Environmental factors such as wind gusts, sustained winds, precipitation, and reduced visibility have a direct impact on UAV performance. Rotary-wing platforms, in particular, are highly sensitive to external disturbances, which may compromise flight stability and significantly increase power consumption. Adverse weather conditions can therefore reduce effective endurance, constrain the operational flight envelope, and introduce additional uncertainty in mission execution.
- **Electromagnetic interference:** External electromagnetic sources may degrade the performance of onboard sensing systems, particularly GNSS, affecting navigation accuracy and positioning reliability. This is especially critical in urban or industrial environments, where signal multipath effects, interference, and partial signal obstruction are more prevalent.
- **Communication limitations:** The quality and reliability of command, control, and telemetry links can be affected by obstacles, terrain, distance, and spectrum congestion. These factors may lead to increased latency, reduced link robustness, or even temporary loss of communication, particularly in BVLOS operations.

In addition, practical airspace usage may introduce further operational constraints not explicitly captured in formal regulations. These include temporary restrictions due to other air traffic, dynamic airspace management measures, or localized operational limitations that must be addressed during mission planning, particularly in complex or congested environments.

From an engineering perspective, these non-regulatory constraints have direct implications for both system design and operation. UAVs intended to operate in demanding environments, such as regions with frequent strong winds or high electromagnetic activity, may require more stringent design specifications than those operating under benign conditions. This can translate into increased robustness of the platform, enhanced sensing and navigation capabilities, and more resilient communication systems.

Moreover, these constraints directly affect mission planning, requiring the incorporation of operational margins, contingency procedures, and more conservative performance assumptions. They also impact the effective energy budget, as non-ideal operating conditions typically lead to higher power consumption and reduced endurance.

Consequently, these constraints, in combination with formal regulatory requirements, define the real operational environment of UAV systems. Their consideration is essential for properly dimensioning system architecture and ensuring reliable and safe mission execution under realistic conditions.

4.7 Applications and use cases

4.7.1 Overview of UAV application domains

Unlike more specialized systems, UAVs are highly versatile platforms that can be adapted to a wide range of applications with relatively minor modifications. A given UAV configuration can therefore be deployed across different operational domains by adjusting elements such as payload, onboard sensors, or mission parameters. This inherent flexibility largely explains the widespread adoption of UAVs across diverse sectors.

It is useful to classify UAV applications in order to better understand their operational diversity. Application domains can be categorized according to different criteria. The classification adopted throughout Section 4.7 is based on the type of mission and illustrated in Fig. 4.7. However, alternative classifications may be more appropriate depending on the context. For instance, from a functional perspective, UAV applications can be grouped into:

- **Sensing applications:** Focused on observation and data acquisition tasks such as monitoring, inspection, and mapping, where sensor capability and data quality are the primary drivers.
- **Transport applications:** Involving the delivery of goods, where payload capacity and endurance become the dominant performance constraints.
- **Intervention applications:** Characterized by direct interaction with the environment, such as agricultural spraying or certain emergency response operations, often requiring specialized payloads or actuation systems.

Although UAVs are inherently flexible platforms, specific application domains impose distinct design constraints. In this sense, an application-driven design approach can be identified, in which mission requirements and operational objectives directly determine the main engineering decisions [7].

4.7.2 Military and defense applications

Historically, military and defense applications have been the primary driver behind the development of UAVs (see Section 4.2), and they continue to represent one of their most demanding operational domains. Their adoption has been largely motivated by the need to extend operational capabilities while minimizing risk to human personnel, a factor that still underpins their widespread use today and continues to shape technological evolution.

Within this context, UAVs are predominantly employed in Inspection, Surveillance and Reconnaissance (ISR) missions, where they provide real-time or near-real-time situational awareness through continuous monitoring of areas of interest. These platforms support both tactical operations, such as battlefield reconnaissance, and

Figure 4.7 Overview of UAV application domains considered in this section, structured according to primary mission categories.

higher-level strategic intelligence activities. In addition, UAVs are frequently used as communication relays, extending connectivity in complex or infrastructure-limited environments, and enabling coordinated operations across distributed assets.

A key advantage of UAVs in military applications is their persistence, defined as the ability to remain airborne for extended periods without direct human intervention. This capability is particularly relevant for long-endurance platforms, allowing sustained surveillance and reducing temporal gaps in data acquisition. As a result, mission effectiveness is significantly enhanced while simultaneously reducing the exposure of human operators to hazardous or contested environments.

From a system design perspective, military UAVs require high levels of autonomy, robustness, and communication reliability. Operations often take place in degraded environments, where GNSS signals may be denied and communication links disrupted. Consequently, these systems incorporate resilient communication architectures, redundancy in critical subsystems, and advanced sensing and navigation capabilities to maintain operational continuity.

Overall, military and defense applications impose stringent technical requirements that have historically driven innovation in UAV technologies, many of which have later been transferred to civil and commercial domains.

4.7.3 Industrial and commercial applications

UAVs have experienced significant penetration in industrial and commercial applications in recent years. This transition from an originally military-focused domain to predominantly civil use has been driven by their ability to improve operational efficiency, reduce costs, and enhance safety across a wide range of activities. In contrast

Table 4.6 Summary of UAV platforms and key requirements across selected industrial and commercial applications.

Application	Key Requirements	Typical UAV Platform
Infrastructure Inspection	Flight stability, high-resolution sensing, precise positioning	Rotary-wing UAVs
Agriculture	Endurance, coverage efficiency, precision in sensing and actuation	Fixed-wing UAVs (monitoring); rotary-wing UAVs (spraying)
Mining and Construction	Advanced sensing, data acquisition and processing capabilities, area coverage	Fixed-wing UAVs (large areas); rotary-wing UAVs (localized inspection)

to military applications (see Section 4.7.2), these use cases are strongly constrained by requirements related to precision, repeatability, and economic viability, which directly influence both system design and platform selection.

A wide variety of industrial and commercial applications have emerged, each characterized by specific operational requirements. Consequently, different UAV configurations are preferred depending on the mission profile, as summarized in Table 4.6. This diversity reflects a strong coupling between application requirements and technological choices, particularly in terms of airframe configuration, sensing capabilities, navigation accuracy, and energy management. In many cases, system integration and data processing capabilities are as critical as the flight platform itself.

One of the most widespread applications is infrastructure inspection, including assets such as power lines, pipelines, bridges, and wind turbines. In these scenarios, rotary-wing UAVs, particularly multirotor platforms, are predominantly used due to their ability to hover and maneuver with high precision in close proximity to complex structures. These UAVs are typically equipped with payloads such as high-resolution optical or thermal cameras, and increasingly LiDAR systems, enabling detailed inspection and defect detection without direct human intervention in potentially hazardous environments.

In the field of agriculture, UAVs are employed for both monitoring and intervention tasks, such as spraying. For large-scale crop monitoring, fixed-wing UAVs are generally preferred due to their higher endurance and ability to efficiently cover extensive areas. These platforms are often equipped with multispectral or hyperspectral sensors for crop health assessment and yield optimization. Conversely, rotary-wing UAVs are more suitable for intervention tasks requiring high precision and low-altitude operation, such as targeted spraying, where maneuverability and payload integration are critical, albeit fixed-wing UAVs are frequently used for large-scale crop spraying when precision is not critical.

In sectors such as mining and construction, UAVs are mainly used for mapping and progress monitoring. These applications often require advanced sensing systems and data processing capabilities, particularly for the generation of high-resolution or three-dimensional maps. Platform selection depends on mission requirements, with fixed-wing UAVs favored for large-area coverage and rotary-wing UAVs better suited for localized inspection tasks in confined or complex environments. Mission planning, data acquisition strategies, and post-processing workflows play a central role in overall system performance.

Overall, industrial and commercial applications clearly illustrate the concept of application-driven design. The diversity of mission requirements leads to highly specialized UAV configurations, where propulsion systems, sensor payloads, and energy management strategies are tailored from early design stages to meet specific operational needs, reinforcing the strong interdependence between platform design and mission objectives.

4.7.4 Logistics and urban air mobility (UAM) use cases

Logistics is one of the domains in which UAVs are currently experiencing a significant expansion in practical use. In particular, UAVs are especially effective for so-called last-mile logistics, where goods are distributed from logistics centers to final delivery points. In this context, UAVs enable the autonomous or semi-autonomous delivery of goods in densely populated urban areas, where conventional transport alternatives often face substantial operational constraints, such as traffic congestion or limited accessibility. From an engineering perspective, these operations are typically characterized by short-range missions with relatively low payload fractions, where system responsiveness and deployment flexibility are prioritized over overall transport efficiency.

Within the logistics domain, especially in high-density urban environments, rotary-wing UAVs are generally preferred over fixed-wing platforms. This is mainly due to their VTOL capability, which allows operation without the need for auxiliary infrastructure such as launch systems or runways. However, this operational flexibility comes at the cost of lower aerodynamic efficiency and higher specific energy consumption, particularly during hover, which directly limits achievable endurance and range for a given battery capacity.

UAVs are also gaining importance in the transportation of small, high-value or time-critical payloads, such as medical supplies. Their ability to provide rapid delivery to hard-to-reach locations, including rural or isolated areas, makes them particularly advantageous compared to conventional mobility solutions. In such missions, the trade-off between payload mass and available energy becomes critical, as even small increases in payload can significantly impact flight time and mission feasibility.

Finally, an emerging and increasingly relevant field is Urban Air Mobility (UAM). This paradigm envisions the integration of UAVs into urban transportation systems, potentially enabling new forms of aerial mobility, including passenger transport. However, such applications face significant technical challenges, including limited endurance, low payload fraction relative to total vehicle mass, and stringent safety and regulatory requirements associated with human transport. As a result, the large-scale deployment of UAM systems remains strongly dependent on advances in energy density, power management, and overall system reliability.

4.7.5 Emergency, public safety, and environmental applications

Finally, it is worth highlighting the role of UAVs in missions related to emergency response, disaster management, and environmental monitoring. In these domains, UAVs are particularly effective due to their rapid deployment capability and their ability to access hard-to-reach areas, providing a significant operational advantage over conventional mobility solutions. A more detailed description of these applications is provided below:

- In emergency operations, UAVs provide the capability to rapidly survey large areas while enabling real-time data transmission. This data transmission capability is especially relevant given the variety of information that can be collected, ranging from visual imagery to the detection of potentially hazardous environmental conditions. Within this domain, UAVs are increasingly used in search and rescue missions, where timely and accurate situational awareness is critical for effective response.
- In disaster management, UAVs provide valuable support in both prevention and response phases for events such as wildfires, earthquakes, and floods. In this context, UAVs can effectively assist in damage assessment, monitoring the evolution of the event, and supporting decision-making processes. Their ability to operate without pre-existing infrastructure makes them particularly suitable in scenarios where conventional systems are disrupted or unavailable.
- The wide range of sensors that UAVs can integrate, together with their ability to acquire high spatial resolution data, makes them an increasingly important tool in environmental monitoring. In this context, UAVs are used for applications such as air quality monitoring, early fire detection, and ecosystem tracking across large or remote areas.

4.8 Role of UAVs in the electric mobility ecosystem

4.8.1 UAVs as an extension of electric mobility

UAVs cannot be understood as an isolated technology outside the electric mobility ecosystem. In fact, UAVs share a common technological foundation with most EVs, including reliance on battery-based energy storage systems, the use of power electronics for energy conversion and management, and electric propulsion architectures enabling precise and efficient system control. Nevertheless, UAVs exhibit important specificities with respect to ground EVs, which are summarized in Table 4.7 and further discussed below.

The most evident and significant difference lies in the operating environment. While ground EVs operate in contact with the surface, UAVs operate in the aerial domain. Although this distinction may appear straightforward, it has major engineering implications. Ground vehicles benefit from passive weight support, whereas UAVs must continuously generate lift. As a result, UAVs are inherently more sensitive to both energy availability and system mass, which directly constrain

Table 4.7 Conceptual comparison between UAVs and ground electric vehicles.

Aspect	UAVs	Ground EVs
Operating environment	Aerial domain	Ground-based
Mass sensitivity	Very high	Moderate
Energy constraints	Critical	More manageable
Energy architecture	Highly optimized	More flexible
Operational margins	Narrow	Wider
Dependence on efficiency	Critical	Important
Control complexity	High	Lower

endurance and payload capacity, leading to narrower operational margins in many applications.

This strong dependence on mass and energy efficiency imposes more stringent design requirements. Consequently, UAV architectures tend to be highly optimized, particularly regarding the coupling between the energy system and propulsion system. In addition, control strategies become more complex and tightly integrated than in most ground EVs, with the primary objective of maximizing energy efficiency and maintaining stable operation under varying flight conditions.

4.8.2 System integration and digital ecosystems

The relevance of UAVs within the electric mobility ecosystem lies in their integration into larger and more complex systems, where multiple technologies and actors operate in a coordinated manner, exchanging information and resources. In this context, the operational value of UAVs becomes significantly more pronounced when they are deployed within digitalized urban environments or distributed logistics systems, where their flexible deployment and ability to access hard-to-reach areas provide clear advantages.

In this sense, UAVs exhibit a stronger dependence on modern digital infrastructures than more conventional EVs. This includes data-driven mission planning, real-time monitoring of both vehicle state and environmental conditions, as well as the capability to dynamically adapt to changing operational scenarios. Within this framework, UAVs are often treated as nodes in large-scale data exchange networks, naturally integrating operators and other mobility technologies. These interactions are typically supported by cloud-based platforms and edge computing solutions, enabling low-latency data processing and more responsive system behavior.

The properties of these digital and information ecosystems enable multiple UAVs to coordinate their actions, leading to the concept of fleet management (see [8,9] for further details). Within this paradigm, several UAVs can operate collaboratively within a shared mission. This cooperation extends the operational range of the fleet, for instance by enabling the coverage of larger areas in monitoring tasks, thereby partially mitigating inherent technical limitations. However, such coordination significantly increases system-level complexity, particularly in terms of control, communication, and decision-making, often requiring advanced algorithms for task allocation, path planning, and conflict avoidance.

4.8.3 Technical and system-level limitations

Similarly to other EVs, the limited energy density of batteries constitutes one of the main constraints for UAVs. As discussed in Chapter 3, the energy density of current lithium-based battery technologies is orders of magnitude lower than that of conventional fossil fuels such as gasoline, which imposes significant limitations on UAV endurance and mission duration (see Section 4.5.1).

Although energy storage limitations are common across the electric mobility ecosystem, payload fraction becomes particularly critical in UAVs. This parameter, defined as the ratio between payload mass and total vehicle mass, is significantly more constrained in UAVs than in conventional terrestrial EVs. This is primarily because a substantial portion of the total vehicle mass must be allocated to energy storage and propulsion systems. Typically, between 30–60% of the total UAV mass is devoted to batteries and propulsion, whereas this range is usually reduced to 15–30% in terrestrial EVs. As a result, the type and capacity of payloads that can be transported are notably limited, directly constraining mission flexibility and operational capability across different use cases.

Regarding integration within larger and more complex digital systems, UAV limitations are largely driven by constraints in communication systems. Effective range, latency, and robustness to interference directly impact both remote control capabilities and data transmission, particularly in complex operations or dense urban environments. These constraints become even more critical in scenarios involving multi-UAV coordination or fleet-level operations.

Finally, most of the aforementioned limitations are further reinforced by a stricter regulatory framework compared to other terrestrial EVs (see Section 4.6), often imposing additional constraints on the operational space and usage conditions, and further narrowing the feasible operational envelope of UAV systems.

4.8.4 Future trends and outlook

The future development of UAVs, both in terms of technological progress and expansion of application domains, is expected to be primarily driven by incremental advancements rather than radical disruptions in the short term. This gradual evolution is consistent with the current maturity level of key enabling technologies within the electric mobility ecosystem (see Fig. 4.8).

In the field of energy storage, progress is expected to remain largely based on lithium-based technologies, with incremental improvements rather than the widespread adoption of disruptive alternatives in the near term. In this context, gradual gains in both cost reduction and energy density are anticipated, enabling progressive increases in flight endurance and mission duration. Although these improvements

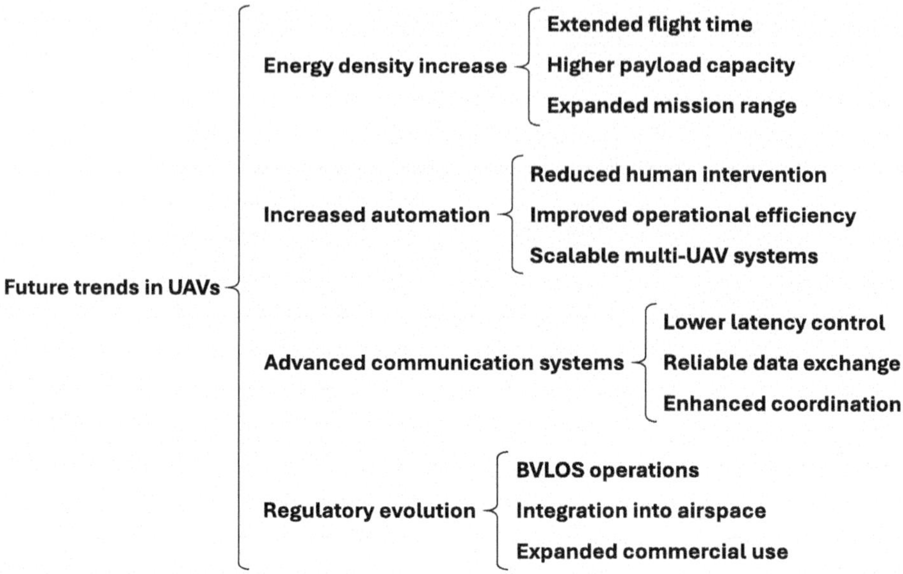

Figure 4.8 Future trends in UAVs.

may appear modest individually, their cumulative impact can significantly enhance operational capabilities over time.

In parallel, continuous advancements in control systems and automation are expected. These developments will enable improved operational efficiency and higher levels of autonomy, reducing the need for human intervention in certain applications. Such progress will likely be reinforced by improvements in communication systems, including lower latency, more reliable data exchange, and enhanced coordination capabilities. These aspects are particularly relevant in fleet management scenarios [8,9], where multiple UAVs operate in a coordinated manner.

Finally, a continued expansion in UAV applications is anticipated, with particular emphasis on logistics, urban mobility, and infrastructure inspection. The progressive deployment of UAVs in both commercial and industrial contexts, including potential passenger transport, will depend not only on technological maturity but also on the evolution of the regulatory framework. In this sense, large-scale adoption is expected to be strongly conditioned by regulatory development alongside technological progress.

References

[1] Market Growth Reports, UAV drones market size, share, growth, and industry analysis, by type (fixed wing, vtol, stuas, male, Hale, others), by application (military, commercial, others), regional insights and forecast to 2035, last updated: December 1, 2025, https://www.marketgrowthreports.com/market-reports/uav-drones-market-101839, 2025. (Accessed 7 April 2026).

[2] N. Elmeseiry, N. Alshaer, T. Ismail, A detailed survey and future directions of unmanned aerial vehicles (UAVs) with potential applications, Aerospace 8 (12) (2021), https://doi.org/10.3390/aerospace8120363, https://www.mdpi.com/2226-4310/8/12/363.

[3] Wikipedia contributors, Unmanned aerial vehicle, https://en.wikipedia.org/wiki/Unmanned_aerial_vehicle, 2026. (Accessed 8 April 2026).

[4] R. Austin, Unmanned Aircraft Systems UAVs Design, Development and Deployment, Wiley, 2010.

[5] M.S. Grewal, L.R. Weill, A.P. Andrews, Global Navigation Satellite Systems, Inertial Navigation, and Integration, Wiley, 2013.

[6] K. Li, Y. Wu, A. Bakar, S. Wang, Y. Li, D. Wen, Energy system optimization and simulation for low-altitude solar-powered UAVs, Aerospace 9 (6) (2022) 331, https://doi.org/10.3390/aerospace9060331, https://www.mdpi.com/2226-4310/9/6/331.

[7] M.H. Sadraey, Design of Unmanned Aerial Systems, Wiley, Chichester, UK, 2020.

[8] T. Lagkas, V. Argyriou, S. Bibi, P. Sarigiannidis, UAV IoT framework views and challenges: towards protecting drones as “things”, Sensors 18 (11) (2018), https://doi.org/10.3390/s18114015, https://www.mdpi.com/1424-8220/18/11/4015.

[9] Margarida Silva, et al., A mission planning framework for fleets of connected UAVs, Journal of Intelligent & Robotic Systems 108 (2) (2023), https://doi.org/10.1007/s10846-023-01820-7, https://link.springer.com/article/10.1007/s10846-023-01820-7.

Other EVs

5

5.1 Electrification in rail transport

5.1.1 Overview of rail electrification

Unlike other EVs such as ECs or UAVs, the electrification of rail transport represents one of the earliest large-scale applications of electric mobility. Its development dates back to the late nineteenth century, coinciding with the emergence of electric motors (see Chapter 1), and it has experienced continuous development and increasing penetration throughout the twentieth century. In the railway sector, electric solutions progressively replaced early steam-based technologies, eventually becoming the dominant form of traction (see Fig. 5.1). This transition led to the establishment of extensive and well-developed electrified rail networks, which today constitute fundamental transport arteries in many regions worldwide.

The primary reason for the consolidation of electric rail transport compared to other EVs lies in its energy supply model. Indeed, vehicles such as ECs or UAVs rely on onboard energy storage systems, typically batteries, whose energy density remains significantly lower than that of conventional fuels (see Chapters 3 and 4). This limitation directly constrains their range and performance when compared to ICE-based vehicles. In contrast, railway systems are far less dependent on onboard energy storage, as they can benefit from a continuous external power supply delivered through infrastructure such as overhead catenary systems or third-rail configurations (see Section 5.1.3).

This reduced dependence on onboard energy storage has important technical implications. For instance, electric trains can operate at high power levels over extended periods without being constrained by range limitations, which is particularly relevant in high-demand corridors. Furthermore, the lower requirement for onboard

Figure 5.1 Milwaukee Road Class ES-2 electric switcher locomotive, featuring a distinctive camelback-style cab, designed for heavy-duty operations on electrified high-traffic railways. The rapid adoption of electric traction during the early twentieth century led to the progressive displacement of steam locomotives, establishing electric railways as a dominant solution for efficient and reliable railway transport. (Image courtesy of Wikimedia Commons.)

Electric Mobility Systems. https://doi.org/10.1016/B978-0-44-349397-3.00013-6

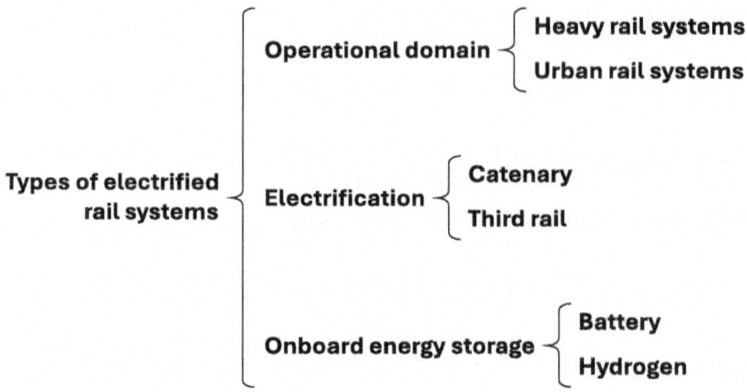

Figure 5.2 Classification of electrified rail systems.

energy storage contributes to a more optimized vehicle mass, which is especially advantageous in applications involving the transport of passengers or goods, particularly when compared to systems such as UAVs (see Chapter 4). In addition, this architecture simplifies certain aspects of energy management at the vehicle level, shifting complexity toward the infrastructure side.

For these reasons, electric rail transport has become firmly established throughout the twentieth century as a predominant solution within the broader mobility landscape, particularly in the logistics domain. Its widespread adoption has acted as a catalyst for continuous technological development over several decades, ultimately positioning electrified rail systems as a highly mature and reliable technology within the transport sector, with a well-established operational and industrial ecosystem.

5.1.2 Types of electrified rail systems

Electric railway systems can be classified according to different criteria. One of the most widely adopted approaches by international agencies such as the International Association of Public Transport (UITP)[1] and the IEA[2] is based on their operational characteristics. This criterion focuses on the primary application of the system, allowing a distinction between heavy rail and urban/light rail transport. Although this is the most common classification approach and largely determines the energy architecture of the train, other criteria can also be relevant. In this sense, and as shown in Fig. 5.2, rail vehicles can also be classified according to their electrical supply system or the type of onboard energy storage.

Regarding heavy rail transport, it typically consists of vehicles designed for high speeds and high transport capacity, both for passengers and freight, over long distances. Their energy systems are adapted to relatively stable demand profiles, characterized by continuous power supply with limited transient peaks, as they usually

[1] https://www.uitp.org/.
[2] https://www.iea.org/.

Table 5.1 Typical characteristics of heavy rail and urban rail systems based on operational requirements.

Characteristic	Heavy Rail	Urban Rail
Primary application	Interurban / freight	Urban passenger transport
Typical distance	Long distance	Short to medium distance
Operational speed	High	Low to medium
Stop frequency	Low	High
Transport capacity	Very high	High (metro) to medium (tram)
Infrastructure segregation	High	Variable
Energy supply system	Overhead catenary	Third rail / catenary
Power demand	High and sustained	Dynamic, with frequent peaks
Operational flexibility	Low	Medium
Integration with environment	Limited	High (especially tramways)

operate on routes with few stops. In this context, heavy rail systems generally rely on robust and efficient external power supply methods such as overhead catenary systems, while the use of onboard energy storage remains limited. This reflects the strong dependence of such systems on continuous external energy supply, which enables sustained high-power operation.

In contrast, urban and light rail systems operate in densely populated environments and are characterized by highly dynamic power demand profiles. These profiles include frequent power peaks associated with repeated acceleration and braking cycles, typical of routes with short inter-station distances. Within this category, metro systems represent one of the most characteristic applications and are among the most widely deployed public transport solutions in high-density urban areas. Urban metro systems typically operate at moderate average speeds and frequently employ third rail power supply systems. This solution provides a compact power delivery method with low visual impact, making it well suited for integration into underground or otherwise controlled urban environments (see Section 5.1.3).

Unlike metro systems, which often operate on segregated or underground tracks, tramways are directly integrated into the urban transport environment, frequently sharing infrastructure with other road vehicles. This operational characteristic limits their achievable speeds compared to metro systems but significantly enhances accessibility and urban integration. On the other hand, light rail systems can be regarded as an intermediate solution between tramways and metro systems, combining partial infrastructure segregation with increased operational flexibility. As a summary, Table 5.1 illustrates the main characteristics of the different types of electric rail vehicles according to their operational profile.

Another important classification criterion is based on the electrification system. In general, overhead catenary systems remain the dominant solution in railway electrification, particularly in heavy rail applications, due to their ability to efficiently deliver high power levels over long distances. In urban environments, however, overhead lines may be replaced or complemented by third rail systems, which offer advantages in terms of compactness and reduced visual impact. This makes them particularly suitable for metro systems operating in controlled environments. Nevertheless, third rail systems present significant safety challenges due to the exposure of energized components, requiring additional protection measures and restricting their use to segregated infrastructures.

Finally, electric rail vehicles can also be characterized according to their onboard energy systems. These solutions are relatively recent and include battery-based trains, known as Battery Electric Multiple Units (BEMUs), in which onboard battery systems provide energy to the vehicle in the absence of external power supply, such as overhead lines [1]. In addition, these batteries can be recharged during operation through catenary or third rail systems in electrified sections. Although this capability allows railway systems to be less sensitive to onboard energy storage limitations compared to other EVs such as UAVs, the relatively low energy density of current battery technologies remains a key limitation, significantly constraining operational range in fully battery-powered segments.

Alternatively, hydrogen-based energy storage and conversion technologies can be employed in railway transport [2,3]. These systems operate under principles similar to FCECs (see Chapter 2), incorporating hydrogen storage (typically in pressurized tanks) and conversion into electricity through fuel cells (see Chapter 8). While hydrogen offers higher energy density compared to batteries and therefore enables greater potential autonomy, it presents drawbacks in terms of lower overall energy conversion efficiency. Furthermore, unlike batteries, hydrogen systems cannot be recharged in transit unless an onboard electricity-to-hydrogen conversion system (electrolyzer) is included, which represents a highly inefficient solution.

5.1.3 Energy supply and system architecture

In contrast to other EVs, in which energy is supplied from onboard storage systems and energy transfer from external sources only occurs at specific moments (charging), electrified railway transport is characterized by a continuous energy supply from the electrical grid to the traction system. In this sense, railway systems can be understood as open energy systems, in which energy flows continuously from outside the vehicle.

Accordingly, the energy flow in electrified railway systems follows a structure similar to that of conventional large-scale electrical consumption. Electrical energy is generated in centralized power plants and delivered to the transmission grid, which in turn connects to the railway system through dedicated substations. These substations play a fundamental role within the overall railway energy system. They are typically purpose-built installations whose primary function is to adapt the electrical energy from the grid to the appropriate type (AC or DC) and voltage levels required by the railway system. In addition, they may incorporate transformation, rectification,

Figure 5.3 Two FS Class E.464 locomotives at Formia-Gaeta station. These electric trains receive power from an overhead catenary system, where electrical energy is transmitted through suspended contact wires and collected by a pantograph mounted on the train roof. (Image courtesy of Wikimedia Commons.)

and protection systems to ensure reliable and stable operation under varying load conditions [4].

The transfer of energy from infrastructure to the train is mainly achieved through two technologies: overhead catenary systems and third rail systems. The catenary consists of overhead electrical conductors (typically bare wires) that establish electrical contact with the train via a pantograph (see Fig. 5.3). This system enables efficient transmission of high power levels and typically operates at kilovolt ranges, both in DC systems (1–3 kV) and AC systems (in the order of tens of kilovolts, e.g., 15–25 kV). In AC systems, onboard transformers are required to step down the voltage before further conversion stages.

In contrast, third rail systems deliver energy through direct contact with an electrified conductor located at track level (see Fig. 5.4). These systems operate almost exclusively in DC and at lower voltage levels than catenary-based systems, typically in the range of 600–750 V. As a result, they are more commonly used in urban transport, where they offer advantages in terms of compactness and reduced visual impact. However, due to the accessibility of the energized conductor, third rail systems require stricter safety measures, particularly in environments with human presence.

The internal architecture of an electric train is composed of subsystems that share similarities with those described for ECs (Chapter 2) and UAVs (Chapter 4). Once energy is transferred from the infrastructure, it is distributed among the functional subsystems of the vehicle using power electronics to adapt voltage and current levels to the requirements of each subsystem. Typically, this involves multiple conversion stages, including AC/DC rectification, DC/DC conversion, and DC/AC inversion for

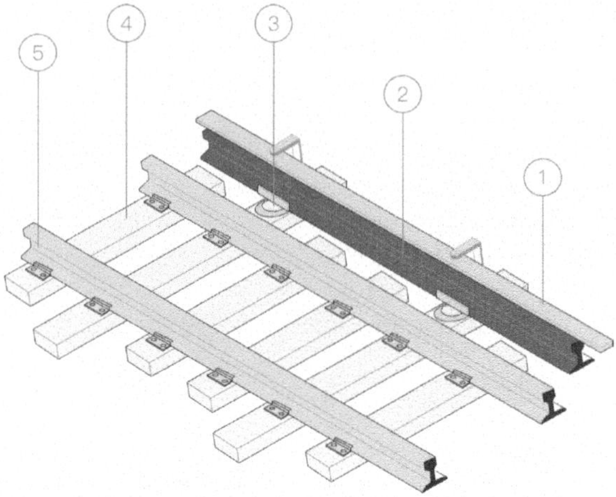

Figure 5.4 Third rail layout: 1: coverboard 2: power rail 3: insulator 4: sleeper 5: rail. (Image courtesy of Wikimedia Commons.)

traction motor control [5]. The energy is primarily allocated to traction systems, which are responsible for train propulsion, and auxiliary systems, including climate control, lighting, control, and communication systems (for further details, see [6,7]).

Another important aspect of the onboard energy architecture is the typical demand profile. In this context, energy demand profiles vary significantly depending on the operational characteristics of the vehicle. Heavy rail systems operating over long distances are generally associated with sustained high-power demand profiles, whereas urban rail systems exhibit more dynamic behavior, with frequent power peaks during acceleration between stations. A key feature of urban rail transport is its ability to recover braking energy through regenerative braking. During deceleration phases, significant amounts of energy can be fed back and, in some cases, stored in onboard energy storage systems, improving overall system efficiency (see results in [1]).

Therefore, the main difference between electrified railway transport and other EVs in terms of energy architecture lies in the reduced dependence of the former on onboard energy storage systems. The continuous external energy supply allows rail transport to decouple operation from storage capacity, reducing the need for large onboard batteries and enabling sustained high-power operation. However, this same characteristic makes railway systems highly dependent on electrical infrastructure, limiting their suitability in regions where the power grid is weak or prone to frequent disruptions [8].

5.1.4 Performance characteristics and representative metrics

The characteristics and performance of electrified railway systems are strongly influenced by the environment in which they operate, the regulatory framework, and the characteristics of the electrical system. This results in significant differences in

Table 5.2 Typical operating speeds and electrification systems in representative railway networks.

Country	System	Typical Speed	Electrification
Spain	High-speed	250–300 km/h	25 kV AC
	Conventional	120–200 km/h	3 kV DC
Japan	High-speed	240–320 km/h	25 kV AC
	Conventional	100–160 km/h	1.5 kV DC
Germany	High-speed	250–300 km/h	15 kV AC, 16.7 Hz
	Conventional	120–200 km/h	15 kV AC, 16.7 Hz
France	High-speed	270–320 km/h	25 kV AC
	Conventional	120–200 km/h	1.5 kV DC / 25 kV AC
Italy	High-speed	250–300 km/h	25 kV AC
	Conventional	120–200 km/h	3 kV DC
United States	High-speed[a]	200–240 km/h	25 kV AC / 12.5 kV AC
	Conventional	80–120 km/h	Limited electrification

[a] Limited high-speed operation; infrastructure is not fully dedicated and maximum speeds are only reached in specific segments.

terms of speed, capacity, and energy demand across the various types of railway transport.

In terms of speed, railway systems exhibit a wide range of operating values, largely determined by their operational context. At the upper end are high-speed rail systems, which typically operate on dedicated infrastructure and can reach speeds of up to 300 km/h. These systems are primarily used for long-distance passenger transport. At the other extreme, urban tramway systems operate at much lower speeds, typically in the range of 15–25 km/h. This limitation is mainly due to their integration within urban transport networks, where they coexist with pedestrians and other vehicles. Due to a higher degree of infrastructure segregation, metro systems can achieve higher speeds than tramways, typically operating in the range of 30–80 km/h.

From an energy perspective, railway systems exhibit a high degree of heterogeneity depending on the geographical region. This variability is largely driven by infrastructure design choices and electrification strategies. As shown in Table 5.2, high-speed systems are typically based on high-voltage AC electrification (on the order of tens of kilovolts), enabling efficient transmission of the large power levels required for high-speed operation. In contrast, conventional railway systems display a wider variety of electrification schemes, including low-voltage DC systems as well as high-voltage AC systems, such as the 15 kV, 16.7 Hz system used in Germany. In this context, the case of the United States is particularly relevant, where conventional rail transport remains strongly dependent on diesel traction, while high-speed services often operate on non-dedicated infrastructure [9].

Another important aspect is the capacity of the vehicle, understood as the number of passengers or the volume of goods that can be transported. Tramway systems typically have limited capacity, transporting on the order of tens to a few hundred passengers

Table 5.3 Typical order of magnitude of installed traction power in railway systems.

System	Typical Power Range
High-speed rail	8–16 MW
Conventional rail (long-distance)	3–8 MW
Metro systems	1–4 MW
Tramway systems	0.2–1 MW

per vehicle, whereas metro systems can accommodate hundreds to thousands of passengers per train, particularly in high-density configurations. At the other end of the spectrum, heavy rail transport not only provides high passenger capacity but is also widely used for large-volume freight transport, making it a key component of global logistics chains [10].

From an energy standpoint, railway transport exhibits high overall efficiency compared to other EVs when evaluated in terms of energy per passenger-kilometer or ton-kilometer. This is largely due to the low rolling resistance associated with the wheel–rail contact (steel-on-steel), which significantly reduces friction losses compared to other land-based transport modes [11]. In addition, the continuous external power supply enables effective energy management, allowing railway systems to operate efficiently over extended periods. This is further enhanced by the ability to recover energy through regenerative braking, particularly in urban systems with frequent stop-and-go operation.

Regarding power levels, railway systems present different demand profiles depending on their application. As shown in Table 5.3, urban systems typically exhibit moderate power demand, except during acceleration phases, whereas heavy rail and high-speed systems require significantly higher power levels. In high-speed applications, traction power can reach several megawatts per train, enabling sustained high-speed operation over long distances.

5.1.5 Limitations and operational constraints

Within the context of electric mobility, rail transport represents one of the most technologically mature sectors. Nevertheless, electrified railway systems present a number of structural and operational limitations that constrain their adoption in certain applications.

Firstly, although the external power supply system based on catenary or third rail (see Section 5.1.3) offers clear advantages in terms of energy efficiency and reduced dependence on onboard storage technologies, the strong coupling between railway transport and the electrical grid makes it particularly sensitive to weak or underdeveloped power systems. In addition, disturbances in the electrical network, such as voltage drops or supply interruptions, can directly impact railway operations over large sections of the network. Furthermore, electric rail transport typically relies on dedicated electrical infrastructure, including specific protection, control, and conversion systems, which increases the overall complexity of its energy architecture compared to other EVs.

From an economic perspective, this infrastructure-intensive approach results in high costs, particularly in terms of capital expenditures (CAPEX). The installation of electrification systems, substations, and associated grid reinforcements represents a substantial upfront investment. In addition, railway infrastructure often entails significant operational expenditures (OPEX), including maintenance of both electrical and mechanical assets, which can reduce its economic competitiveness relative to other transport networks such as road-based systems. This economic barrier limits the deployment of railway lines in low-demand corridors, making them less attractive for sparsely populated regions [12]. Consequently, many railway networks exhibit uneven levels of electrification, with non-electrified sections persisting where full deployment is not economically justified.

In terms of operation, rail transport is inherently constrained to fixed routes, which restricts its ability to dynamically adapt to demand fluctuations or network disruptions. This rigidity also limits route diversification and makes rerouting complex in case of incidents. Moreover, due to the low adhesion between wheel and rail, railway vehicles face limitations in terms of maximum gradient and acceleration capabilities, which in turn influence route design and operational flexibility. Finally, it is important to highlight system-specific constraints. Tram systems are subject to strict speed limitations due to their direct interaction with pedestrians and other road users (see Section 5.1.4), while high-speed rail requires highly specialized infrastructure with stringent geometric tolerances, dedicated tracks, and advanced signaling systems, further increasing system complexity and cost.

5.2 Electric maritime transport

5.2.1 Overview and operational context

It is estimated that currently more than 80% of global trade by volume is transported by sea [13]. This figure highlights the critical role of maritime transport in the global economy, establishing it as one of the main pillars of the worldwide logistics system. Unlike other transport segments, maritime operations are characterized by long transport cycles, ranging from hours in coastal shipping to several weeks in intercontinental routes. In addition, they exhibit a high degree of operational inertia, requiring complex route planning, extended port calls, and the management of large cargo volumes. These factors result in highly optimized but relatively inflexible operational frameworks.

Regarding electrification, this transition is primarily driven by increasing regulatory pressure, which demands a progressive decarbonization of heavy transport. This is further reinforced by the need to reduce environmental impact at a global scale, particularly in coastal and port areas, where emissions and noise levels have significant local implications. In this context, electrification is also seen as a key enabler for improving local air quality and meeting increasingly stringent environmental standards.

In contrast to other EVs, the electrification of maritime transport requires the deployment of energy systems with demanding specifications, capable of sustaining high power levels over extended periods of operation. This results in the need for large-capacity energy storage systems, often associated with significant volume and integration constraints. At the same time, maritime EVs are generally less sensitive to mass compared to other platforms such as UAVs (see Chapter 4), allowing for greater flexibility in system architecture and technology integration.

These characteristics make the maritime sector one of the most challenging domains for electrification. In terms of technological maturity and deployment, maritime transport is currently in an intermediate transition phase, positioned between the well-established rail sector and the still emerging aviation sector in terms of electrification readiness and adoption. Consequently, a wide range of technological solutions coexists, each adapted to specific operational profiles and mission requirements.

5.2.2 Types of electrified maritime systems

In maritime transport, electrification does not rely on a single technological solution; instead, multiple configurations coexist, whose applicability strongly depends on the energy requirements of the mission. In this context, it is useful to classify maritime vehicles according to their level of electrification and energy architecture. This leads to three main categories, which are outlined below:

- **Fully electric vessels:** In this type of configuration, propulsion is entirely based on electric systems, including electric motors and onboard energy storage, typically batteries. These vessels are inherently limited in terms of range, primarily due to the relatively low energy density of current electrical energy storage technologies. As a result, their application is mainly restricted to short and predictable routes, such as ferry services (see Fig. 5.5) or inland waterway transport. Their operation is typically associated with fixed routes and the availability of dedicated charging infrastructure.
- **Hybrid-electric ships:** This configuration currently represents the most widespread solution. In hybrid-electric vessels, electric propulsion systems are combined with conventional fuel-based systems, enabling multiple operating modes. This versatility provides increased operational flexibility, allowing different propulsion modes to be used depending on mission requirements, such as low-emission operation in ports and conventional propulsion during cruising. Hybridization also enables improved energy management and partial reduction of fuel consumption.
- **Diesel-electric architectures:** In this configuration, an ICE is integrated with an electrical system. However, the ICE is not mechanically coupled to the propulsion system. Instead, it is used exclusively to generate electrical power, which is then supplied to electric propulsion motors. This architecture allows for optimized operation of the ICE and facilitates the integration of additional energy sources or

Figure 5.5 MF Tycho Brahe, a Danish battery-electric car-ferry owned by Øresundslinjen that operates on the 5 km (3.1 mi) HH Ferry route. (Image courtesy of Wikimedia Commons.)

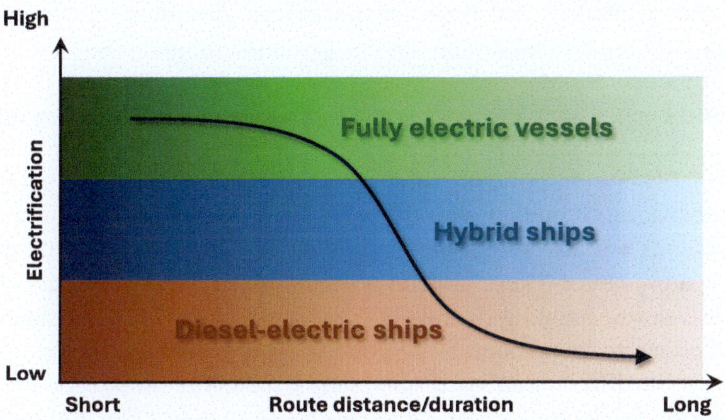

Figure 5.6 Conceptual relationship between the level of electrification and the operational profile in maritime transport. The figure illustrates the inverse correlation between electrification and route distance or duration, where fully electric vessels are predominantly suitable for short-range operations, hybrid-electric configurations enable intermediate flexibility, and diesel-electric architectures remain dominant in long-distance applications due to current energy storage limitations.

storage systems. It can be considered a transitional solution towards higher levels of electrification, analogous to ERECs (see Chapter 2).

From an operational perspective, it is useful to establish a relationship between the level of electrification and the maximum route distance or duration. This relationship follows a clear inverse trend, as illustrated in Fig. 5.6. Fully electric vessels are therefore best suited for short-range operations, while diesel-electric architectures are more appropriate for long-distance missions, with hybrid-electric ships occupying an intermediate position.

It is also important to note that electrification in maritime transport is not limited to propulsion systems. In this context, the term "hotel load" refers to the set of non-propulsive electrical loads onboard a vessel, including systems required for habitability, operation, and onboard services, but not directly contributing to propulsion [14]. These loads may represent a significant share of the total energy consumption, particularly in passenger vessels, and are typically fully electrified even in conventional ships.

5.2.3 Propulsion and energy system architecture

In contrast to rail transport, where the vehicle receives energy continuously from the electrical grid, electric maritime vehicles must be understood as energetically closed systems, in which all processes of energy storage, generation, conversion, distribution, and consumption take place onboard. This fundamental characteristic directly constrains the design of the energy system, as well as the associated strategies for energy management and optimization. In particular, the need to ensure energy availability over long mission durations introduces strict requirements on storage capacity, system reliability, and power management. Additionally, redundancy and fault tolerance become critical design drivers, given the limited possibility of external assistance during operation.

The onboard energy architecture is typically organized into three main functional blocks. First, the energy storage and/or generation block, which in fully electric vessels is primarily composed of high-capacity battery systems and, to a lesser extent, hydrogen-based storage coupled with FCs [15]. In hybrid configurations, ICE-based generation systems are also integrated, supplying electrical power to the propulsion and auxiliary subsystems. These generation units are generally decoupled from the mechanical drivetrain, enabling flexible operation and optimized load management.

Second, the energy conversion stage conditions the electrical power to meet the requirements of the different onboard subsystems, including propulsion and auxiliary loads (hotel loads, see Section 5.2.2). This stage relies on power electronics converters, which not only adapt voltage and current levels but also actively control power flows within the onboard electrical network. Advanced control strategies are often implemented to optimize efficiency, ensure system stability, and coordinate multiple energy sources and loads.

Finally, the power distribution system interconnects all onboard subsystems, enabling the controlled transfer of energy from storage or generation units to propulsion and auxiliary consumers. Depending on the vessel design, distribution architectures may be based on AC or DC systems. In recent years, DC distribution has gained increasing attention due to its higher efficiency in certain operating conditions, reduced conversion stages, and improved compatibility with energy storage systems and power electronic interfaces [16].

The electric propulsion system itself typically relies on high-power electric machines designed for continuous operation under demanding conditions. These motors are mechanically coupled to propulsion devices such as conventional propellers or more advanced solutions like electric pods, which integrate the motor and

Figure 5.7 The MS *Tûranor PlanetSolar* is the largest solar-powered boat in the world and circumnavigated the globe solely with the aid of solar power in 2010. (Image courtesy of Wikimedia Commons.)

propulsor into a single steerable unit. Podded propulsion systems provide enhanced maneuverability, improved hydrodynamic efficiency, and greater flexibility in vessel design, particularly for dynamic positioning and low-speed operations.

It is also worth noting the emergence of vessels incorporating onboard renewable energy generation, primarily through PV systems. Representative examples include the solar-powered catamaran *Tûranor PlanetSolar* (see Fig. 5.7), which demonstrated the feasibility of intercontinental navigation powered exclusively by solar energy, and the RA66 *Helio*, a smaller-scale vessel designed for inland waterway operation. However, the overall contribution of onboard generation remains limited compared to the total energy demand of most maritime missions. This limitation is mainly due to constraints in available surface area and the inherent variability of solar irradiance. As a result, onboard generation systems currently act as supplementary sources supporting the primary energy storage system rather than as standalone solutions [17].

5.2.4 Energy profile and performance characteristics

Electrified maritime transport is characterized by sustained high power demand profiles over extended periods of operation. In this sense, maritime applications exhibit demand patterns that differ significantly from other EVs with more dynamic profiles, such as tramways, and show greater similarity to heavy rail systems. This particularity implies that, although the power demand of electrified vessels is high, it remains relatively predictable during most operational phases.

Operational cycles are primarily determined by the route to be covered by the vessel, ranging from several hours in short-range applications to days or even weeks in intercontinental ocean operations. This behavior results in large cumulative energy consumption, typically in the order of megawatt-hours and potentially reaching gigawatt-hours in high-capacity vessels. Such energy requirements position the

Table 5.4 Representative examples of electrified maritime vessels and their operational energy profiles.

Vessel	Architecture	Typical Route / Range	Battery Capacity
MF Ampere (Norway)	Fully electric	~5.7 km (per crossing)	~1.0 MWh
Tycho Brahe / Aurora	Fully electric	~4 km (per crossing)	~4.1 MWh
Candela P-12	Fully electric	up to ~90 km	~0.25 MWh
Baleària fast ferry (Tarifa–Tangier project)	Fully electric with backup	~30 km	~11–12 MWh
Incat Hull 096 (Buquebus)	Fully electric	~200 km (estimated)	>40 MWh

maritime sector as one of the most demanding within the electrification landscape, often requiring very high-capacity onboard energy storage systems.

This variability in the energy profile strongly influences vessel design and system characteristics, which must be tailored to the specific application. In this context, Table 5.4 presents representative examples of electrified vessels in the industry, covering a range of operational scenarios. As shown, battery capacity requirements span from below one megawatt-hour to several tens of megawatt-hours, depending on the intended application and required autonomy.

In contrast to rail transport, where the wheel–rail interface results in relatively low rolling resistance, energy losses in maritime transport are dominated by hydrodynamic resistance. Although electric propulsion systems are inherently efficient, vessel motion through water involves significant drag forces with a strong dependence on speed, which can become a limiting factor. As a result, electrified vessels are highly sensitive to speed variations, and moderate increases in cruising speed can lead to disproportionately large increases in power demand. Consequently, maintaining stable cruising speeds is essential for efficient operation.

On the other hand, unlike other EVs such as UAVs, the relative weight of the energy storage system is a less critical constraint in maritime transport. This provides additional flexibility in system design, allowing the integration of large-scale energy storage solutions. Finally, while the energy profile of vessels is generally stable during cruising phases, it can become more variable during port maneuvers. This introduces some degree of uncertainty in energy management strategies, although its overall impact remains more limited compared to other electric mobility sectors.

5.2.5 Limitations, infrastructure and applications

As in other electric mobility sectors, the energy density of current battery technologies represents the main barrier to the widespread adoption of fully electrified solutions in maritime transport. This limitation requires the integration of large-scale battery systems, adding significant weight and volume to the vessel, which directly impacts its design as well as the payload capacity it can accommodate.

As a result, the primary challenge for maritime electrification remains the scalability of fully electric solutions. In this context, while fully electric vessels are increasingly deployed in short-range applications such as ferry routes or inland waterways, hybrid solutions are still necessary to ensure sufficient autonomy in longer missions, including intercontinental routes. The rapid increase in energy demand with range makes full electrification progressively less feasible with current storage technologies, particularly for large vessels requiring continuous operation over extended periods.

Another major challenge is the supporting infrastructure required for electrified maritime transport. Electrified vessels rely on the availability of electrified ports and high-power charging systems (shore power), which can reach hundreds of kilowatts or even megawatt-scale levels depending on the application. This high power demand introduces significant challenges for integration with the electrical grid, complicating infrastructure planning and energy management, particularly in ports with high traffic density and simultaneous charging requirements.

Charging time is also a critical operational constraint. Although high-power charging systems are available, charging durations remain substantial due to the large energy capacity of onboard storage systems. Even in lightweight and highly efficient vessels such as the Candela P-12, fast charging times are on the order of one hour under high-power DC conditions [18]. For larger vessels with significantly higher battery capacities, these charging times increase considerably, reinforcing the need for robust high-power charging infrastructure and optimized operational scheduling.

Despite these challenges, electrification has already achieved practical deployment in several well-defined applications. Fully electric vessels have been successfully used for years in short-distance, high-frequency ferry operations, coastal transport, and inland waterways. These applications are characterized by controlled mission profiles, reduced autonomy requirements, and continuous proximity to port infrastructure, which facilitates both charging operations and overall system integration. Consequently, maritime electrification is currently most viable in these niches, while large-scale, long-distance applications remain constrained by the limitations of existing energy storage technologies.

5.3 Electric aviation

5.3.1 Overview and sector context

Within the aviation domain, it is essential to clearly differentiate between UAVs and other electrified aerial transport systems. Although UAVs can be considered a particular subset of electric aviation, they exhibit distinct characteristics in terms of deployment level, scale, applications, and regulatory requirements. In this context, the present section focuses on crewed aircraft, primarily intended for passenger transport or large-scale cargo operations. These applications inherently impose much stricter requirements in terms of safety, redundancy, and reliability. Such differences have strongly influenced the pace of technological adoption, with UAVs already relying predominantly on electrified solutions, whereas electrification in crewed aviation remains significantly less widespread.

The aviation sector represents a domain with substantial decarbonization potential. At present, aviation constitutes the most effective transport mode for covering long distances within short timeframes, playing a key role in global connectivity. However, this capability is largely sustained by the intensive use of fossil fuels with high energy density, which are associated with significant environmental impact. The combination of high fuel consumption and the strategic importance of aviation in global transport systems positions this sector as a major target for greenhouse gas emission reduction. Consequently, ongoing regulatory developments continue to push towards progressive electrification and the adoption of low-emission technologies within the aviation industry.

As in other EV domains, the energy density of battery systems represents the primary limitation for the electrification of air transport. Nevertheless, this constraint becomes significantly more critical in aviation due to two main factors. First, aircraft operates as fully closed energy systems, where all the energy required for a given mission must be stored onboard, with no possibility of in-mission external energy supply. Second, similar to UAVs, aircraft performance and design are highly sensitive to vehicle mass. Even relatively small increases in system weight can result in substantial reductions in range, payload capacity, or overall efficiency. This strong coupling between mass and performance directly constrains the sizing and feasibility of onboard energy storage systems.

In addition, air transport imposes stringent power requirements, particularly during critical flight phases such as take-off and climb. These phases demand high peak power levels, requiring electrical systems capable of delivering significant power while maintaining low mass and high reliability. This trade-off between power capability and weight further complicates the design of fully electric aircraft and represents a key technical bottleneck.

For these reasons, the electrification of the aviation sector has progressed at a considerably slower pace compared to other transport modes, currently being limited to a relatively narrow set of applications. Nevertheless, its potential impact in terms of emission reduction remains substantial, making electric aviation a strategic area of development for achieving global decarbonization targets, strongly dependent on future advancements in energy storage technologies.

5.3.2 Types of electric and hybrid aircraft

In contrast to other electric mobility sectors, where a dominant technological solution is progressively emerging and leading the electrification process, electric aviation is characterized by the coexistence of multiple aircraft configurations. This results in a wide range of solutions that differ significantly in terms of mission profile, energy requirements, and technological maturity. The literature does not provide a universally accepted classification of electric aircraft. However, several approaches converge

Table 5.5 Comparison of main electric and hybrid aircraft categories based on mission profile, range, passenger capacity, and energy source.

Parameter	eVTOL	Regional Electric Aircraft	Hybrid-Electric Aircraft
Typical mission profile	Urban / short-range	Short- to medium-range (regional)	Medium- to long-range
Range (order of magnitude)	~10–50 km	~100–500 km	>500 km (depending on configuration)
Passenger capacity	2–6	9–50	50+ (potentially)
Energy source	Battery-electric	Battery-electric	Hybrid (fuel + electric)

towards a synthesis based on three main categories [19,20], whose key characteristics are summarized in Table 5.5 and further described below:

- **Electric Vertical Take-Off and Landing (eVTOL) aircraft:** These represent one of the most prominent configurations in electric aviation. They are typically small-scale aircraft with limited passenger capacity. Their VTOL capability makes them particularly suitable for operation in urban environments, as they do not require conventional airport infrastructure for take-off and landing (see Fig. 5.8). However, this operational flexibility comes at the expense of very high power requirements, especially during take-off and hover phases. These peak power demands, combined with the current limitations of battery energy density, significantly constrain both range and payload capacity, making eVTOLs primarily suitable for short-range missions.
- **Regional electric aircraft:** These correspond to medium-capacity aircraft, typically designed to transport between 9 and 50 passengers. Their primary objective is to electrify regional routes over short to medium distances that are compatible with current battery technologies. In contrast to eVTOL platforms, which rely on rotor-based configurations, regional electric aircraft generally adopt conventional fixed-wing aerodynamic designs, which are inherently more energy-efficient (see Chapter 4). This results in lower overall energy consumption during cruise and more stable power demand profiles, enabling increased range and making them suitable for longer routes compared to urban air mobility platforms.
- **Hybrid-electric aircraft:** These configurations represent a transitional solution, where propulsion is primarily based on ICE powered by conventional fuels, complemented by onboard electric systems. These electrical subsystems are typically used to supply peak power demands or to optimize specific phases of flight, such as take-off or climb. As a result, hybrid-electric architectures improve overall system efficiency while mitigating the limitations imposed by battery energy density. This enables operation over longer distances and with higher passenger capacities (potentially exceeding 50 passengers). Due to these

Figure 5.8 Joby Aviation eVTOL, representative of emerging electric aircraft designed for urban air mobility applications. (Image courtesy of Wikimedia Commons.)

characteristics, hybrid-electric aircraft are widely regarded as the most viable medium-term solution for the electrification of the aviation sector [19].

The fact that a transitional solution such as hybrid-electric aircraft represents the most viable option in the medium term highlights the relatively early stage of electrification in aviation compared to other transport sectors. In this context, the coexistence of different aircraft configurations enables a gradual adoption of fully electric solutions in applications where the technology is already viable, such as short-range and low-capacity transport, while awaiting further technological advancements that allow scaling purely electric configurations to more demanding use cases.

5.3.3 Energy and propulsion system architecture

In electric aviation, the energy and propulsion system plays a central role, strongly influencing overall aircraft design. This is primarily due to the fact that, unlike other transport systems such as rail, the aircraft operates as a closed energy system, where all the energy required for propulsion must be generated or stored onboard. This constraint imposes stringent and often competing requirements, including high power delivery capability, elevated efficiency, low weight, and strict safety standards. As a result, the design of the energy and propulsion system becomes a key driver in the overall aircraft configuration.

It is useful to structure the onboard energy system into three main functional blocks: energy storage, power conversion and conditioning through power electronics, and propulsion. In fully electric configurations, this architecture typically follows a relatively simple unidirectional energy flow, where energy is stored in batteries and delivered to the propulsion system after appropriate conditioning. This configuration is inherently dependent on the autonomy provided by the battery system, which remains limited due to the relatively low specific energy of current battery technologies.

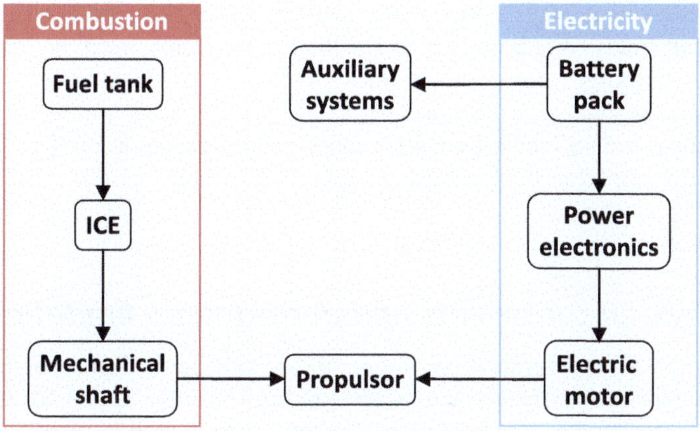

Figure 5.9 Simplified parallel hybrid-electric propulsion architecture, illustrating the direct mechanical coupling of internal combustion and electric motors for combined thrust generation.

Consequently, the achievable range and payload capacity are significantly constrained in purely electric aircraft.

To mitigate these limitations, hybrid-electric architectures have been proposed, where electric propulsion is combined with a combustion-based system relying on ICEs. This approach enables an extension of the operational range while retaining some of the efficiency benefits of electric propulsion. Although different hybrid configurations exist, including series architectures analogous to those used in ECs (see Chapter 2), parallel architectures (such as the one shown in Fig. 5.9) are more commonly adopted in aviation. This is mainly because they allow both propulsion systems, electric and combustion-based, to operate simultaneously. In such configurations, the electric system typically provides supplementary power during high-demand phases such as take-off or climb, while the combustion system sustains the aircraft during cruise and provides the bulk of the energy required for long-range operation.

The propulsion system itself must meet demanding requirements in terms of power-to-weight ratio and efficiency across varying load conditions. Permanent magnet synchronous motors are currently the most widely used solution in electric aviation, owing to their high efficiency, compactness, and high power density. Induction motors are also considered in some designs due to their robustness and simpler construction, although they generally exhibit lower efficiency and power density compared to permanent magnet synchronous motors.

Electric propulsion units can be arranged in conventional centralized configurations or in more advanced layouts such as distributed electric propulsion (DEP). In DEP architectures, multiple propulsion units are distributed along the airframe rather than concentrated in a few locations. This approach offers potential advantages in terms of aerodynamic efficiency, boundary layer control, and design flexibility. However, it also introduces additional complexity in terms of system integration, control, and reliability.

Finally, redundancy plays a critical role in aeronautical systems in order to comply with stringent safety requirements. Redundant subsystems must be incorporated across energy storage, power electronics, and propulsion units, ensuring fault tolerance and continued safe operation in the event of component failures. While this enhances safety, it also increases system mass and architectural complexity, further reinforcing the importance of careful system-level design optimization in electric aviation.

5.3.4 Physical constraints and performance limitations

In electric aviation, the physical constraints of the vehicle itself have a significantly greater impact than in other electric transport modes. These limitations act as dominant constraints on system design, operational feasibility, and overall performance. Among them, four factors stand out due to their critical influence on aircraft performance: the specific energy of the energy storage system, mass sensitivity, the power-to-weight ratio, and the flight operational profile.

- **Specific energy of the energy storage system:** Current battery technologies exhibit an energy density several orders of magnitude lower than that of conventional fossil fuels. As a result, achieving even moderate flight autonomy requires the installation of relatively heavy energy storage systems. As previously discussed for UAVs (see Chapter 4), the weight of the storage system becomes a critical design parameter, requiring a trade-off between energy capacity and overall mass. Alternative solutions, such as hydrogen-based storage systems, present additional challenges in the context of aviation. These include the weight and complexity of associated components, such as FCs, as well as safety concerns related to high-pressure hydrogen storage (see Chapter 8). Consequently, these limitations restrict the current operational domain of electric aircraft primarily to short- and medium-range missions.
- **Mass sensitivity:** This aspect is closely linked to the specific energy of the storage system and is considerably more critical than in ground or maritime transport. In aviation, any increase in system mass directly translates into a higher energy requirement to sustain flight. This leads to a cascading effect, negatively impacting range, efficiency, and payload capacity. As a result, the sizing of the energy storage system becomes a highly constrained optimization problem, where structural weight, propulsion requirements, and energy capacity must be carefully balanced to ensure viable operation.
- **Power-to-weight ratio:** This limitation is particularly relevant during specific phases of flight, such as take-off, where high power peaks are required over short time intervals. Meeting these demands necessitates electrical systems capable of delivering high instantaneous power, including batteries, power electronics, and propulsion units. However, designing for such peak power conditions often leads to oversizing of components, increasing system mass and further penalizing range and payload. This reinforces the need for optimized design strategies that balance performance requirements with strict weight constraints.

- **Flight operational profile:** Unlike other transport modes such as heavy rail or maritime systems, which typically operate under relatively stable energy demand conditions, aviation is characterized by a highly dynamic operational profile. Different phases of flight impose distinct energy and power requirements: take-off demands maximum power, climb and cruise involve sustained high energy consumption, and descent is associated with reduced power demand. This variability complicates onboard energy management and limits the effectiveness of conventional optimization strategies. As a result, the design of energy systems must account for a wide range of operating conditions while maintaining safety and performance.

Overall, these constraints highlight that, in electric aviation, fundamental physical limitations (particularly those related to energy storage and mass) constitute the primary barriers to performance improvement. Unlike other transport sectors, where infrastructure or system integration may dominate, the feasibility of electric aircraft is largely dictated by these intrinsic physical factors, which strongly constrain their scalability and operational envelope.

5.3.5 Certification and operational constraints

Electric aviation is not only constrained by intrinsic physical limitations, but also by a significantly more demanding regulatory framework than other transport sectors. Indeed, this mode of transport is subject to certification and safety requirements that are considerably more stringent than those applied in ground or maritime transport. As an illustrative example, Table 5.6 compares selected requirements imposed by the European Union Aviation Safety Agency (EASA) and the Federal Aviation Administration (FAA) on electric aircraft prior to their deployment. It is worth noting that these two agencies operate under bilateral agreements, such as the Bilateral Aviation Safety Agreement (BASA), which enable mutual recognition of certifications and avoid duplication of approval processes [21].

All aircraft must undergo a certification process before entering service, which results in long development and deployment timelines. This process typically involves extensive testing, validation, and compliance demonstration to ensure airworthiness, often extending over several years depending on system complexity. Both EASA and FAA apply largely common certification frameworks to conventional and electric aircraft, requiring new designs to demonstrate a level of safety equivalent to or higher than that of established technologies. This requirement imposes strict constraints on system design, leading to the integration of redundancy, fault-tolerant architectures, and advanced monitoring systems. In addition, the introduction of novel subsystems such as high-voltage electrical architectures and battery systems further increases certification complexity, as these technologies require dedicated validation procedures and safety assessments.

Safety constitutes the fundamental pillar governing operational margins in electric aviation. The need to ensure sufficient onboard energy throughout the mission significantly reduces operational flexibility. In addition, regulatory requirements

Table 5.6 Comparison of key certification and safety requirements for electric aviation in major regulatory regions.

Aspect	Europe (EASA)	United States (FAA)
Regulatory authority	Centralized agency at EU level	National authority (United States)
Type certification (Type Certificate)	Mandatory for entry into service	Mandatory for entry into service
Certification approach	Compliance-based with organizational oversight	Compliance-based with product-focused certification
Safety requirements	Very stringent, in some cases more conservative	Very stringent, risk-based approach
Redundancy requirements	High requirements for critical systems	High requirements with focus on failure mitigation

impose mandatory energy reserves to address contingencies such as diversions or holding patterns. These reserves further reduce the effective usable energy for the primary mission, thereby limiting the achievable range of the aircraft. Furthermore, unlike fuel-based systems, this reserved energy must be carried as constant mass throughout the flight, which amplifies its impact on overall efficiency and payload capacity. Consequently, fully electric aircraft are currently best suited to missions with predictable profiles and well-defined operational envelopes, restricting their applicability in more flexible or long-range scenarios.

5.3.6 Technological maturity

From a technological perspective, electric aviation remains far less mature than other electric transport sectors such as rail. Most developments are still in an emerging stage, with commercial applications largely limited to eVTOL platforms and short-range aircraft. Although numerous prototypes and demonstrators exist [22], large-scale commercial deployment has yet to be achieved. In this context, hybrid-electric configurations are currently considered the most viable solution for extending operational range without being fully constrained by the limitations of onboard energy storage. In parallel, ongoing research efforts are focused on improving battery performance, thermal management, and system integration, which are all critical to enabling future scalability.

These factors collectively introduce a high degree of technological uncertainty. Unlike more mature sectors, where development trajectories are well established, electric aviation remains strongly dependent on future advances in key enabling technologies, particularly energy storage systems and power electronics. This uncertainty directly affects investment decisions, certification strategies, and the timeline for large-scale deployment beyond UAV applications. Moreover, the lack of fully standardized certification pathways for emerging propulsion concepts further contributes to this uncertainty.

5.4 Integration within the EV ecosystem

5.4.1 Role of energy storage and energy carriers

Within electric mobility, the concept of an energy carrier (understood as the means by which energy is transported, stored, and ultimately consumed) plays a fundamental role. In this context, electricity as an energy carrier presents important differences with respect to conventional fuels such as diesel or gasoline. While traditional fuels inherently combine energy storage and transport within a single medium, electric systems rely on a separation between energy generation, transmission, storage, and end use. This separation enables the integration of multiple technological solutions, each characterized by specific operational and physical properties.

Along the energy chain, energy storage becomes a critical element, particularly in transport systems that operate as energetically closed systems, such as ECs, UAVs, maritime transport, and electric aviation. In these cases, onboard energy storage directly determines vehicle autonomy, operational flexibility, and overall performance. As a result, the selection and integration of suitable storage technologies constitute a central design consideration.

At present, batteries represent by far the most widely deployed energy storage solution in electric mobility, mainly due to their relatively high efficiency, technological maturity, and ease of integration. However, hydrogen-based energy storage systems emerge as a relevant alternative in applications where higher energy density is required, particularly when extended range or reduced refuelling time becomes critical. In parallel, hybrid configurations combining different storage and conversion technologies are frequently adopted as transitional solutions, enabling a balance between performance, flexibility, and technological constraints.

Despite these alternatives, all energy storage technologies are subject to fundamental physical limitations. Among these, energy density and specific power are especially relevant, as they impose inherent trade-offs between stored energy, system mass, and power delivery capability. These constraints ultimately influence vehicle design, mission profiles, and system-level performance. Further details regarding battery and hydrogen-based energy storage technologies are provided in Chapters 7 and 8, respectively. Nevertheless, from a systemic perspective, it is evident that energy storage constitutes the key element governing the architecture, design, and performance of any electric mobility system.

5.4.2 Energy supply paradigms and infrastructure integration

Different energy supply alternatives exist in electric mobility, defined by the manner in which energy is delivered to the vehicle and by the degree of dependence on external infrastructures. These supply paradigms have a direct impact on the overall energy architecture of the vehicle and strongly influence its operational capabilities. In general terms, energy supply solutions in electric mobility can be classified into four main categories:

- **Continuous external supply:** this paradigm is characteristic of transport systems that operate as energetically open systems, such as rail transport. In this configuration, the vehicle continuously receives energy from external infrastructure through systems that enable uninterrupted current flow, such as overhead catenary lines or third-rail systems. This approach effectively decouples vehicle autonomy from onboard energy storage capacity, thereby reducing dependence on storage technologies and enabling sustained operation without significant onboard energy reserves.
- **Distributed charging:** in this case, the vehicle operates as an energetically closed system during most of its operation but receives energy from external sources at specific points. This paradigm is typical of ECs and certain rail-based solutions. The vehicle completes the majority of its mission relying on onboard stored energy, while periodically accessing external charging infrastructure to replenish its energy reserves. This approach requires a distributed charging network to ensure energy availability across the system and implies a strong integration with the electrical grid and user-dependent operational patterns.
- **Closed energy systems:** this paradigm is characteristic of electrified maritime transport and aviation, where all the energy required for operation is stored onboard prior to the mission. External energy supply is typically limited to pre-operation phases, such as refuelling or charging at ports or airports. As a result, vehicle performance and operational feasibility are highly dependent on the capabilities and limitations of onboard energy storage technologies.
- **Hybrid systems:** in some vehicles, such as ERECs (see Chapter 2), onboard energy generation systems are integrated alongside energy storage technologies. These systems, often based on conventional fuel-powered generators, allow for the partial replenishment of onboard energy during operation. Hybrid configurations provide increased operational flexibility and are frequently adopted as transitional solutions when energy storage technologies alone are not sufficient to meet performance requirements.

5.4.3 Common architectural building blocks

From a functional perspective, the architecture of EVs, regardless of the transport mode, is structured around a set of common building blocks. These elements define a modular framework upon which different technological solutions are implemented, ultimately determining how energy is managed, distributed, and utilized within the vehicle.

First, control systems constitute the supervisory and decision-making layer of the architecture. These systems monitor key variables through sensing and data acquisition subsystems, and generate reference signals for quantities such as torque or rotational speed. To accomplish this, control systems implement energy management

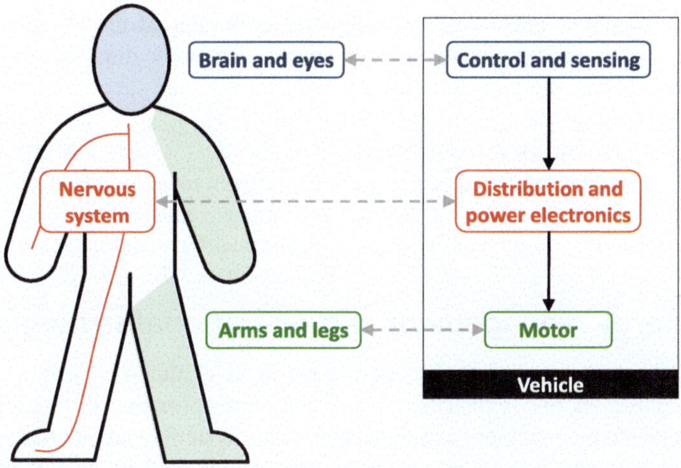

Figure 5.10 Conceptual analogy between the main functional blocks of an EV and the human body.

strategies, often supported by optimization-based approaches, with the objective of maximizing overall system efficiency while ensuring safe and reliable operation under varying operating conditions.

Once the reference signals are generated, they are propagated to the different functional blocks through energy distribution systems. These systems are closely coupled with power electronics devices, which play a critical role in adapting voltage and current levels between subsystems. Components such as inverters, rectifiers, and DC-DC converters enable efficient energy conversion and precise control of internal power flows. In addition, power electronics provide the necessary interface between energy storage systems, propulsion units, and auxiliary loads, ensuring compatibility and operational flexibility across the system.

Finally, electric machines are responsible for converting electrical energy into mechanical motion, which constitutes the final usable form of energy for vehicle propulsion. In electric mobility applications, these machines operate under demanding conditions in terms of power density, efficiency, and thermal constraints. Consequently, achieving a high power-to-weight ratio is a key design requirement to avoid excessive mass penalties. Among the most widely adopted technologies are permanent magnet synchronous machines and induction machines, both offering suitable performance characteristics for a broad range of applications.

From a conceptual standpoint, these three functional blocks can be interpreted through an analogy with the human body, as illustrated in Fig. 5.10. In this analogy, the control system corresponds to the brain, responsible for decision-making processes; the energy distribution and power electronics systems resemble the nervous system, transmitting signals and regulating energy flows; and the electric machines act as the limbs, executing motion based on the received commands.

The design process of EVs also exhibits common principles across different transport sectors. In this context, modularity and scalability are key concepts, facilitating component reuse and enabling the adaptation of the system to diverse applications. These principles also allow the architecture to be dimensioned according to specific power and energy requirements. Furthermore, the specific implementation of these functional blocks is ultimately determined by factors such as the power-to-energy ratio, redundancy requirements, and the desired level of structural and functional integration, which directly impact performance, reliability, and system complexity.

5.4.4 Energy management and operational strategies

Energy management constitutes the operational layer of the vehicle that determines how the available energy is distributed, utilized, and optimized throughout its duty cycle. Appropriate management strategies enable the system to adapt efficiently to different operating conditions, thereby maximizing overall performance. Consequently, the implementation of suitable energy management approaches has direct implications on vehicle design, influencing component sizing, system integration, and overall efficiency.

Energy management in EVs is structured around key variables such as the power–energy relationship and the characterization of the load profile. The power–energy trade-off plays a central role in defining vehicle performance. While energy is directly linked to vehicle range or mission duration, power is more closely associated with dynamic response and transient capability. Achieving an appropriate balance between these two variables enables an optimized sizing of system components, contributing to reductions in weight and volume while maintaining the required performance levels.

The load profile, defined as the temporal evolution of energy demand, represents another critical aspect. Accurate characterization of this profile allows for the effective implementation of demand management strategies. Depending on the application, load profiles may vary significantly, ranging from quasi-steady profiles characterized by sustained power demand over long periods (heavy rail or maritime transport), to highly dynamic profiles with frequent power peaks followed by lower demand phases (light rail or aviation). These differences directly impact the design of control strategies and the required responsiveness of the system.

A proper understanding of the load profile also enables mission planning, introducing a strategic dimension to energy management. In this context, peak power management strategies can be implemented to limit transient loads, reducing thermal stress and potentially allowing for smaller component sizing. In parallel, efficiency-oriented strategies based on optimization techniques can be applied to minimize energy losses over the operational cycle.

The complexity of these strategies depends on the nature of the load profile. Systems with relatively constant demand may rely on simpler control approaches, whereas highly dynamic applications require more advanced control algorithms capable of real-time adaptation. This highlights the importance of not only accurately characterizing the load profile, but also implementing an energy management strategy tailored to the specific operational requirements of the vehicle.

Table 5.7 Functional roles of different transport modes within the electric mobility ecosystem.

Transport Mode	Primary Role
Rail transport	Mass transportation
Electric road vehicles	Distributed and flexible mobility
Maritime transport	Global logistics
Electric aviation	Rapid long-travel mobility
UAVs	Specialized operational capabilities

5.4.5 System-level integration within the transport ecosystem

The integration of different electric mobility modes within the transport sector cannot be approached in isolation, but rather from a systemic perspective in which each transport niche fulfills a differentiated role. In this context, it is essential to exploit the intrinsic capabilities of each transport mode in order to optimize overall system performance. As shown in Table 5.7, each electric mobility solution can be assigned a specific function within the system, aligned with its technical characteristics. This enables different mobility segments to complement each other, providing capabilities that may be limited or inefficient in other technologies.

Such an intermodal approach requires the adoption of a global system-level design perspective, where measures are implemented to ensure the effective integration of each vehicle within the broader transport ecosystem. Importantly, this framework extends beyond the operational coordination of transport modes, encompassing also energy management and infrastructure-level integration.

From this standpoint, the design of the ecosystem must address the joint optimization of both electrical and logistical infrastructures, which are inherently coupled in electrified transport systems. On the one hand, connection to the electrical grid provides the necessary charging points that are critical for certain transport modes requiring either continuous or in-route energy supply, as is the case for rail systems. On the other hand, this dependency introduces new constraints in system planning, particularly in urban environments.

Consequently, urban and transport planning must incorporate the electrical network as an active component of the mobility system, rather than as a passive support infrastructure. This integrated approach enables the simultaneous optimization of energy and traffic flows, reducing inefficiencies associated with poorly managed modal transitions or underutilized infrastructure. Ultimately, such coordination is key to achieving a coherent and efficient electric mobility ecosystem at system level.

5.4.6 Technological convergence and future trends

The electric mobility ecosystem is evolving towards a progressive technological convergence across transport segments that have historically been highly differentiated.

In this context, modes traditionally based on fossil fuels, such as road transport, as well as others historically confined to niche applications, such as UAVs, are gradually transitioning towards partial or fully electrified architectures, alongside an expansion into a broader range of operational domains.

This convergence is supported by two fundamental pillars: the adoption of common technological building blocks (see Section 5.4.3) and the increasing standardization of energy architectures. Together, these trends enable economies of scale and facilitate knowledge transfer across applications. However, the transition towards electrification remains heterogeneous across transport sectors. Mature technologies can be identified where electrification has been historically established, such as rail transport, while other sectors, including road and maritime transport, are currently in an intermediate stage. In contrast, aviation remains at an early stage of electrification, primarily constrained by fundamental physical limitations.

The shift towards an electrified technological base is not only driven by economic and societal trends, but also by increasing regulatory pressure and the need for decarbonization, often influenced by geopolitical considerations [23]. In parallel, advancements in transversal technologies such as digitalization and automation are expected to introduce additional layers of system-level optimization and control. These technologies enable enhanced coordination not only of traffic flows but of the entire transport ecosystem. Concepts such as smart grids and the Internet of Things have emerged as key enablers for integrated system management, allowing EVs to be incorporated within a holistic control framework.

From a technological perspective, although subject to uncertainty, significant improvements in energy storage technologies are anticipated, particularly in terms of battery energy density. Such advancements are expected to enhance the performance of transport modes that are highly sensitive to mass, especially aviation, thereby enabling the progressive expansion of electrification into domains that are currently constrained. Ultimately, the evolution of the electric mobility ecosystem will depend on the combined progress in energy storage, system integration, and cross-sector technological convergence.

5.5 Cross-sector comparative analysis

5.5.1 Comparative framework and evaluation criteria

The objective of this section is to establish a systematic comparison of the different transport modes analyzed in Chapters 2, 4, and 5. This comparison is motivated by the differentiated characteristics of each transport mode, previously synthesized in Section 5.4, which have led to a clearer definition of their primary roles within the electric mobility ecosystem (as previously outlined in Table 5.7), as well as to notable differences in their technological maturity and level of deployment.

In order to define a consistent and system-level comparative framework, this section evaluates the main transport modes (road transport, maritime, rail, UAVs, and aviation) based on five fundamental criteria:

Table 5.8 Qualitative comparison of transport modes based on key evaluation criteria.

Criterion	Road	Rail	Maritime	Aviation	UAVs
Energy Requirements	Medium	High	Very High	High	Low
Mass Sensitivity	Medium	Low	Low	Very High	Very High
Infr. Dependence	Medium	Very High	Variable	High	Low
Operational Flexibility	High	Low	Medium	Medium	High
Techn. Maturity	High	High	Medium	Low	Medium–High

- **Energy requirements**: primarily determined by the operational consumption profile, including aspects such as the power-to-energy ratio and the total energy demand associated with each system.
- **Mass sensitivity**: referring to the influence of the total vehicle mass on key performance parameters such as range, payload capacity, and overall efficiency.
- **Infrastructure dependence**: capturing the extent to which each transport mode relies on dedicated energy supply infrastructures, such as electrified networks or charging systems.
- **Operational flexibility**: describing the ability of each transport mode to adapt to mission profiles beyond its primary role.
- **Technological maturity**: providing an indication of the current level of development and deployment, as well as the expected trajectory of each technology.

To facilitate a structured comparison, Table 5.8 summarizes the qualitative assessment of each transport mode according to the defined evaluation criteria.

5.5.2 Energy requirements and power profiles

Energy requirements and power profiles play a fundamental role in determining the feasibility of electrification across different transport modes, thus constituting a key criterion for the classification of EVs. In this context, it is essential to consider not only the total energy demand of a vehicle, but also additional factors such as the power-to-energy ratio and the temporal and spatial distribution of energy consumption.

Regarding total energy demand, significant differences can be observed across sectors, spanning several orders of magnitude. Maritime transport typically represents the highest energy demand, with consumption levels in the order of gigawatts-hour for large vessels. Aviation, in contrast, generally operates in the megawatts-hour range, particularly in regional applications. In the case of rail transport, a distinction must be made between heavy and light rail systems, which often occupy an intermediate position between aviation and maritime transport. Finally, road transport (particularly

ECs) and UAVs are characterized by comparatively lower energy demands, typically within the kilowatts-hour range, which facilitates the adoption of onboard energy storage solutions.

From the perspective of the power-to-energy ratio, air transport (including both UAVs and electric aircraft) requires high power capabilities, especially during critical phases such as takeoff and climb. In contrast, maritime transport is characterized by relatively stable and continuous power demand profiles associated with cruise operation. Road transport can be considered an intermediate case, with moderate power requirements and variable operating conditions. In rail transport, it is again necessary to distinguish between system types: heavy rail systems typically exhibit more stable load profiles, whereas light rail systems are characterized by frequent power peaks associated with repeated acceleration and braking cycles between stations.

Finally, the consumption profile introduces an additional dimension of analysis. Road-based EVs generally exhibit distributed and decentralized consumption patterns, reflecting their diverse and user-dependent operation. In contrast, rail transport is characterized by more centralized energy consumption, closely linked to fixed infrastructure. For aviation and UAVs, energy consumption is strongly dependent on predefined mission profiles, with clearly differentiated phases in terms of power demand.

5.5.3 Mass sensitivity and its impact on system design

In electric mobility, mass sensitivity has a direct impact on vehicle design and ultimately affects range, payload capacity, and overall performance. As such, it plays a critical role in determining the feasibility of electrification across different transport modes. However, mass sensitivity is not uniform across the sector, but rather highly heterogeneous, reflecting the distinct physical constraints and operational requirements of each system.

At one extreme, UAVs and electric aviation exhibit very high sensitivity to mass. In these systems, vehicle design is strongly constrained by weight considerations, primarily due to the continuous requirement to generate lift. In addition, high power demands during critical phases such as takeoff and climb further exacerbate this sensitivity. As a result, any increase in mass has a direct and significant impact on performance, reducing range and payload while increasing energy consumption. This imposes strict limitations on the amount of onboard energy storage that can be integrated, thereby constraining the achievable autonomy of fully electric configurations.

Road transport occupies an intermediate position. Although ECs are sensitive to mass, its impact is less critical than in aviation. Increased vehicle weight leads to higher energy consumption and reduced efficiency. However, the absence of lift requirements allows for greater design flexibility. This enables the integration of larger battery systems, albeit with trade-offs in terms of efficiency, cost, and overall vehicle performance.

At the opposite end of the spectrum, rail and maritime transport exhibit comparatively low sensitivity to mass. In rail systems, the low rolling resistance associated with steel-on-steel contact, combined with the availability of continuous external energy supply, significantly reduces the influence of vehicle mass on performance. In maritime transport, while mass contributes to increased hydrodynamic resistance, its relative impact remains secondary compared to other dominant factors such as vessel speed, operating profile, and hull design.

5.5.4 Infrastructure dependence

The different segments within electric mobility exhibit varying degrees of dependence on external infrastructure, including charging systems, continuous power supply, and other installations required for vehicle operation. This parameter not only affects the deployment of electric solutions across the transport sector, but also directly impacts their scalability.

In this context, electric rail transport represents the mobility segment with the highest dependence on external infrastructure. Its electrification is intrinsically linked to the availability of continuous power supply systems such as overhead catenary lines or third-rail configurations (see Section 5.1.3). The deployment of such infrastructure, together with its integration into the electrical grid, entails significant CAPEX, which ultimately constrains large-scale expansion despite enabling highly reliable and continuous energy delivery.

The aviation sector also exhibits a high level of infrastructure dependence, although of a fundamentally different nature compared to rail. Certain aircraft configurations require dedicated take-off infrastructure, such as launch systems, particularly in specific operational concepts. In addition, fully electric aircraft require high-power charging points, while hybrid-electric architectures typically rely on adapted energy supply systems. These requirements, combined with strict operational and safety constraints, limit infrastructure flexibility and standardization.

Road transport occupies an intermediate position in terms of infrastructure dependence. Although it requires the deployment of an adequate network of charging points, this infrastructure can be implemented progressively and in a decentralized manner, often leveraging the existing electrical grid. This characteristic reduces its dependence relative to highly centralized systems such as rail, while enabling scalable adoption, albeit with challenges related to grid integration and demand management.

UAVs, particularly at small scales, present minimal infrastructure requirements. While fixed-wing UAVs may require specific take-off infrastructure (see Chapter 4), such systems are typically compact and less complex. Charging operations are generally performed using conventional means, further reducing infrastructure constraints. This low dependence contributes to the high operational flexibility of UAV systems.

In the case of maritime transport, infrastructure dependence is strongly application-dependent. Medium- and long-distance operations rely primarily on onboard energy generation or storage systems, resulting in reduced dependence on external infrastructure during operation. In contrast, short-range applications such as ferries require shore-based charging installations, including electrified port infrastructure.

5.5.5 Operational flexibility and mission profiles

Operational flexibility refers to the ability of a transport system to adapt to different mission requirements, support a wide range of applications, and integrate effectively within complex transport ecosystems. This parameter plays a critical role in the adoption of electric mobility solutions and is strongly influenced by factors such as infrastructure dependence, energy requirements, and the capability to accommodate diverse routes and operational profiles.

Within the electric mobility landscape, road transport and UAVs exhibit the highest degree of operational flexibility. Electric road vehicles benefit from an extensively developed road network combined with a progressively expanding and decentralized charging infrastructure. This enables high adaptability in terms of routing, scheduling, and usage conditions. UAVs, on the other hand, offer an even greater degree of flexibility due to their minimal reliance on physical infrastructure and their ability to operate in three-dimensional space. This allows for highly dynamic mission profiles and access to otherwise unreachable areas. However, their operational flexibility can be constrained by regulatory frameworks, airspace restrictions, and safety requirements, which must be considered in practical deployments.

At the opposite end of the spectrum, rail transport presents the lowest operational flexibility. Its operation is inherently tied to dedicated infrastructure, with movement strictly constrained to predefined tracks. This results in limited adaptability to changes in demand or operational conditions. Maritime transport occupies an intermediate position. While vessels are not bound to fixed physical tracks, their operation is influenced by established shipping routes, port infrastructure, and logistical constraints. This allows for a greater degree of flexibility compared to rail, although still limited, particularly in long-distance and large-scale operations.

Finally, aviation exhibits a conditionally flexible operational profile. Although aircraft can traverse the airspace with a high degree of freedom, their operation is constrained by regulatory requirements, air traffic management, and safety considerations. Additionally, dependence on airport infrastructure, certification processes, and structured flight planning introduces further limitations, particularly in the context of electric and hybrid-electric aircraft.

5.5.6 Technological maturity and deployment readiness

Technological maturity and deployment readiness determine not only the technical feasibility of each solution, but also the rate at which it can be adopted at commercial scale. Within this framework, rail and road transport represent the most mature segments of electric mobility. In particular, electric rail systems benefit from decades of continuous development, resulting in highly standardized infrastructure, well-established operational models, and proven system architectures. This long-term evolution enables reliable operation and facilitates large-scale deployment, albeit within the constraints imposed by infrastructure dependence.

Electric road vehicles, while historically more recent than rail electrification, have undergone rapid development throughout the twenty-first century. This progress has been largely driven by advances in lithium-based battery technologies, significant cost reductions, and increasing regulatory pressure aimed at decarbonization (see Chapter 1). As a result, ECs have reached a high level of technological maturity, supported by a progressively expanding and increasingly decentralized charging infrastructure.

UAVs occupy an intermediate position. They exhibit a relatively high level of maturity in specific applications, particularly in short-range and mission-specific operations. However, their large-scale deployment remains constrained by regulatory frameworks, airspace integration challenges, and operational limitations, which collectively hinder broader adoption.

Similarly, electrified maritime transport presents a heterogeneous maturity landscape. Fully electric solutions are well established in short-range applications such as ferries and inland waterways. In contrast, long-distance maritime transport remains at a comparatively early stage of electrification, primarily due to the limitations imposed by current energy storage technologies and the high energy demands associated with deep-sea operations.

Finally, aviation is generally characterized as an emerging sector in the context of electrification. Despite significant progress at the prototype level, commercial deployment remains limited. This relative immaturity is mainly driven by stringent certification requirements, high sensitivity to mass, and current constraints in battery energy density, all of which impose significant limitations on scalability and operational viability.

References

[1] M. Tostado-Véliz, P. Arévalo, F. Jurado, An optimization framework for planning wayside and on-board hybrid storage systems for tramway applications, Journal of Energy Storage 43 (2021) 103207, https://doi.org/10.1016/j.est.2021.103207.

[2] P. Garcia, L.M. Fernandez, C.A. Garcia, F. Jurado, Energy management system of fuel-cell-battery hybrid tramway, IEEE Transactions on Industrial Electronics 57 (12) (2010) 4013–4023, https://doi.org/10.1109/TIE.2009.2034173, https://ieeexplore.ieee.org/abstract/document/5289980.

[3] J.P. Torreglosa, P. García, L.M. Fernández, F. Jurado, Predictive control for the energy management of a fuel-cell–battery–supercapacitor tramway, IEEE Transactions on Industrial Informatics 10 (1) (2014) 276–285, https://doi.org/10.1109/TII.2013.2245140, https://ieeexplore.ieee.org/abstract/document/6450094.

[4] R.D. White, Ac/dc railway electrification and protection, in: IET Professional Development Course on Electric Traction Systems, Institution of Engineering and Technology, 2008.

[5] R.J. Hill, Traction drives and converters, in: 3rd IET Professional Development Course on Railway Electrification Infrastructure and Systems, Institution of Engineering and Technology, 2007.

[6] N. Rodrigues, J. Cunha, V. Monteiro, J.L. Afonso, Railway auxiliary power supply system: a modular multilevel converter approach, Transportation Research Procedia 72 (2023) 3182–3189, https://doi.org/10.1016/j.trpro.2023.11.877, in: tRA Lisbon 2022 Conference Proceedings Transport Research Arena (TRA Lisbon 2022), 4th-17th November 2022, Lisboa, Portugal, 2022.
[7] H. Hu, Y. Liu, Y. Li, Z. He, S. Gao, X. Zhu, H. Tao, Traction power systems for electrified railways: evolution, state of the art, and future trends, Railway Engineering Science 32 (2024) 1–19, https://doi.org/10.1007/s40534-023-00320-6.
[8] F.J. Bris-Peñalver, R. Verdecia-Peña, J.I. Alonso, A survey of AI-enabled predictive maintenance for railway infrastructure: models, data sources, and research challenges, Sensors 26 (3) (2026), https://doi.org/10.3390/s26030906.
[9] International Energy Agency, The future of rail: Opportunities for energy and the environment, Tech. Rep., International Energy Agency, 2019, https://www.iea.org/reports/the-future-of-rail.
[10] International Union of Railways, Railway statistics, Tech. Rep., International Union of Railways, 2024, https://uic.org/statistics.
[11] M. Chester, A. Horvath, Life-cycle energy and emissions inventories for motorcycles, diesel automobiles, school buses, electric buses, Chicago rail, and New York city rail, Tech. Rep., Center for Future Urban Transport: A Volvo Center of Excellence, UC Berkeley, 2009, https://escholarship.org/uc/item/6z37f2jr.
[12] J. Campos, G. de Rus, Some stylized facts about high-speed rail: a review of hsr experiences around the world, Transport Policy 16 (1) (2009) 19–28, https://doi.org/10.1016/j.tranpol.2009.02.008.
[13] United Nations Conference on Trade and Development (UNCTAD), Review of maritime transport 2025, Tech. Rep., United Nations, Geneva, 2025, https://unctad.org/publication/review-maritime-transport-2025.
[14] R. Borstlap, H. ten Katen, Ship Electrical Systems, 2nd edition, Dokmar Maritime Publishers B.V., 2021.
[15] W. Guan, L. Chen, Z. Wang, J. Chen, Q. Ye, H. Fan, A 500 kw hydrogen fuel cell-powered vessel: from concept to sailing, International Journal of Hydrogen Energy 89 (2024) 1466–1481, https://doi.org/10.1016/j.ijhydene.2024.09.418.
[16] European Commission, Scalable, plug & play and modular dc power systems for low-emission large vessels (all-dc-ships), horizon Europe Project, Grant Agreement No. 101192732, https://doi.org/10.3030/101192732, https://cordis.europa.eu/project/id/101192732, 2024.
[17] N. Vahabzad, M. Jadidbonab, B. Mohammadi-Ivatloo, S. Tohidi, A. Anvari-Moghaddam, Energy management strategy for a short-route hybrid cruise ship: an igdt-based approach, IET Renewable Power Generation 14 (10) (2020) 1755–1763, https://doi.org/10.1049/iet-rpg.2019.0882.
[18] Candela Technology AB, Candela p-12 completes world's longest electric sea journey, press release. Retrieved February 2, 2026, https://www.mynewsdesk.com/candela-speedboat-ab/pressreleases/candela-p-12-completes-worlds-longest-electric-sea-journey-3430126, 2026.
[19] B.J. Brelje, J.R. Martins, Electric, hybrid, and turboelectric fixed-wing aircraft: a review of concepts, models, and design approaches, Progress in Aerospace Sciences 104 (2019) 1–19, https://doi.org/10.1016/j.paerosci.2018.06.004.
[20] International Civil Aviation Organization (ICAO), Electric and hybrid aircraft platform for innovation (e-hapi), https://www2023.icao.int/environmental-protection/Pages/electric-aircraft.aspx. (Accessed 13 April 2026).

[21] AirlineFYI, How aircraft are certified: faa and easa type certification, https://airlinefyi.com/guide/safety-and-standards/how-aircraft-are-certified/. (Accessed 13 April 2026).
[22] European Alternative Fuels Observatory, Electric aircraft, https://alternative-fuels-observatory.ec.europa.eu/transport-mode/aviation/alternative-fuelled-aircrafts, 2024. (Accessed 13 April 2026).
[23] M. Giuli, S. Oberthür, Assessing the eu's evolving position in energy geopolitics under decarbonisation, The International Spectator 58 (3) (2023) 152–170, https://doi.org/10.1080/03932729.2023.2199648.

Charging technologies

6

6.1 Fundamentals of charging systems

6.1.1 Power vs energy in EV charging

In the charging process of EVs, it is essential to distinguish between the concepts of power and energy, as the main characteristics of the charging process are governed by these two parameters. In this context, power represents the instantaneous rate at which energy is transferred from the charging infrastructure to the vehicle, whereas energy corresponds to the total amount of energy delivered over a given time interval. In other words, power can be interpreted as the "speed" at which energy is transferred.

An intuitive analogy can be drawn with filling a water tank. The total volume of water to be stored represents the energy, while the flow rate at which water is supplied corresponds to the power. A higher flow rate allows the tank to be filled more quickly, just as a higher charging power enables faster energy transfer to the vehicle. Therefore, increasing the charging power reduces the time required to recharge a given amount of energy, although practical limitations prevent this relationship from being strictly linear.

In a simplified scenario where the charging power remains constant throughout the entire process, the relationship between energy, power, and time can be expressed through Eq. (6.1). Under this assumption, the charging time is inversely proportional to the available power. However, in real charging processes, this condition is rarely satisfied, as the power varies significantly over time

$$E = P \cdot t. \tag{6.1}$$

For this reason, a more accurate representation of the transferred energy is given by Eq. (6.2), where power is considered a time-dependent variable. This formulation reflects the fact that the charging process is inherently dynamic and influenced by several operating conditions

$$E = \int_0^t P(t)\,dt. \tag{6.2}$$

In practice, the charging power depends on multiple factors, most notably the SoC of the energy storage system, defined in Eq. (6.3). EVs can typically sustain high charging power levels only at low or intermediate SoC values. As the SoC approaches its upper limit, the charging power is progressively reduced in a phenomenon known as *tapering*, leading to non-linear charging curves in which the final stage of the process is significantly slower

$$\text{SoC} = \frac{E}{C}. \tag{6.3}$$

Electric Mobility Systems. https://doi.org/10.1016/B978-0-44-349397-3.00014-8

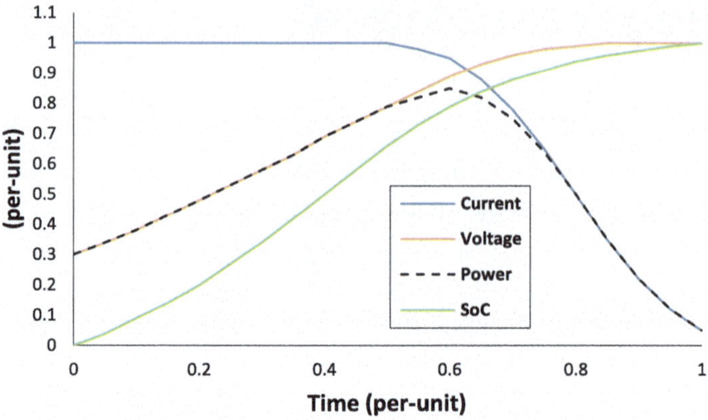

Figure 6.1 Typical CC–CV charging profile, showing the evolution of current, voltage, power, and SoC as a function of normalized time.

As a consequence, the total charging time cannot be accurately estimated solely from the nominal power of the charger. Instead, it is more appropriate to approximate it using an average charging power $\left(P_{\text{avg}}\right)$, as reflected in Eq. (6.4), which represents the effective power delivered over the entire process. This average power is generally lower than the nominal power, due to the variable nature of real charging profiles and system constraints

$$t \approx \frac{E}{P_{\text{avg}}}. \tag{6.4}$$

This distinction between nominal and effective power has direct implications for both user experience and the design of charging infrastructure. From the user perspective, it determines the actual time required to recharge the vehicle. From the infrastructure perspective, it affects the utilization of installed capacity, peak demand conditions, and the ability to meet charging demand under realistic operating scenarios.

6.1.2 Charging profiles and system constraints

The charging process of an EV is characterized by a dynamic power profile, in which the applied power varies throughout the entire charging period. A charging profile can therefore be defined as the temporal evolution of the power delivered to the vehicle during the charging process. In practice, several characteristic profiles exist, although the most widely adopted is the constant current–constant voltage (CC–CV) profile, illustrated in Fig. 6.1.

The CC–CV profile divides the charging process into two clearly differentiated phases. The initial stage corresponds to the constant current phase, during which the current remains approximately constant. In this phase, the voltage increases progressively, which leads to a corresponding increase in the transferred power and the SoC. As a result, a significant amount of energy is delivered over a relatively short period of time, making this stage particularly effective in terms of energy transfer.

As the charging process progresses, the system transitions to the constant voltage phase. At this point, the voltage is maintained at an approximately constant value, while the current begins to decrease. This reduction in current occurs more rapidly than the increase in voltage observed during the previous phase. Consequently, the charging power decreases progressively, giving rise to the phenomenon known as *tapering*. This effect explains why the charging rate is significantly reduced as the energy storage system approaches full charge (SoC $\approx$ 1), as shown in Fig. 6.1.

Although the CC–CV profile is the most commonly used in practical applications, alternative approaches have been proposed. Among these, multistage charging profiles [1] adjust the current and voltage in discrete steps according to the instantaneous value of SoC and operating conditions. When properly designed, these strategies can contribute to extending the lifetime of the energy storage system and reducing thermal stress. Additionally, pulse charging methods [2] are employed in specific applications, where current is applied intermittently in the form of pulses. This approach can help mitigate internal polarization effects and, under certain conditions, improve charging efficiency, although it introduces additional complexity in terms of control and power electronics implementation.

The design of the charging profile has important implications from the perspective of charging system design. As illustrated in Fig. 6.1, maintaining high power levels in order to reduce charging time would require sustaining high current levels throughout the process, which in turn leads to increased thermal and electrical stress on system components. These factors must be carefully considered when dimensioning charging equipment, as they impose practical limits on parameters such as the maximum current supplied by the charger and the maximum allowable voltage of the energy storage system.

In this context, it is important to emphasize that the energy storage system of the vehicle ultimately determines the power that can be accepted at each instant. The charging process is therefore governed by a dynamic interaction between the charger and the vehicle, in which the delivered power is continuously adapted to the internal conditions of the system. As a result, a charging system cannot be considered in isolation, but must be properly matched to the characteristics of the vehicle, giving rise to different charging technologies and operational strategies.

6.1.3 Efficiency and thermal limitations

As in any energy transfer process, EV charging is inherently non-ideal from an energetic standpoint. A fraction of the supplied power is unavoidably lost due to different physical phenomena occurring within the charging system, meaning that not all the transferred energy reaches the vehicle and part of it is dissipated as heat. From a system perspective, these losses not only reduce the overall efficiency of the charging process but also impose fundamental operational limits.

From an analytical standpoint, it is useful to classify these losses into three main categories. First, there are losses associated with energy conversion processes. These occur both in the charging infrastructure and within the vehicle itself, depending on the charging architecture. They originate from the operation of power electronic converters, such as AC/DC conversion in AC charging systems or DC/DC conversion in certain subsystems. Although modern converters achieve high efficiency levels, these losses become increasingly relevant at high power levels, where even small efficiency deviations translate into significant absolute power dissipation [3].

Second, resistive losses arise due to current flowing through cables, conductors, and electrical connections with non-negligible resistance. These losses can be expressed as:

$$P_{\text{loss}} = I^2 R. \tag{6.5}$$

As shown in (6.5), resistive losses scale with the square of the current. This quadratic dependence is particularly critical in EV charging applications. Increasing the charging power without a proportional increase in voltage leads to higher current levels, which in turn significantly increase power losses. As a result, high-current operation introduces not only efficiency penalties but also severe thermal constraints.

Finally, additional losses are associated with the switching processes of power electronic devices. These switching losses are linked to the operation of semiconductor materials and become more pronounced under demanding operating conditions, such as high switching frequencies or elevated current levels. Together with conduction losses, they contribute to the overall efficiency degradation of the system.

In practice, AC chargers typically operate at lower power levels and under less demanding thermal conditions, resulting in relatively high efficiency values, usually in the range of 90–95% [4]. In contrast, DC chargers are generally used for fast charging applications at significantly higher power levels. Under these conditions, increased current and switching activity lead to higher losses, and the overall efficiency typically falls within the range of 85–95%, depending on the operating point and system design [4].

The power losses described above are ultimately converted into heat, which must be properly managed to ensure safe and reliable operation. This is particularly critical in high-power applications, where components such as cables and connectors can reach elevated temperatures. Excessive thermal stress may lead to accelerated component degradation or, in extreme cases, safety hazards. For this reason, thermal management plays a fundamental role in EV charging systems. Different cooling strategies are employed depending on the power level, ranging from air cooling in low- and medium-power applications to liquid cooling solutions in high-power charging systems, where thermal constraints become significantly more demanding.

6.2 Charging technologies for ECs

6.2.1 AC charging

Nowadays, AC charging systems constitute the most widely deployed solution for charging ECs, mainly due to their ease of integration into existing electrical grids, which results in relatively low installation costs. Table 6.1 summarizes some of the

Table 6.1 Comparison between AC and DC charging technologies for ECs.

Characteristic	AC Charging	DC Charging
Energy conversion location	Onboard	Off-board
Typical power levels	Up to ~22 kW	50 kW to >350 kW
Typical voltage levels	230–400 V	400–800 V (up to ~1000 V)
Charging time	Hours	Minutes to ~1 hour
Infrastructure complexity	Low	High
Cost of infrastructure	Relatively low	High
Efficiency (typical)	90–95%	85–95%
Current levels	Moderate	Very high
Cooling requirements	Passive or air cooling	Often liquid cooling

main characteristics of this charging approach. One of the distinguishing elements of AC charging is the need for an onboard charger (OBC). This device is an AC/DC converter responsible for converting the AC signal supplied by the charging point into DC so that it can be used by the batteries and other electrical systems of the EC.

From a functional perspective, the OBC represents the main limitation in terms of the power that can be delivered in AC charging. Indeed, the maximum charging power must be constrained by the nominal rating of the OBC in order to avoid degradation or damage to this component. This limitation explains why AC charging power levels are typically low to moderate, ranging from a few kilowatts in residential applications up to approximately 22 kW in more advanced configurations. In practice, the effective charging power may also depend on grid conditions and installation constraints, which can further limit the achievable performance.

In terms of electrical configuration, AC charging can be implemented using either single-phase or three-phase networks. Single-phase systems are typical in residential environments, where the existing electrical infrastructure imposes constraints that limit the adoption of higher-power configurations. Single-phase AC charging is therefore usually restricted to relatively low power levels in order to avoid excessive current, which results in longer charging times. In contrast, three-phase charging systems can accommodate higher power levels due to improved current distribution, leading to significantly reduced charging times. This distinction is particularly relevant in public or semi-public charging points, where higher utilization rates justify increased power levels.

AC charging systems are often easily integrated into existing distribution networks, typically requiring only connection points equipped with appropriate control and protection devices. This simplicity constitutes a key advantage of AC charging and translates into lower infrastructure costs compared to DC charging, as shown in Fig. 6.2. These cost reductions arise not only from the straightforward integration into the grid, but also from the absence of large-scale power electronic conversion equipment at the charging station level (see Fig. 6.2a). Furthermore, due to the

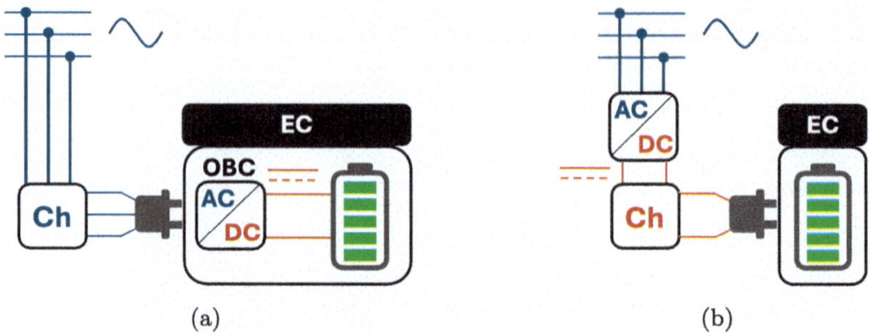

Figure 6.2 Typical (a) AC and (b) DC charging configurations for ECs. Ch: charger.

moderate power levels involved, AC charging systems generally do not require advanced cooling mechanisms, with natural or, in some cases, forced air ventilation being sufficient.

Owing to their relatively low power levels, AC charging systems exhibit high efficiency, typically in the range of 90–95% [4]. In addition, the associated losses tend to be distributed across different components and circuits, particularly within the OBC. This distributed loss profile allows for more effective management, as efficiency improvements can be addressed at the individual component level rather than requiring system-wide optimization. However, efficiency can still be affected by operating conditions such as partial load operation or voltage deviations.

The main limitation of AC charging is therefore its restricted power capability, which is typically imposed by the current limits of the OBC. As a result, charging times are generally on the order of several hours, especially in single-phase residential applications. This limitation makes AC charging less suitable for in-transit charging scenarios, where significantly shorter charging times are required. Consequently, AC charging is primarily focused on applications where charging time is not a critical factor, such as overnight residential charging, workplace charging, or long-duration parking. In these scenarios, the availability of long dwell times compensates for the lower charging power.

In summary, AC charging currently forms the backbone of EC charging infrastructure due to its ease of deployment and relatively low cost. However, its limited power capability constrains its applicability to everyday charging scenarios, significantly restricting its use in applications requiring fast energy replenishment. Despite these limitations, its scalability and compatibility with existing grids ensure its continued relevance within the overall charging ecosystem.

6.2.2 DC charging

DC charging of ECs represents a paradigm shift with respect to the AC charging approach described in Section 6.2.1. In contrast to AC charging, where the AC/DC conversion is performed onboard the vehicle through the OBC, in DC charging this function is carried out by an AC/DC converter installed off-board. This arrangement effectively decouples the charging power from the maximum rating of the OBC, enabling significantly higher power levels and, consequently, reduced charging times.

Fig. 6.2b illustrates this configuration in detail. As shown, DC charging relies on a high-power AC/DC converter located outside the vehicle. This converter draws energy directly from the distribution grid and delivers it in DC form to the vehicle interface. In practice, this architecture bypasses the OBC during DC charging operation, as its functionality is assumed by the external converter. However, this does not imply that the OBC is eliminated from the vehicle; it remains installed to support AC charging modes but is bypassed when the EC is connected to a DC charger.

The external AC/DC converter is designed to handle high power levels. In practical implementations, charging power typically ranges from approximately 50 kW in standard applications up to 350 kW in high power charging (HPC) systems [5]. These values contrast sharply with the more moderate power levels associated with AC charging, as summarized in Table 6.1. From an architectural standpoint, this converter incorporates AC/DC rectification stages followed by DC/DC conversion stages, which allow dynamic adjustment of voltage and current levels at the DC bus to match the instantaneous requirements of the charging process. This multi-stage structure also facilitates grid compliance, controlled power delivery, and operational flexibility under varying conditions, particularly under fluctuating grid conditions or variable load demands.

Regarding the charging profile, DC charging follows a progression similar to the CC–CV profile described in Fig. 6.1. Initially, a quasi-constant current phase dominates, during which both power and SoC increase rapidly. This is followed by a phase in which current and power progressively decrease as the SoC rises. In DC charging, this behavior is particularly relevant, as it results in a charging profile where the maximum power is sustained only during a limited portion of the overall process. Consequently, special care must be taken when estimating $\left(P_{\text{avg}}\right)$ in Eq. (6.4), since the discrepancy between peak power and average power can be substantial, especially in HPC scenarios.

Despite its advantages in reducing charging time, DC charging is subject to several important physical limitations. First, while removing the OBC constraint enables higher power operation, even when higher voltage levels are employed (see Table 6.1), delivering such power inherently involves large current levels. This leads to significant resistive losses, as described by Eq. (6.5). These losses increase quadratically with current, making them a critical factor in high-power systems.

As discussed in Section 6.1.3, elevated resistive losses can result in severe thermal stresses on system components, potentially leading to accelerated degradation or safety concerns. To mitigate these effects, DC charging systems require more advanced thermal management solutions compared to AC charging. In particular, liquid-cooled cables are commonly employed in HPC applications [6], allowing higher current densities to be sustained while maintaining acceptable operating temperatures. This requirement significantly increases system complexity relative to AC charging, where passive or air-based cooling is typically sufficient.

A second key limitation arises from the interaction between the charger and the vehicle. Although the external charger may be capable of delivering high power, the actual charging power can be constrained by the vehicle itself. Internal protection mechanisms within the EC may limit the accepted power based on variables such as temperature or SoC. These dynamic constraints are especially relevant in HPC scenarios, where the nominal charger rating may not reflect the effective charging power experienced in practice.

Another important aspect concerns communication. Unlike AC charging, where the communication architecture is relatively simple, DC charging requires continuous information exchange between the charger and the vehicle. This communication enables real-time adjustment of charging parameters to ensure safe and efficient operation, as discussed in more detail in Section 6.2.3. The need for such interaction adds an additional layer of system complexity but is essential for managing high-power operation under varying conditions and ensuring interoperability across different charging systems.

In terms of efficiency, DC charging typically exhibits slightly lower efficiency compared to AC charging, as indicated in Table 6.1. This is primarily due to increased losses associated with high current conduction and the presence of multiple power conversion stages within the external charger. Nevertheless, this reduction in efficiency is often offset by the substantial decrease in charging time, which is the primary advantage of DC charging.

Indeed, the ability to significantly reduce charging time is the defining feature of DC charging systems. The high power levels achievable in this mode enable the energy required for a typical EC to be replenished within minutes rather than hours. This characteristic makes DC charging particularly suitable for in-route charging and, more generally, for applications where time is a critical factor, such as high-utilization vehicle fleets.

However, these benefits come at the cost of increased infrastructure complexity. DC charging stations require high-capacity grid connections, advanced power electronics, and sophisticated monitoring, control, and protection systems. Additionally, the integration of active cooling solutions further increases system complexity. As a result, the deployment cost of DC charging infrastructure is significantly higher than that of AC charging, which limits its widespread adoption to specific use cases.

In conclusion, DC charging plays a central role in the EC charging ecosystem by overcoming key limitations associated with AC charging, particularly in terms of power and charging time. Nevertheless, its higher cost and technical complexity lead to a scenario in which both AC and DC charging technologies coexist, each serving distinct applications within the broader charging infrastructure.

6.2.3 Connectors, standards and communication

The electric mobility ecosystem is evolving towards increased standardization, ultimately enabling greater interoperability and technological convergence. In this context, implemented standards define both the physical connection interfaces and

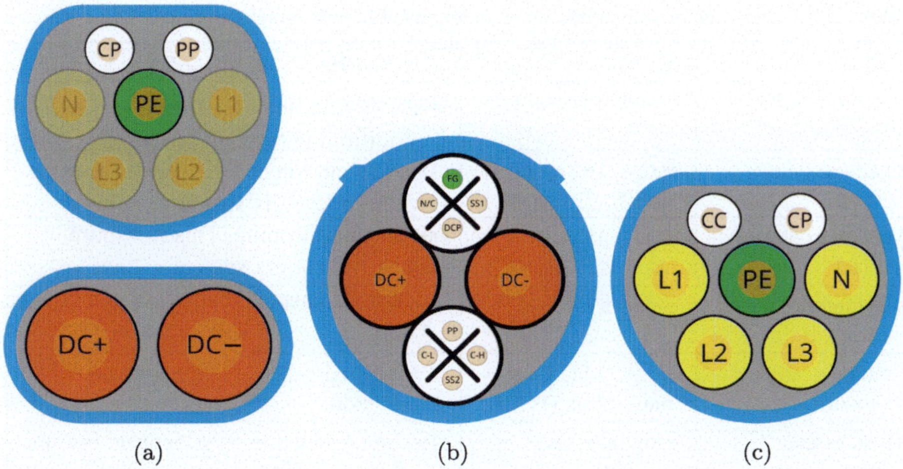

Figure 6.3 Overview of the main EC charging connectors: (a) CCS connector, (b) CHAdeMO connector and (c) GB/T connector. These connectors represent the dominant standards across different global regions and illustrate the diversity of physical interfaces currently used in electric vehicle charging infrastructure. (Image courtesy of Wikimedia Commons.)

the communication protocols between the vehicle and the charging infrastructure. This progressive harmonization is essential not only for ensuring compatibility between different systems, but also for facilitating large-scale deployment of charging infrastructure and improving user experience across regions. Moreover, it contributes to reducing technological fragmentation and supporting economies of scale in both vehicle manufacturing and infrastructure development.

Connectors represent the primary physical interface between the vehicle and the external charging system, and therefore constitute one of the critical elements determining the technical and operational feasibility of charging systems. Significant efforts are being made towards global standardization. However, substantial heterogeneity still exists, largely determined by regional deployment. Today, three main standards coexist, whose physical characteristics are illustrated in Fig. 6.3 and described below.

- **CCS**: this is the most widely deployed standard in Europe and the United States. Its distinguishing feature is the integration of AC and DC connections within a single connector assembly, facilitating operation and enabling multiple charging modes through a unified interface. Currently, it supports the highest power levels, up to 350 kW, which makes it the reference standard for HPC applications. Its modular design also simplifies vehicle-side integration and supports future scalability.
- **CHAdeMO**: this standard was developed in Japan and was one of the first to enable DC fast charging. In contrast to CCS, it supports only DC charging, which limits its flexibility. It achieved significant adoption in the early stages of electric mobility development; therefore, despite its declining use compared

Table 6.2 Main characteristics of the most widely used EC charging connectors.

Feature	CCS	CHAdeMO	GB/T
Region of main adoption	Europe, USA	Japan (legacy)	China
Charging type	AC + DC (combined)	DC only	AC and DC (separate)
Maximum power (typical)	Up to 350 kW	Up to 100–200 kW	Up to 250 kW
Maximum power (projected)	>500 kW	Up to ~400 kW	>500 kW
Communication protocol	Power Line Communication (PLC)	CAN-based communication	CAN-based communication
Market trend	Dominant and expanding	Declining adoption	Dominant in China

to other standards, it remains highly relevant for legacy EC models and existing infrastructure, particularly where early deployment was concentrated.

- **GB/T**: this is the dominant standard in China, and its adoption is strongly driven by the national regulatory framework. Its main characteristic is that it supports both AC and DC charging, but unlike CCS, it implements these modes using separate connectors, leading to a different system architecture at vehicle level. This approach provides design flexibility but may increase system complexity in certain implementations.

It is worth noting that all three standards are evolving towards significantly higher power capabilities. In all cases, future versions converge towards HPC levels in DC charging. Specifically, CCS is moving towards next-generation HPC solutions,[1] while GB/T is also expected to evolve towards new versions targeting similar performance, both aiming at power levels around 500 kW. In parallel, CHAdeMO 3.0 reaches up to 400 kW.[2] This convergence towards higher power levels is driven by the need to reduce charging times and to enable operational conditions comparable to conventional refuelling processes, particularly in long-distance and high-utilization scenarios.

In order to synthesize the characteristics of each standard and provide a structured comparison, Table 6.2. As shown, although there is some convergence in certain aspects, notable disparities still exist in connector standards, particularly in terms of integration approach, communication architecture, and regional deployment strategies.

Standardization has also progressed in the communication protocols between the vehicle and the charger. These protocols define how the EC and the charging system

[1] https://www.charin.global/technology/.

[2] https://www.chademo.com/.

exchange information in order to ensure a safe and efficient charging process. In addition, they enable the implementation of advanced functionalities such as session management, user authentication, and real-time monitoring, which are particularly relevant in public charging environments and large-scale infrastructure networks.

As discussed in Section 6.2.2, communication becomes more critical in DC charging. This is because, unlike AC charging (where the vehicle, through the OBC, controls the process) in DC charging the charger assumes control of power conversion and energy delivery. As a result, a larger volume of information must be exchanged between the vehicle and the charger, with the charger dynamically adjusting parameters such as voltage and current in real time based on vehicle-side variables such as SoC and temperature. In both charging modes, the process begins with verification that the physical connection between the vehicle and the charger has been correctly established, in a procedure known as "handshake", during which the required conditions are validated before initiating energy transfer. This initial synchronization step is essential to ensure safe operation and to prevent faults during high-power energy transfer.

As shown in Table 6.2, communication standards are still far from being fully unified across connectors. Specifically, the CCS system employs Power Line Communication (PLC), whereby data is transmitted through the same conductors used for electrical power transfer. This approach provides high data transmission capacity and is particularly suitable for fast charging applications and integration within more complex systems. In contrast, CHAdeMO and GB/T rely on Controller Area Network (CAN)-based communication. This technology is widely used in the automotive sector due to its robustness and reliability, and it provides adequate performance for charging control. However, CAN-based communication often presents limitations in terms of data throughput and flexibility when compared to PLC, which may restrict the implementation of more advanced functionalities and future scalability requirements.

6.3 Wireless charging technologies

6.3.1 Static wireless charging

Wireless charging is based on the principle of electromagnetic induction, by means of which energy can be transferred in the form of a magnetic field. This enables power transfer without physical contact between the emitting element (charger) and the receiving element (vehicle). In this sense, wireless charging represents a paradigm shift with respect to conventional conductive charging, where a physical connection between the charger and the vehicle is strictly required to establish energy transfer.

Fig. 6.4 illustrates a schematic representation of the system. A wireless charging system for ECs is essentially composed of two coils. On the one hand, the primary coil assumes the role of the charger and is supplied with high-frequency AC, which is typically used to reduce the effective impedance of the coil and improve power transfer capability. As a result, a time-varying magnetic field is generated. This primary coil is generally embedded in a fixed platform or directly integrated into the ground.

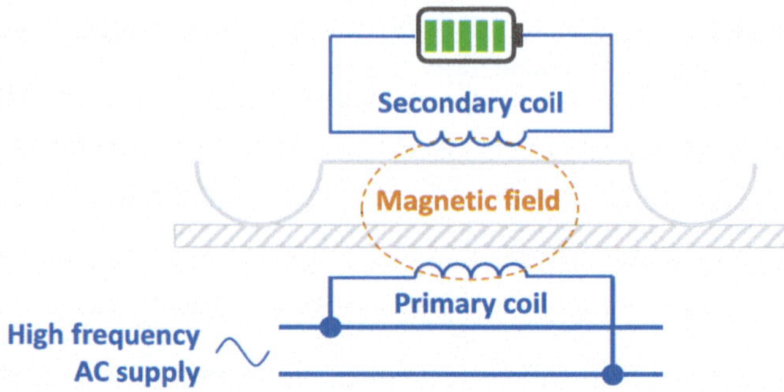

Figure 6.4 Schematic representation of a wireless charging system.

On the other hand, the secondary coil is installed in the vehicle itself, typically on its underside and positioned at a relatively short distance from the primary coil. The secondary coil concatenates the time-varying magnetic field generated by the primary coil. According to Faraday's Law,[3] the concatenated magnetic flux induces an electric current in the secondary coil, which can then be rectified and conditioned to charge the onboard batteries. In this way, energy transfer from the primary to the secondary coil is achieved without any direct electrical contact between them.

The efficiency of a wireless charging system depends on several factors, including the alignment between coils, the air gap separating them, and the operating frequency. Under ideal conditions, with optimal alignment and minimal separation, such systems can reach efficiency values comparable to conductive charging in moderate power applications (90–93% [7]). However, under more realistic conditions, where misalignment and non-ideal coupling are present, efficiency may decrease to values around 85%. This places wireless charging systems below conventional conductive solutions in terms of overall efficiency.

Despite this limitation, the absence of physical connectors provides significant advantages in terms of usability, automation, and reduced mechanical wear. Consequently, wireless charging is particularly well suited for applications where user convenience is a key factor, such as overnight charging in residential environments. In such scenarios, this technology eliminates the need for manual cable connection and simplifies the charging process, potentially enhancing user acceptance and facilitating integration into everyday mobility routines.

6.3.2 *Dynamic wireless charging*

In contrast to static charging, where the charging process is carried out while the vehicle is stationary, dynamic charging enables energy transfer while the vehicle is in motion. In this way, dynamic charging systems allow continuous and contactless power transfer to the vehicle, thereby extending the wireless charging paradigm beyond stationary applications.

[3] Faraday's law of induction (Wikipedia).

Regarding components and operating principles, dynamic charging relies on concepts and hardware similar to those used in static wireless charging (see Section 6.3.1). The main difference between both approaches lies in the spatial distribution of the primary coil. In static charging, the primary coil (i.e., the charging element) is concentrated at a single location, whereas in dynamic charging it is distributed along the path followed by the vehicle. To avoid excessive energy consumption and unnecessary activation, the primary infrastructure is typically divided into multiple segments that are selectively energized as the vehicle passes over them. This segmented operation requires more advanced sensing, control, and automation systems than those used in static charging.

From a performance perspective, dynamic wireless charging generally exhibits lower efficiency than static wireless charging and significantly lower efficiency than conventional conductive charging. This is primarily due to the non-stationary coupling conditions between coils, which make it more difficult to maintain optimal alignment during operation. In addition, extra losses arise from the sequential activation and deactivation of infrastructure segments, as well as from variations in coupling due to vehicle speed and position. As a result, overall system efficiency under realistic operating conditions is typically around 75% [7], although higher values may be achieved under controlled conditions. It should be noted that these efficiency figures are highly application-dependent and can vary significantly depending on factors such as vehicle speed, lateral misalignment, air gap variations, and environmental conditions.

The main advantage of dynamic charging is its ability to provide continuous energy supply to the vehicle while it is in operation. This feature reduces the dependence on the limited energy density of onboard batteries and opens the possibility of designing vehicles with smaller battery capacities. Such an approach can lead to reductions in vehicle mass, cost, and overall energy consumption. This characteristic is particularly attractive for applications involving predefined routes, such as public transport systems or logistics fleets, where the infrastructure can be strategically deployed along frequently used paths.

However, the implementation of dynamic charging presents significant technical, economic, and operational challenges. On the one hand, its effectiveness is strongly linked to use cases involving predictable trajectories, which limits its applicability for general-purpose ECs with highly variable routes. On the other hand, the integration of charging infrastructure into roadways entails substantial investment and maintenance costs, typically much higher than those associated with conventional conductive charging systems. These challenges currently constrain the large-scale deployment of dynamic wireless charging technologies, despite their strong conceptual potential.

6.3.3 Limitations and comparison

Wireless charging presents significant potential from a conceptual standpoint. However, its practical implementation involves a number of challenges and limitations that explain its currently limited deployment compared to conventional conductive charging systems. From an analytical perspective, these limitations can be broadly categorized into performance-related constraints and economic considerations.

Both static and dynamic wireless charging systems generally achieve lower efficiency than conductive solutions. Moreover, the performance of wireless systems is highly dependent on operating conditions and environmental factors. In particular, efficient energy transfer requires proper alignment between the primary and secondary coils in order to ensure a sufficiently high coupling coefficient. Achieving optimal alignment, however, can be challenging in practice. This issue is especially critical in dynamic charging, where factors such as vehicle speed and lateral displacement make it difficult to maintain stable and optimal coupling conditions. As discussed in Sections 6.3.1 and 6.3.2, the efficiency values typically achieved in wireless charging are significantly lower than those observed in conductive systems, which becomes particularly relevant in high-power applications where losses have a stronger impact on overall system performance.

From an economic standpoint, wireless charging systems involve substantially higher implementation and maintenance costs than conductive charging. These increased costs arise from several factors, including the need to integrate coil systems both in the infrastructure and within the vehicle. This effect is even more pronounced in dynamic charging, where the primary coil must be deployed along extended sections of roadway. In addition, wireless charging systems require power electronic stages capable of generating high-frequency AC in order to enable efficient inductive coupling and reduce transfer losses. This requirement contrasts with conductive systems, which can be more easily integrated into existing electrical distribution networks, particularly in the case of AC charging (see Section 6.2.1).

Finally, it is important to note that wireless power transfer inherently relies on the generation of a time-varying magnetic field to induce current in the secondary coil. As a result, these systems are intrinsically linked to AC-based operation, which introduces additional complexity in power conversion and may limit their practical scalability to very high power levels.

Overall, while wireless charging offers clear advantages in terms of usability, automation, and reduced mechanical wear, its lower efficiency, higher cost, and greater technical complexity currently limit its widespread adoption. Nevertheless, it may represent a viable solution in specific applications where these drawbacks are offset by operational or user-experience benefits.

6.4 Charging infrastructure and grid interaction

6.4.1 Grid impact and physical constraints

The large-scale deployment of EV charging infrastructure introduces significant challenges regarding its integration into the electrical grid. This impact is expected to be more pronounced at the lower layers of the power system, particularly in distribution networks, which are closer to end users [8]. From an operational perspective, EV charging can exhibit high and highly variable power demand peaks, especially in HPC applications, thereby introducing new challenges for both grid planning and operation.

A direct consequence of transport electrification is the overall increase in electricity demand. However, it is important to emphasize that the demand associated with EV charging may exhibit usage patterns that are particularly stressing for the electrical infrastructure. Indeed, EV charging demand tends to concentrate both geographically and temporally, often coinciding with periods of high activity such as the return from workplaces in areas with a high density of charging points, particularly residential zones. These characteristics introduce important challenges, as they may lead to overload conditions in critical grid components and reduce the available operational margins.

In addition to the increase in total demand, EV charging can significantly alter conventional temporal demand profiles [9]. Unlike most traditional electrical loads, whose behavior is relatively stable and well characterized, EV charging demand depends on highly variable factors such as the initial SoC of the batteries or user mobility patterns. As a result, demand profiles that have historically been predictable and stable over long periods may be substantially modified. This introduces additional uncertainty in both system operation and planning, making demand forecasting more complex.

Moreover, EV charging may introduce steep power gradients over short time intervals. These gradients are typically associated with the simultaneous initiation of multiple charging processes, particularly in scenarios with high penetration of EVs. This dynamic behavior can lead to local voltage fluctuations and increased stress on distribution equipment, further complicating grid management.

The increase in demand also has a direct impact on key operational parameters of the grid, especially in areas with a high concentration of charging points. In such environments, the individual effect of each charging process aggregates with others connected to the same network, resulting in a collective behavior that may differ significantly from that of isolated loads. This aggregation effect can produce demand peaks for which the existing infrastructure is not dimensioned, potentially leading to operational issues such as network congestion or voltage drops.

As a consequence of the above, the capability of the existing electrical infrastructure may be compromised. A critical element within the energy distribution chain is the power transformer. These devices are typically designed to operate under relatively predictable load profiles. However, EV charging can introduce more aggressive demand patterns, characterized by high power peaks and pronounced demand gradients, for which existing transformers may not be adequately sized. This can lead to operation near or beyond nominal limits, increasing the risk of failures or supply interruptions. Similarly, distribution lines may experience higher current levels, resulting in increased resistive losses (see Eq. (6.5)) and associated thermal constraints.

As a direct consequence of these operational challenges, there is an increasing need to reinforce and upgrade the existing electrical infrastructure. Such actions may include the replacement of transformers, the expansion of distribution lines, or the installation of additional infrastructure elements. However, these solutions typically involve high investment costs and may face technical constraints, particularly related to space availability. In addition, their deployment often requires long implementation times, which limits their effectiveness as short-term corrective measures. It should also be noted that the variability and geographical dispersion of EV charging demand, together with its uncertain future evolution, introduce additional complexity in planning decisions related to infrastructure investment.

Therefore, the impact of EV charging on the electrical grid does not depend solely on the total aggregated energy demand, but also on its temporal and spatial distribution. These aspects position EVs as a key factor in the future development of the electrical system. Their integration will strongly influence infrastructure planning and investment decisions, as well as the need for advanced demand management solutions, as introduced in Sections 6.4.3 and 6.4.4.

6.4.2 Impact of EV charging on electricity markets

The impact of EV charging is not limited to its physical effects on grid infrastructure, but also extends to the operation of wholesale electricity markets. One of the most immediate consequences of the aggregated demand associated with EV charging is an upward pressure on electricity prices.

To understand this effect, it is necessary to briefly introduce the operation of marginal electricity markets, which are widely adopted in liberalized power systems. In such markets, generation units submit bids specifying the quantity of energy they are willing to supply at a given price for each time interval, typically on an hourly basis. These bids are sorted in ascending order of price, forming the so-called supply curve (red curve in Fig. 6.5). On the demand side, consumers submit bids reflecting their willingness to consume electricity at different price levels. These bids are arranged in descending order of price, forming the demand curve (blue curve in Fig. 6.5). For further details on wholesale electricity market design and alternative approaches, the reader is referred to [10,11].

In marginal markets, a single clearing price is determined by the intersection of the supply and demand curves. Referring to Fig. 6.5, consider an initial scenario without EV demand, represented by the solid blue demand curve. In this case, the market equilibrium results in a price p_1 and an energy level E_1. With a significant penetration of EVs, the demand curve shifts to the right due to the increased energy requirement, resulting in the dashed blue curve. As illustrated, the increase in demand from E_1 to E_2 leads to a new equilibrium point, raising the electricity price from p_1 to p_2. This price increase can be particularly pronounced when the additional demand pushes the system into regions of the supply curve characterized by steeper slopes, typically associated with higher-cost generation technologies.

Mitigating this price increase can be approached by acting on the supply side. In particular, investment in generation technologies with low marginal cost, such as renewable energy sources, can exert downward pressure on electricity prices, partially offsetting the effect of increased demand due to EV charging. Similarly, policies promoting self-consumption can reduce the total energy required from the

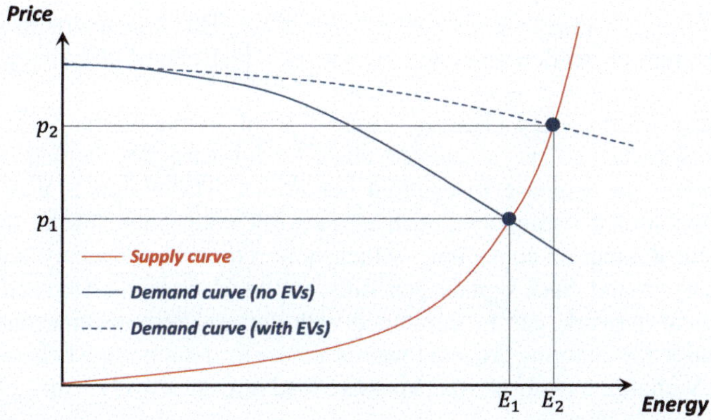

Figure 6.5 Market-clearing mechanism illustrating the impact of EV charging on electricity prices. The increase in demand due to EV integration shifts the demand curve from E_1 to E_2, moving the equilibrium point along the supply curve and raising the market price from p_1 to p_2. This effect highlights how additional flexible demand can influence price formation depending on system conditions.

market, effectively shifting the demand curve to the left in Fig. 6.5. Furthermore, EVs themselves can contribute to improving the integration of renewable generation. Due to their inherent flexibility, EV charging demand can be shifted to periods with high penetration of PV generation, enabling the absorption of surplus energy that would otherwise be curtailed, thus improving the overall efficiency of the system.

In addition, the temporal concentration of EV charging can amplify price volatility in short-term markets. If large numbers of users initiate charging simultaneously (particularly during existing peak demand periods) the resulting demand spikes may lead to sharp, short-lived price increases. Conversely, coordinated or incentivized charging during off-peak periods can smooth demand profiles and reduce stress on the market.

Finally, it is important to note that the integration of EVs may introduce additional variability and uncertainty in electricity prices, as charging demand depends on user behavior and mobility patterns, which are inherently difficult to predict (see Section 6.4.1). Consequently, the effective integration of EVs into electricity markets requires coordinated actions across multiple domains, including the promotion of self-consumption, the deployment of renewable generation, and the implementation of demand-side management strategies (see Section 6.4.3).

6.4.3 Smart charging

Part of the challenges associated with EV charging and its integration into the power grid and electricity markets, as discussed in Sections 6.4.1 and 6.4.2, can be mitigated by optimally exploiting the intrinsic flexibility of EV charging demand. Indeed, EV charging constitutes a flexible load capable of dynamically adjusting its instantaneous power as well as shifting its consumption over time. This characteristic introduces additional degrees of freedom in system operation, which can be effectively leveraged through the concept known as smart charging.

The adaptation of the EV charging demand profile is driven by signals received by both the vehicle and the associated charging infrastructure, which govern the charging process in response to external conditions. These signals may originate from grid operational parameters, such as network congestion levels, transformer loading, or local capacity constraints, which may require a temporary reduction or modulation of demand. Such signals, generated as part of the monitoring and operation of the electrical network, can be combined with demand response mechanisms that ultimately guide the charging process towards a more grid-supportive behavior [12]. In this sense, EV charging evolves from a passive load into an actively managed resource within the power system.

In addition to grid-driven signals, the control of the smart charging process can also be based on economic incentives. In this context, time-of-use (TOU) tariffs have proven to be particularly effective in shaping EV charging demand [13]. Specifically, TOU schemes rely on time-varying electricity prices, thereby directly influencing charging behavior by penalizing undesirable demand profiles and incentivizing load shifting towards more favorable periods, such as those with high PV generation or generally lower system demand. This economic layer reinforces user participation without requiring direct technical intervention.

The primary objectives of smart charging are therefore closely aligned with mitigating adverse impacts on the power system, including network congestion and voltage stability issues (see Chapter 11). When properly implemented, smart charging strategies can significantly reduce power peaks associated with the simultaneous charging of multiple EVs, improve the overall load factor of the system by enabling a more uniform utilization of grid assets, and facilitate the integration of renewable energy sources. As a result, smart charging emerges as a key enabler for enhancing grid operation while potentially reducing the need for costly infrastructure reinforcements required to accommodate large-scale EV deployment. Furthermore, it contributes to a more coordinated interaction between end-users and system operators, reinforcing the role of EVs as active elements within future energy systems.

6.4.4 Bidirectional charging

During conventional charging operation, the EV behaves as a purely consumptive load, drawing energy from the grid. However, this mode of operation only exploits part of the capabilities of the onboard battery system. Bidirectional charging enables not only the charging of the EV, but also the reversal of the energy flow, allowing the onboard batteries to operate in discharge mode and export energy outside the vehicle.

The concept of bidirectional charging is integrated within the broader V2X paradigm (see Chapter 10), through which the EV can act as an energy provider for different applications. Within the V2X framework, several operating modes can be distinguished depending on the final use of the exported energy:

- **Vehicle-to-Grid (V2G):** in which the exported energy is injected directly into the electrical grid, potentially contributing to grid support and system operation [14].
- **Vehicle-to-Home (V2H):** in which the exported energy is used to supply domestic loads, increasing local energy autonomy [13].
- **Vehicle-to-Vehicle (V2V):** where different EVs can exchange energy among themselves, which may be useful in specific scenarios such as energy assistance or fleet-level energy management [15].

From a system perspective, bidirectional charging can contribute to improving grid operation, resilience, and overall system security. This paradigm enables the deployment of distributed energy storage resources, allowing for a more efficient and coordinated management of energy flows within the power system. As a result, it can help mitigate adverse effects such as line congestion, steep demand gradients, and challenges associated with renewable energy integration.

Furthermore, bidirectional charging introduces a significant economic dimension, as it enables EVs to participate as active agents in energy markets through energy arbitrage. In this context, energy can be stored in the vehicle during periods of low electricity prices or high PV penetration, and later exported back to the grid during periods of high demand, when electricity prices are typically higher (see Section 6.4.2), thereby generating potential economic benefits [14]. This capability reinforces the role of EVs as flexible and economically valuable assets within modern power systems.

6.5 Charging in other electric mobility sectors

6.5.1 Rail

Charging strategies in rail transport differ significantly from the paradigms typically employed in ECs. Unlike road vehicles, rail systems are characterized by the use of continuous power supply schemes, primarily based on dedicated infrastructure rather than discrete charging processes.

The most widely adopted system for electrified rail transport is based on overhead catenary lines. In this configuration, an energized overhead conductor is supplied directly from the electrical grid and deployed along the railway track. Energy transfer is achieved through physical contact between the catenary and the vehicle via a pantograph. This arrangement enables continuous power delivery from the grid to the train, supporting the high and sustained power levels required for railway operation.

In scenarios where the deployment of overhead catenary is impractical, such as in certain urban or underground environments, third rail systems are often employed. In this case, the electrical supply is provided through a conductor located at track level, and energy is collected in the vehicle using sliding contact shoes. While this approach eliminates the need for overhead infrastructure, it introduces additional safety requirements due to the accessibility of the energized conductor and typically presents limitations in terms of maximum transferable power.

Finally, in cases where continuous electrification is not feasible, such as on partially electrified lines or isolated segments, hybrid solutions can be implemented. In these systems, the external power supply is complemented by onboard energy storage systems, typically based on batteries or other storage technologies [16]. These hybrid configurations enable trains to operate autonomously over non-electrified sections while recharging during operation under catenary or third rail systems, thereby increasing operational flexibility without requiring full infrastructure deployment.

6.5.2 Maritime

The charging of electrified maritime transport presents notable differences compared to rail-based systems. In most cases, the charging process takes place while vessels are docked at port facilities, where suitable electrical infrastructure is available. This process is commonly referred to as *cold ironing*. Due to the intrinsic operational characteristics of maritime transport, vessels cannot rely on continuous power supply systems such as catenary-based solutions. As a result, energy replenishment is typically limited to docking periods.

Cold ironing infrastructure must be designed to handle high power levels, which can reach several megawatts for large vessels such as cruise ships or container carriers. Consequently, charging operations are generally performed at medium or high voltage levels, often requiring dedicated electrical infrastructure, including substations and reinforced grid connections, to ensure safe and reliable operation.

In addition, high-power conductive charging systems have been developed for smaller vessels such as ferries [17]. These systems enable fast partial recharging during short docking intervals, allowing vessels operating on predefined routes to maintain high utilization rates. This operational concept is analogous to opportunity charging in tramway systems (see Chapter 5). In such applications, automated connection mechanisms are essential to simplify the charging process and reduce turnaround times.

It is also important to highlight that maritime charging infrastructure must be designed to withstand harsh environmental conditions. Equipment is required to operate reliably under high humidity, salt exposure, and mechanical stress, which imposes stringent requirements on materials, insulation, and system robustness.

Electrified maritime charging faces significant challenges. The deployment cost is considerably higher than in other transport sectors, mainly due to the high power demand and the need for dedicated infrastructure. Furthermore, the current lack of global standardization limits interoperability and slows down large-scale deployment, posing additional barriers to technological development in this field.

6.5.3 Aviation

In contrast to other transport sectors, aviation does not allow for in-route energy replenishment, meaning that charging processes must be carried out exclusively on the ground, in what is commonly referred to as *ground charging*. This constraint is compounded by intrinsic limitations of the aviation sector, such as the strong dependence on the energy density of onboard storage systems, which stems from

Figure 6.6 Ground charging process of an electric aircraft using a conductive high-power connection at an airport facility. (Image courtesy of Wikimedia Commons.)

the high sensitivity of aircraft to mass, as well as the substantial energy demands associated with flight operations (see Chapter 5).

In this context, charging infrastructure must be deployed within airport facilities (see Fig. 6.6). The required power levels vary significantly depending on the application, ranging from low-power systems for light aircraft to higher power levels for larger platforms. Furthermore, the integration of charging systems into airport environments introduces an additional electrical load to infrastructures that already exhibit high energy demand due to other operations. This often necessitates substantial investments in grid reinforcement and dedicated electrical installations.

However, the main challenge associated with electric aircraft charging lies in the limited amount of energy that can be stored onboard, primarily due to weight constraints. The high energy requirements of flight frequently result in long charging times, even when high-power infrastructure is available. This can conflict with operational requirements, particularly in applications where short turnaround times are critical.

Additionally, charging installations in airport environments must comply with stringent safety and reliability standards, consistent with the highly regulated nature of the aviation sector. Charging processes must be performed under tightly controlled conditions and carefully coordinated with flight operations to avoid interference with mission execution or other critical airport activities.

Overall, these constraints currently limit the applicability of electrified aviation to small aircraft and short-range operations, although ongoing technological developments may progressively expand its feasibility.

6.5.4 UAVs

Charging systems for UAVs must be adapted to the specific characteristics of this type of vehicle, which typically features reduced size and weight and operates in short, repetitive mission cycles. As a result, UAV charging systems are generally designed to be fast and highly efficient. In this context, two main paradigms can be identified for UAV energy replenishment: direct charging and *battery swapping*.

Direct charging is performed through conventional conductive connections, usually at relatively low power levels, and is suitable for applications where operational downtime is not critical. However, charging times can become a significant limitation in scenarios requiring continuous operation, such as surveillance or monitoring tasks.

In contrast, *battery swapping* enables the replacement of a depleted battery with a fully charged one in a very short time, often within seconds. This approach also facilitates the deployment of automated charging platforms capable of performing the process without human intervention, making it particularly suitable for UAV fleets and fully automated operations. Nevertheless, this solution requires a certain degree of standardization in battery design, as well as efficient management of the energy inventory.

Fast charging has also been explored as a means to reduce charging times. However, this approach presents notable limitations in UAV applications due to increased thermal stress associated with higher power levels and elevated losses. This is particularly critical in UAVs, where thermal constraints are more restrictive (see Chapter 4).

In summary, the choice between direct charging and *battery swapping* largely depends on the mission profile. Compared to other transport sectors, UAV charging offers greater flexibility in terms of energy management strategies, but it is also subject to more stringent constraints related to energy availability and operational requirements inherent to these platforms.

6.6 Global deployment of EC charging infrastructure

6.6.1 Global overview

The number of charging points has experienced sustained growth in recent years, closely linked to the expansion of the global EC fleet [4]. However, this growth has not been homogeneous in terms of the technologies deployed. Instead, an increasing diversification in EC charging technologies can be observed. This trend reflects the need to adapt charging solutions to different use cases and operational profiles, ranging from low-power, long-duration residential charging to high-power, short-duration in-route charging.

As shown in Fig. 6.7, the expansion of charging infrastructure has been primarily driven by a steady increase in the number of AC chargers, typically associated with slow charging systems. This type of charging is characterized by relatively low power levels and ease of integration into existing electrical grids, which has historically favored its widespread deployment. As a result, AC charging infrastructure has become the backbone of the global charging network, particularly in residential and urban environments. Additionally, its lower installation and operational costs have facilitated large-scale roll-out, especially in regions with mature distribution networks.

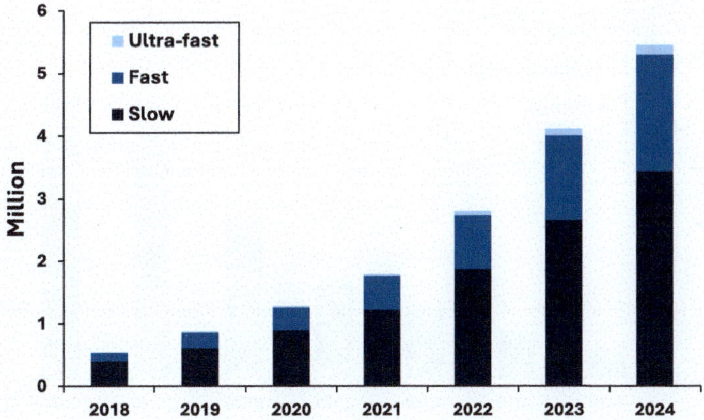

Figure 6.7 Global evolution of publicly accessible EC charging infrastructure, disaggregated into slow (AC, < 22 kW), fast (DC, ≥ 22 kW, < 150 kW), and ultra–fast (DC, ≥ 150 kW) chargers.

In contrast, DC chargers, associated with fast charging systems, have experienced more pronounced growth in recent years, albeit from a smaller installed base. The deployment of these chargers is largely motivated by the need to enable long-distance travel, where reduced charging times are essential. Fast charging infrastructure allows for shorter stops and improved operational flexibility, making it a critical component for the broader adoption of ECs beyond urban settings. In this context, the emergence of ultra-fast chargers represents a further step in performance, specifically targeting in-route charging scenarios rather than long-duration parking applications. These systems are increasingly deployed along highways and strategic transport corridors, where high throughput and reduced dwell times are key requirements.

A combined analysis of these three categories highlights a clear transition towards a multi-technology charging ecosystem. Rather than converging to a single solution, the infrastructure is evolving towards the coexistence of complementary charging approaches, each optimized for specific operational requirements. This paradigm provides greater system flexibility, which is essential to support the large-scale deployment of electric mobility at a global level. Furthermore, the balance between AC and DC infrastructure will continue to evolve as user behavior, vehicle capabilities, and grid constraints develop over time.

6.6.2 Regional disparities

As shown in Fig. 6.8, the number of EC charging points exhibits significant regional disparities, with a strong concentration in specific geographical areas. This uneven distribution can be attributed to multiple factors, including differences in regulatory frameworks, levels of public and private investment, and the varying pace of EC adoption across regions.

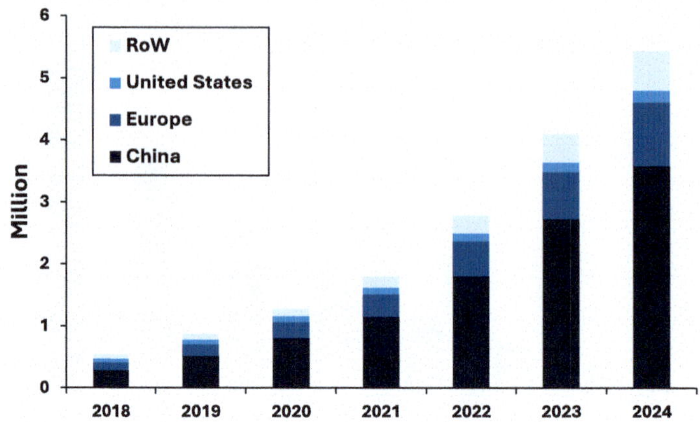

Figure 6.8 Global stock of public charging points by region (RoW = Rest of the World).

China clearly stands out as the global leader in terms of charging infrastructure deployment. The country not only accounts for the largest absolute number of installed chargers, but also demonstrates the highest growth rate over the period considered. This rapid expansion has been driven by a combination of strong central planning, extensive policy support for transport electrification, and the accelerated growth of the domestic EC fleet. As a result, China has developed a highly dense and continuously expanding charging network, which plays a key role in supporting large-scale EC adoption.

Europe occupies an intermediate position, with a steady and sustained increase in the number of charging points over time, although at a slower pace compared to China. In this region, infrastructure deployment has been largely driven by ambitious regulatory targets related to emission reduction and decarbonization of the transport sector. Additionally, coordinated initiatives at both national and supranational levels have contributed to improving infrastructure coverage and interoperability. Nevertheless, it is important to note that significant disparities exist among European countries, with some markets showing much higher levels of infrastructure maturity than others, despite the overall upward trend.

In the case of the United States, the growth in charging infrastructure has historically been more moderate than in both China and Europe. This difference can be largely explained by a lower degree of centralized planning and a stronger reliance on private-sector initiatives. However, recent years have shown a clear acceleration in infrastructure deployment, supported by new policy measures and investment programs aimed at promoting EC adoption and expanding the national charging network.

Finally, other regions, including several emerging economies, have historically contributed a relatively small share of the global charging infrastructure. However, a noticeable increase has been observed in recent years, indicating the early stages of market development in these areas. This trend suggests that countries beyond the main

Table 6.3 Charging modes classification based on charging speed, technology, and typical characteristics.

Mode	AC/DC	Typical power range	Typical charging time
Slow	AC	~1–22 kW	Several hours
Fast	DC	~50–150 kW	Tens of minutes
Ultra-fast	DC	$\gtrsim$150 kW	Few minutes

markets may play an increasingly important role in the future expansion of electric mobility, contributing to the global decarbonization of the transport sector.

6.6.3 Charging modes

EC charging systems can be classified according to the charging mode, which encompasses both technical characteristics, such as power levels, and operational aspects, such as typical charging time under common use cases. Based on this criterion, charging infrastructure can be categorized into slow, fast, and ultra-fast charging. Table 6.3 summarizes the main characteristics of each charging mode, providing a coherent comparison framework based on architecture (AC or DC), power level, and typical charging time.

Within this classification, slow charging is typically associated with AC systems and moderate power levels, resulting in extended charging durations that may span several hours. This mode is well suited for applications where long dwell times are available, such as residential charging or workplace environments. In contrast, fast and ultra-fast charging modes rely on significantly higher power levels and are generally implemented using DC architectures, enabling substantially shorter charging times compared to slow charging.

As shown in Fig. 6.7, slow charging represents the dominant solution in absolute terms at the global level. This predominance is largely driven by its lower deployment cost and its straightforward integration into existing electrical distribution networks. Furthermore, its compatibility with long-duration parking scenarios makes it a structurally essential component of the charging ecosystem.

Conversely, fast charging exhibits the highest relative growth rate among the different modes. This trend reflects an increasing demand for infrastructure capable of supporting long-distance travel, addressing a key limitation of slow charging in scenarios where reduced charging time is critical. The expansion of fast charging infrastructure is therefore closely linked to the broader usability of ECs beyond urban or short-range applications.

Ultra-fast charging, while still representing a relatively small share of the total installed base, has shown a clear upward trend in recent years. Its limited current penetration reflects both its emerging nature and the higher technical and economic requirements associated with its deployment (see Section 6.7.1). Nevertheless, its

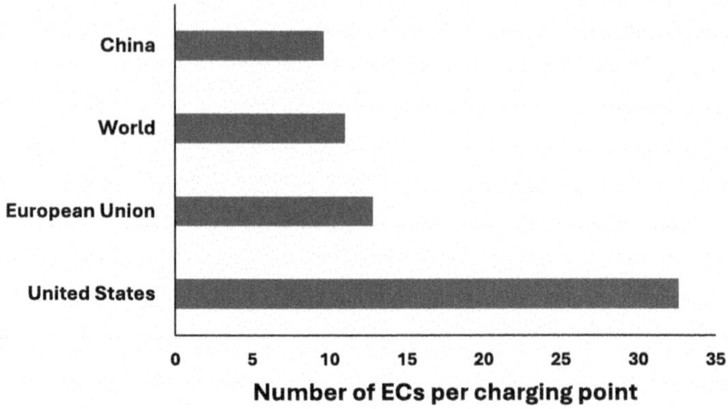

Figure 6.9 Number of electric light-duty vehicles per charging point in different regions in 2024.

role is expected to become increasingly relevant in high-demand scenarios requiring minimal charging interruptions.

In summary, the adoption landscape of charging modes points towards a coexistence paradigm in the near future, where each mode addresses specific operational needs. This multi-layered structure enables a more efficient allocation of resources, aligning infrastructure capabilities with diverse user requirements and usage patterns.

6.6.4 Infrastructure vs EC stock

The absolute number of chargers shown in Fig. 6.8 does not fully capture the actual level of development of charging infrastructure within a given region. In this context, a more representative metric is the ratio between the number of ECs and the available charging points, as this indicator reflects both the effective availability of charging infrastructure per vehicle and the level of saturation of the existing network. This metric is presented for several representative regions in Fig. 6.9.

Among the regions considered, China exhibits the lowest EC-to-charging-point ratio, with approximately 9.6 ECs per public charging point in 2024. This value indicates that, in addition to leading globally in EC adoption, China has achieved a relatively well-synchronized expansion of its charging infrastructure. As a result, the system operates under comparatively low saturation conditions, suggesting that the deployment of charging points has closely followed the growth in vehicle stock.

In contrast, the United States shows a significantly higher ratio, reaching approximately 32.6 ECs per charging point in 2024, well above the global average. This figure reflects a comparatively slower infrastructure rollout, historically more dependent on private initiatives and less centrally coordinated planning. Consequently, this imbalance may lead to increased pressure on existing charging stations, particularly in high-demand scenarios or along long-distance travel corridors, where charging availability becomes a critical constraint.

Europe occupies an intermediate position, with a ratio of around 12.8 ECs per charging point in 2024, slightly above the global average. This suggests a more balanced infrastructure deployment compared to the United States, although still less extensive than in China. It is important to note that this aggregated value conceals significant disparities across European countries, with some regions exhibiting high infrastructure density while others remain underdeveloped. As in the case of EC adoption, the expansion of charging infrastructure in Europe has been largely driven by a stringent and ambitious regulatory framework (see Chapter 3).

From a global perspective, the observed ratios indicate that a misalignment still exists between EC adoption and charging infrastructure deployment, with pronounced regional disparities. This imbalance continues to represent a relevant barrier to the large-scale adoption of ECs, particularly in regions where the number of vehicles per public charging point remains high. In such contexts, users may experience longer waiting times, reduced accessibility, and increased uncertainty in trip planning.

The future evolution of this indicator will be critical in assessing the viability and scalability of electric mobility systems. A progressive decrease in the EC-to-charger ratio would reflect an improvement in infrastructure availability, while increasing values could signal emerging bottlenecks in deployment. However, beyond absolute values, the spatial distribution of charging infrastructure must also be considered. Ensuring an adequate geographic allocation of charging points, particularly in regions where infrastructure is currently underdeveloped relative to EC stock, will be essential. In this sense, a well-planned expansion of charging infrastructure can act as a key enabler for accelerating EC adoption in emerging markets.

6.7 Future trends in EC charging

6.7.1 *Ultra-fast charging*

Ultra-fast charging represents an evolution of fast charging systems, enabling higher power levels and significantly reducing charging times to the order of a few minutes (see Table 6.3). This capability makes it a key enabler for long-distance travel, particularly along highway corridors where minimizing charging times is critical. Although its deployment is no longer negligible (see Fig. 6.7), the total number of ultra-fast chargers remains relatively small compared to other charging modes, positioning it as an emerging technology still in early stages of adoption.

Ultra-fast charging is based on DC systems, allowing the bypass of the OBC and enabling power levels exceeding 150 kW. However, as with other charging modes, the delivered power is not constant throughout the charging process. Instead, it is strongly influenced by factors such as the SoC and thermal conditions, leading to a progressive reduction in power at higher SoC levels.

Thermal constraints represent one of the main limitations of ultra-fast charging. The high current levels required to sustain elevated power result in significant resistive losses (see (6.5)), which increase heat generation in cables, connectors, and battery systems. This not only reduces overall efficiency but also raises concerns regarding component degradation and operational safety, thereby imposing stringent cooling requirements.

In addition, the high power demand associated with ultra-fast charging introduces challenges at the grid level. The deployment of such systems may require reinforcement of the electrical infrastructure or the integration of auxiliary solutions, such as local energy storage, to mitigate peak loads. These factors contribute to increased system complexity and cost, which currently limit large-scale deployment.

6.7.2 Megawatt charging

The Megawatt Charging System (MCS) represents a further step beyond ultra-fast charging, constituting the next evolutionary stage in the electrification of transport. These systems typically operate at power levels in the megawatt range, significantly exceeding ultra-fast charging, which is commonly limited to approximately 350 kW [18]. MCS is primarily targeted at heavy-duty road transport applications, such as electric trucks, where energy demand and required battery capacities are substantially higher than in light-duty vehicles.

MCS architectures are based on high-power DC systems, leading to stringent requirements in terms of electrical design, thermal management, and operational safety. While their fundamental limitations are analogous to those of ultra-fast charging (see Section 6.7.1), they manifest with greater severity due to the substantially higher power levels involved. In particular, the increase in current results in significantly higher resistive losses, leading to pronounced thermal stresses in cables, connectors, and power electronics. As a consequence, advanced cooling solutions (often based on liquid-cooled components) are required to ensure safe and reliable operation.

From a power system perspective, megawatt-scale charging introduces considerable challenges. The concentration of such high-power loads can generate substantial demand peaks at specific nodes of the distribution network, potentially leading to infrastructure overloading or voltage stability issues. Addressing these effects may require grid reinforcements, local energy storage integration, or coordinated load management strategies.

Finally, the deployment cost of MCS infrastructure is significantly higher than that of lower-power charging technologies, particularly when grid upgrades are necessary. These technical and economic constraints currently limit large-scale deployment. Nevertheless, MCS is expected to play a key enabling role in the electrification of long-haul heavy-duty transport, where reducing charging time is critical for operational viability.

6.7.3 Battery swapping

In contrast to conventional charging systems based on energy transfer, battery swapping relies on replacing depleted onboard batteries with pre-charged units, effectively eliminating charging time from the user perspective.

Despite this conceptual advantage, battery swapping requires a high degree of standardization in both battery design and mechanical and electrical interfaces. This represents a major barrier to large-scale deployment, particularly in markets with multiple manufacturers and heterogeneous vehicle platforms. From an infrastructure standpoint, battery swapping stations must incorporate automated and highly reliable systems capable of performing rapid battery exchange, as well as handling the storage and recharging of multiple battery units.

Additionally, the management of depleted batteries introduces further operational complexity. Charging scheduling, inventory control, and battery health monitoring must be coordinated to ensure system availability and reliability, which can significantly impact overall feasibility [19].

At present, the adoption of battery swapping remains limited, primarily due to the lack of standardization and the associated infrastructure costs. Nevertheless, it presents a promising solution for specific use cases, particularly high-utilization fleets, where minimizing downtime is a critical operational requirement.

6.7.4 Charging + renewables

From a system-level perspective, the integration of charging infrastructure with renewable generation units represents a viable approach to mitigate many of the challenges associated with advanced charging technologies, particularly those with high power demand (see Sections 6.7.1 and 6.7.2).

Within this paradigm, charging points can leverage distributed generation to locally supply part of the required energy. This approach reduces peak loads on the grid, alleviating localized congestion and potentially avoiding costly reinforcements of existing transmission and distribution infrastructure.

Moreover, EV charging demand can act as an enabler for renewable integration. In particular, it provides an opportunity to better exploit energy sources with short and concentrated generation windows, such as PV systems, improving their effective utilization and reducing curtailment.

However, the inherent intermittency of most renewable sources constitutes the main challenge in this context. The temporal mismatch between generation and charging demand increases operational complexity and may compromise reliability if not properly managed. As a result, the integration of large-scale energy storage systems often becomes necessary to decouple generation from consumption [20]. Additionally, smart charging strategies can further enhance system flexibility and improve the overall balance between supply and demand (see Chapter 11).

References

[1] M. Usman Tahir, A. Sangwongwanich, D.-I. Stroe, F. Blaabjerg, Overview of multi-stage charging strategies for li-ion batteries, Journal of Energy Chemistry 84 (2023) 228–241, https://doi.org/10.1016/j.jechem.2023.05.023.

[2] J. Liu, X. Wang, Investigating effects of pulse charging on performance of li-ion batteries at low temperature, Journal of Power Sources 574 (2023) 233177, https://doi.org/10.1016/

j.jpowsour.2023.233177.
[3] M. Yilmaz, P.T. Krein, Review of battery charger topologies, charging power levels, and infrastructure for plug-in electric and hybrid vehicles, IEEE Transactions on Power Electronics 28 (5) (2013) 2151–2169, https://doi.org/10.1109/TPEL.2012.2212917.
[4] I.E. Agency, Global ev outlook 2025, Tech. Rep., IEA, Paris, 2023, https://www.iea.org/reports/global-ev-outlook-2025. (Accessed 27 March 2026).
[5] A. Parlikar, M. Schott, K. Godse, D. Kucevic, A. Jossen, H. Hesse, High-power electric vehicle charging: low-carbon grid integration pathways with stationary lithium-ion battery systems and renewable generation, Applied Energy 333 (2023) 120541, https://doi.org/10.1016/j.apenergy.2022.120541.
[6] Y. Li, Q. Zhang, Y. Liao, L. Yang, X. Bai, H. Tao, Thermal management optimization for high power ev charging connectors: a heat pipe-based design approach, Applied Thermal Engineering 279 (2025) 127680, https://doi.org/10.1016/j.applthermaleng.2025.127680.
[7] Z. Bi, T. Kan, C.C. Mi, Y. Zhang, Z. Zhao, G.A. Keoleian, A review of wireless power transfer for electric vehicles: prospects to enhance sustainable mobility, Applied Energy 179 (2016) 413–425, https://doi.org/10.1016/j.apenergy.2016.07.003.
[8] A. González-Garrido, M. González-Pérez, F.J. Asensio, A.F. Cortes-Borray, M. Santos-Mugica, I. Vicente-Figueirido, Hierarchical control for collaborative electric vehicle charging to alleviate network congestion and enhance ev hosting in constrained distribution networks, Renewable Energy 230 (2024) 120823, https://doi.org/10.1016/j.renene.2024.120823.
[9] Y. Mao, X. Yu, F. Wang, J. Zhu, Electric vehicle charging demand forecasting: a data-driven integrated learning approach, Renewable Energy 256 (2026) 124141, https://doi.org/10.1016/j.renene.2025.124141.
[10] S. Prabhakar Karthikeyan, I. Jacob Raglend, D. Kothari, A review on market power in deregulated electricity market, International Journal of Electrical Power & Energy Systems 48 (2013) 139–147, https://doi.org/10.1016/j.ijepes.2012.11.024.
[11] K. Mayer, S. Trück, Electricity markets around the world, Journal of Commodity Markets 9 (2018) 77–100, https://doi.org/10.1016/j.jcomm.2018.02.001.
[12] S.A. Mansouri, E. Nematbakhsh, A.R. Jordehi, M. Marzband, M. Tostado-Véliz, F. Jurado, An interval-based nested optimization framework for deriving flexibility from smart buildings and electric vehicle fleets in the tso-dso coordination, Applied Energy 341 (2023) 121062, https://doi.org/10.1016/j.apenergy.2023.121062.
[13] M. Tostado-Véliz, S. Kamel, H.M. Hasanien, R.A. Turky, F. Jurado, Optimal energy management of cooperative energy communities considering flexible demand, storage and vehicle-to-grid under uncertainties, Sustainable Cities and Society 84 (2022) 104019, https://doi.org/10.1016/j.scs.2022.104019.
[14] M. Tostado-Véliz, H.M. Hasanien, J. Carpio, F. Jurado, Risk-aware strategies for optimal participation of parking lots in day-ahead electricity markets, Energy 322 (2025) 135406, https://doi.org/10.1016/j.energy.2025.135406.
[15] Y. Cao, J. Cui, S. Liu, X. Zhang, L. Xu, H. Sun, A reservation-based hybrid electric vehicle energy supply system via direct and asynchronous vehicle-to-vehicle charging modes, International Journal of Electrical Power & Energy Systems 176 (2026) 111704, https://doi.org/10.1016/j.ijepes.2026.111704.
[16] M. Tostado-Véliz, P. Arévalo, F. Jurado, An optimization framework for planning wayside and on-board hybrid storage systems for tramway applications, Journal of Energy Storage 43 (2021) 103207, https://doi.org/10.1016/j.est.2021.103207.
[17] DNV, Maritime forecast to 2050, https://www.dnv.com/maritime/maritime-forecast/, 2025. (Accessed 27 March 2026).

[18] CharIN e.V., Megawatt charging system (mcs) specification, Tech. Rep., Charging Interface Initiative (CharIN), https://www.charin.global/technology/mcs/, 2025. (Accessed 27 March 2026).

[19] W. Xu, J. Chen, X. Liu, X. Chen, Optimization of charging-discharging strategies and ev battery swapping recommendation algorithm considering aggregation of battery swapping stations, Global Energy Interconnection (2026), https://doi.org/10.1016/j.gloei.2026.01.004.

[20] M. Tostado-Véliz, A. Rezaee Jordehi, Y. Zhou, S.A. Mansouri, F. Jurado, Best-case-aware planning of photovoltaic-battery systems for multi-mode charging stations, Renewable Energy 225 (2024) 120300, https://doi.org/10.1016/j.renene.2024.120300.

Batteries

7

7.1 Fundamentals of electrochemical energy storage

7.1.1 Basic operating principles

A rechargeable battery operates based on the reversible conversion between chemical energy and electrical energy through reduction and oxidation reactions occurring at two electrodes (active materials) separated by an electrolyte. Fig. 7.1 schematically illustrates the operating principle of a battery, distinguishing between its two reversible processes: charging, through which the battery increases its stored energy, and discharging, during which the battery behaves as an energy source, delivering energy to an external system.

From an electrochemical perspective, oxidation and reduction (redox[1]) reactions are defined in terms of electron transfer. Oxidation corresponds to the loss of electrons, whereas reduction involves the gain of electrons. These processes always occur simultaneously and are spatially separated within the battery, enabling controlled energy conversion.

During discharge, the anode undergoes an oxidation reaction, in which the active material releases electrons. These electrons flow through the external electrical circuit, where a load may be connected, allowing useful work to be performed. The electron flow continues toward the cathode, where a reduction reaction takes place, and electrons are accepted by the active material. This electron movement through the external circuit constitutes the electrical current supplied by the battery and defines its ability to power external devices.

During the charging process, the overall mechanism is reversed, as shown in Fig. 7.1. In this case, an external energy source forces electrons to flow in the opposite direction to that observed during discharge. Consequently, the cathode undergoes oxidation, while reduction takes place at the anode. This externally driven process restores the original chemical state of the active materials, thereby recovering the battery energy storage capability and enabling repeated operation over multiple cycles.

In parallel with electron transfer, ionic transport occurs within the cell through the electrolyte, which separates the two electrodes. The electrolyte allows ionic conduction while preventing electron transport, thereby avoiding internal short circuits. During discharge, ions migrate from the anode to the cathode to maintain charge neutrality, whereas during charging, this ionic flow is reversed. This internal ion movement is essential to sustain the electrochemical reactions and ensure continuous operation of the battery under varying operating conditions.

As in any energy conversion system, the electrochemical processes occurring in batteries are not ideal. Energy losses arise due to internal resistances, overpotentials, and non-ideal transport phenomena. These effects reduce the overall efficiency of the

[1] https://en.wikipedia.org/wiki/Redox.

Electric Mobility Systems. https://doi.org/10.1016/B978-0-44-349397-3.00015-X

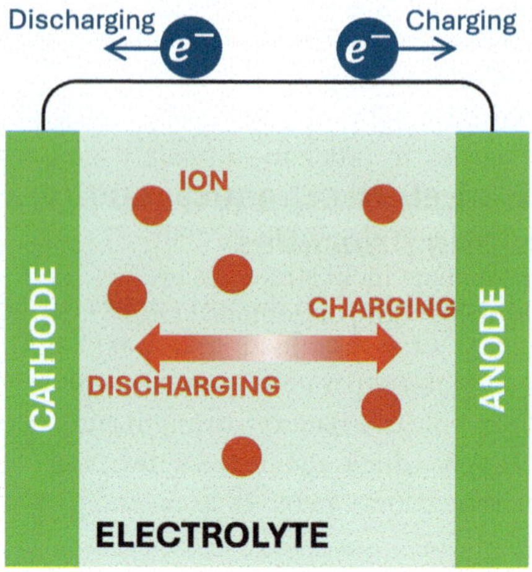

Figure 7.1 Schematic representation of the operating principles of a rechargeable electrochemical cell. During discharge, oxidation at the anode releases electrons that flow through the external circuit to the cathode, where reduction occurs, while ions migrate through the electrolyte to maintain charge balance. During charging, an external power source reverses these processes, forcing electron flow in the opposite direction and driving ions back to their original electrodes.

battery and play a critical role in its performance, thermal behavior, and practical operation over its lifetime.

7.1.2 Electrochemical potential and cell voltage

In the operation of a battery, the concept of electrochemical potential describes the energy associated with the chemical species involved in redox reactions. In this context, each electrode (anode or cathode) exhibits a characteristic potential that depends on the active material (i.e., the material that directly participates in the electrochemical reactions responsible for energy storage and release) as well as on the system conditions. Accordingly, the cell voltage or potential difference is defined as the difference between the potentials of the anode and the cathode, and constitutes the fundamental quantity governing the flow of electrons through the external circuit.

Formally, the voltage of an electrochemical cell can be expressed as a function of the Gibbs free energy of the overall electrochemical reaction:

$$V = -\frac{\Delta G}{nF} \tag{7.1}$$

where ΔG is the variation of Gibbs free energy, n is the number of electrons transferred in the reaction, and F is the Faraday constant ($F = 96{,}485$ C/mol). Eq. (7.1) establishes a direct relationship between the chemical nature of the system and its electrical behavior, as it determines the amount of energy available per unit charge.

From an energy perspective, a higher potential difference (voltage) implies a greater capability to perform electrical work, which translates into a higher specific energy of the system, provided that the charge storage capacity remains constant. This observation highlights the importance of carefully selecting the active materials in a battery. Since the cell voltage is intrinsically linked to the electrochemical potentials of the active materials, their properties directly influence key performance aspects such as overall efficiency and energy output.

The dependence of voltage on electrode materials provides significant flexibility in the design of electrochemical cells, allowing their characteristics to be tailored to specific applications. Different combinations of anode and cathode materials can lead to substantially different voltage levels, which in turn affect parameters such as energy density, power capability, safety, and lifetime. In addition, the choice of materials also influences reaction kinetics and internal resistances, further impacting the effective voltage under real operating conditions.

It is important to distinguish between the nominal voltage and the actual voltage of a battery. The actual voltage varies dynamically during charge and discharge processes. In particular, during discharge, a voltage drop is observed due to internal resistive losses, whereas during charging, overpotentials may arise as a result of polarization phenomena. By contrast, the nominal voltage refers to a representative average value observed during operation and is commonly used to characterize and compare different battery technologies.

Furthermore, while the nominal voltage is treated as a characteristic parameter of the device, the actual voltage is influenced by several operating conditions, including the SoC, temperature, and applied current. These factors can cause significant deviations from the nominal value, and therefore must be considered when analyzing the real performance of electrochemical energy storage systems. Moreover, transient effects during dynamic load conditions can introduce additional voltage fluctuations that are relevant in practical applications.

7.1.3 Energy and power concepts in batteries

Although the concepts of energy and power have already been introduced in Chapter 6, it is necessary to particularize them in the context of battery operation, as these two parameters play a fundamental role in defining the operational capabilities of the device. In particular, the energy stored in a battery represents the total useful work that can be delivered over time.

From a physical standpoint, the energy can be approximated as the product of the voltage and the total transferred charge:

$$E = V \cdot q \tag{7.2}$$

where E is the energy, V represents the voltage, and q is the total stored charge. The fundamental unit of energy is the joule (J). However, this unit is often impractical in engineering applications, and therefore watt-hour (Wh) or ampere-hour (Ah) are more commonly used. In particular, the watt-hour arises naturally from the definition of electrical power:

$$P = V \cdot I \tag{7.3}$$

where P is the power (in watts) and I is the electric current (in amperes). In this sense, while energy determines the autonomy of a system, power is directly related to its ability to respond to dynamic demands, such as rapid acceleration or fast charging events. Consequently, energy can be interpreted as the time integral of power, which naturally leads to its expression in watt-hours in practical systems.

On the other hand, the ampere-hour is more closely related to the amount of stored charge and can be particularly useful from an operational perspective, as it directly links capacity to discharge current and operating time. In practice, both units are related through the nominal voltage of the battery, since the energy in watt-hours can be obtained as the product of the capacity in ampere-hours and the nominal voltage. This relationship explains why energy (Wh or kWh) is typically used to compare different battery systems, while capacity (Ah) is often used to describe their electrical behavior in more detail.

From an operational perspective, the ability of a battery to deliver both energy and power is constrained by several physical limitations. In particular, kinetic phenomena associated with electrochemical reactions and ionic transport processes within the electrolyte and the active materials impose significant restrictions. These limitations become more pronounced under high current conditions, such as fast charging or high power discharge. In such scenarios, the underlying processes may not proceed sufficiently fast, which, combined with internal resistances, leads to increased voltage drops and reduced efficiency, ultimately limiting the effective power output of the battery. As a consequence of these constraints, the capabilities to deliver energy and power are inherently coupled and often exhibit a complementary behavior. Batteries optimized to maximize energy storage typically present limited power delivery capabilities, whereas technologies designed for high power output usually achieve this at the expense of total stored energy. This complementarity gives rise to the well-known energy–power trade-off, which depends strongly on the selected active materials, cell design, and operating conditions. Additionally, factors such as electrode thickness, porosity, and electrolyte conductivity also influence this balance.

This trade-off is particularly relevant in electric mobility applications, where EVs require both sufficient energy to ensure adequate driving range and high power to meet dynamic performance demands. For instance, high power capability is essential for acceleration and regenerative braking, while high energy density is critical for extending vehicle autonomy. Consequently, battery design becomes a key aspect in EV development, requiring careful optimization to achieve an appropriate balance between these competing requirements and to ensure overall system performance under realistic driving conditions.

7.1.4 Key performance metrics

In order to evaluate and compare different battery technologies (and energy storage systems in general), it is necessary to introduce a set of key metrics that characterize battery behavior from an energetic, dynamic (response capability), and operational efficiency perspective. These metrics provide a standardized framework to assess performance and to guide design choices depending on the requirements of the intended application. The most relevant metrics are summarized as follows:

- **Energy density:** This parameter relates the total energy that can be stored in the battery to its mass or volume. Accordingly, a distinction is made between gravimetric energy density, expressed in watt-hours per kilogram (Wh/kg), and volumetric energy density, expressed in watt-hours per liter (Wh/L). The former refers to the mass of the device, while the latter accounts for its volume. This metric is particularly critical in electric mobility applications, where weight and space constraints strongly influence system design and feasibility. Batteries with high energy density are generally preferred, as they enable larger amounts of stored energy (and therefore greater driving range) while minimizing mass and volume. However, achieving high energy density often requires compromises in other performance aspects.
- **Power density:** This concept is analogous to energy density but refers instead to the ability to deliver or absorb power per unit mass (W/kg) or volume (W/L). Unlike energy, which is directly related to vehicle range, power is associated with dynamic response capability. Therefore, batteries with high power density are desirable for applications requiring rapid energy exchange, such as vehicle acceleration, take-off in aerial platforms, or high-power charging processes (see Chapter 6). As discussed in Section 7.1.3, energy and power densities typically exhibit a complementary behavior, leading to an inherent trade-off that must be carefully balanced depending on the target application. This trade-off is a central consideration in battery engineering.
- **Efficiency:** This metric provides a global measure of energy losses occurring during a complete charge–discharge cycle. It is commonly defined as the ratio between the energy delivered during discharge and the energy supplied during charge (round-trip efficiency). Efficiency is influenced by several factors, including internal resistive losses, irreversible electrochemical processes, and heat generation. High efficiency values are particularly desirable in applications involving frequent cycling, as they directly impact overall system performance and energy consumption. In addition, efficiency is closely related to thermal management requirements, as energy losses are typically dissipated as heat.
- **Cycle life:** This concept provides an indication of the duration over which a battery can be operated under acceptable conditions without significant degradation of its nominal parameters. It is typically quantified as the number of charge–discharge cycles that can be completed before the battery capacity falls below a predefined threshold (commonly 70–80% of its initial value [1]). Cycle life is strongly influenced by degradation mechanisms such as loss of active material, structural changes in the electrodes, and side reactions within the cell. These phenomena will be analyzed in detail in Section 7.5.2, where the most relevant degradation mechanisms in battery systems are discussed. It is also important to note that cycle life depends on operating conditions such as temperature, depth of discharge (DoD), and current rates.

7.2 Battery architecture and components

7.2.1 Cell components

Within a battery, the electrochemical cell constitutes the fundamental functional unit where the electrochemical reactions responsible for the conversion between chemical and electrical energy take place. Each cell is capable of delivering a certain voltage and storing a limited amount of energy. For this reason, in practical applications, multiple cells are typically interconnected in series and parallel configurations in order to achieve the voltage and capacity levels required by the specific application (see Section 7.2.3). Consequently, the overall performance of a battery is strongly influenced by the design and behavior of its individual cells.

From a constructive point of view, an electrochemical cell is composed of several key components, whose function and properties determine its electrochemical performance:

- **Anode:** The anode is the material that hosts the chemical species responsible for releasing electrons to the external circuit during the discharge process. From a design perspective, a suitable anode material should exhibit a high charge storage capacity, good electronic conductivity, and structural stability under repeated cycling conditions. Additionally, it should possess an appropriate electrochemical potential that contributes to maximizing the overall cell voltage. Reversibility is also a critical requirement, meaning that the material should sustain repeated insertion and extraction of ionic species without significant degradation or loss of performance.
- **Cathode:** The cathode is the material where reduction reactions take place during discharge, acting as the receiver of electrons from the external circuit. An appropriate cathode material should present a high redox potential, good chemical and structural stability, and sufficient capacity to reversibly store and release ionic species. The selection of the cathode material plays a decisive role in determining the cell voltage and overall energy performance. Similarly to the anode, it must also ensure efficient ionic exchange during both charge and discharge processes, while maintaining its structural integrity over time.
- **Electrolyte:** The electrolyte acts as the medium that enables ionic transport between the anode and the cathode, thereby closing the internal circuit of the cell. A suitable electrolyte must exhibit high ionic conductivity while maintaining negligible electronic conductivity in order to prevent internal short-circuit currents. Depending on the technology, the electrolyte may be in liquid, solid, or gel form. Furthermore, it must be chemically and electrochemically stable within the operating voltage window of the cell, compatible with the electrode materials, and safe from both thermal and chemical standpoints.
- **Separator:** The separator is a porous, electrically insulating material placed between the anode and the cathode. Its primary function is to prevent direct physical contact between the electrodes, thereby avoiding internal short circuits, while still allowing the free transport of ions through its porous structure. An effective separator should exhibit adequate mechanical strength, thermal stability,

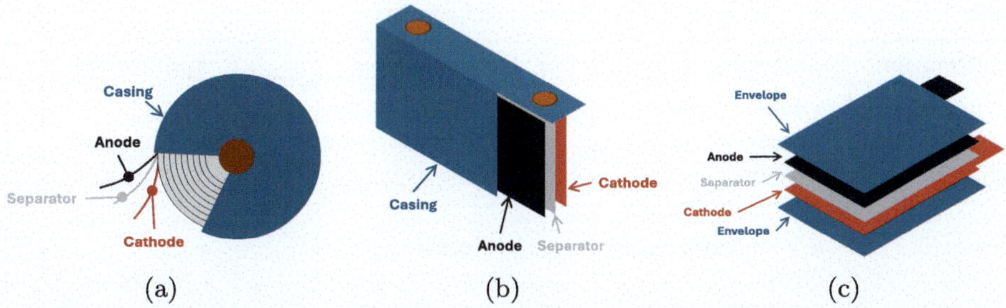

Figure 7.2 Schematic representation of the main electrochemical cell formats: (a) cylindrical, (b) prismatic, and (c) pouch, highlighting their distinct mechanical architectures and internal structural configurations.

and a porosity level that facilitates ionic transport without compromising the safety of the system.

It is important to note that the performance (or efficiency) of a battery does not depend solely on the selection of individual materials, but also on the interactions between them. The long-term behavior of an electrochemical cell is governed by factors such as interface stability, transport phenomena, and associated degradation processes. These aspects must be carefully considered in battery design in order to optimize performance while ensuring critical requirements such as safety and lifetime are not compromised.

7.2.2 Cell formats and mechanical design

The physical format and mechanical design of electrochemical cells have a significant impact on the operational behavior, performance, and application range of battery systems. In particular, the mechanical configuration strongly influences key aspects such as energy density, thermal management, structural robustness, and manufacturability. In industrial and commercial contexts, three main electrochemical cell formats are commonly employed (see Fig. 7.2), each presenting specific advantages and limitations:

- **Cylindrical cells** (Fig. 7.2a): In this configuration, the electrodes and separator are wound into a spiral structure, commonly referred to as a "jelly roll", and enclosed within a rigid cylindrical casing. This architecture provides excellent mechanical robustness and a strong tolerance to internal efforts and thermal cycling. From a manufacturing perspective, cylindrical cells benefit from a high degree of standardization and well-established automated production processes, which contribute to cost reduction and scalability. However, their geometry leads to suboptimal space utilization, resulting in lower volumetric energy density compared to alternative formats. Cylindrical cells are widely used in consumer electronics and in certain EV designs, such as those developed by Tesla.
- **Prismatic cells** (Fig. 7.2b): These cells typically adopt a stacked configuration of electrodes and separator layers, enclosed within a rigid rectangular casing.

Table 7.1 Comparison of main electrochemical cell formats.

Cell Format	Advantages	Disadvantages	Applications
Cylindrical	• Mechanical robustness • Mature manufacturing	• Energy density • Inefficient space utilization	• EVs • Consumer electronics
Prismatic	• Volumetric energy density • Efficient space utilization • Simplified pack integration	• Mechanical complexity • Internal stress and swelling	• EVs • Stationary energy storage
Pouch	• Energy density • Lightweight and flexible format • Design adaptability	• Mechanical robustness • Sensitive to swelling and degradation	• Smartphones and laptops • Aerospace applications

This geometry allows for improved packing efficiency and better utilization of the available volume within battery modules, leading to higher volumetric energy density. As a result, prismatic cells are particularly suitable for applications where space constraints are critical, such as in EV battery packs. Nevertheless, their mechanical design is more complex, requiring careful management of internal stresses and volume changes associated with electrode expansion during operation. These factors can impact long-term reliability if not properly addressed.

- **Pouch cells** (Fig. 7.2c): Pouch cells consist of stacked electrode layers enclosed within a flexible laminated polymer-aluminum envelope. Their main advantage lies in their ability to maximize both gravimetric and volumetric energy density due to the absence of a rigid casing. Additionally, their flexible form factor enables adaptable designs and efficient integration into compact systems. However, this flexibility also implies reduced mechanical robustness, making pouch cells more sensitive to swelling and external mechanical constraints. Consequently, they typically require external support structures within the battery pack. Pouch cells are commonly used in portable electronic devices such as smartphones and laptops, as well as in certain aerospace and high-performance applications.

Table 7.1 summarizes the main advantages, disadvantages, and typical applications of each cell format. As can be observed, these three configurations are largely complementary, enabling their coexistence across different technological domains depending on the specific design requirements and operational constraints.

7.2.3 Battery pack structure

In order to facilitate integration, scalability, and system management, battery-based energy storage systems in electric mobility applications are organized according to a

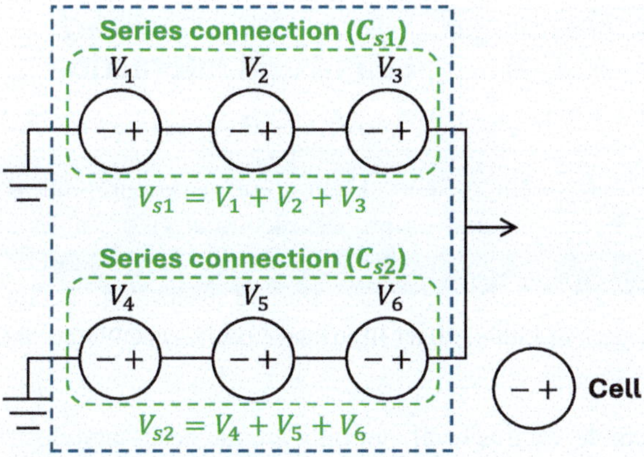

Figure 7.3 Example of a battery module configuration combining series and parallel interconnections. In each series string (e.g., C_{s1} and C_{s2}), the total voltage is obtained as $V_s = \sum_i V_i$, while the capacity remains unchanged. The parallel combination of series strings leads to an equivalent capacity $C_p = C_{s1} + C_{s2}$, with the voltage remaining equal to that of each individual series string.

hierarchical structure, typically consisting of three levels: cell, module, and battery pack. The purpose of this hierarchical arrangement is to progressively increase the voltage and capacity of the storage system, thereby adapting it to the operational requirements of the vehicle.

At the lowest level of this hierarchy are the electrochemical cells, where the reduction and oxidation reactions take place, enabling the flow of electrons between their terminals. Individual cells exhibit relatively low voltage and capacity values, far below those required for electric mobility applications. To overcome these limitations, cells are combined using series and parallel configurations, as illustrated in Fig. 7.3. As shown in the figure, the series connection of cells with individual voltage V_i results in a total voltage given by $V_s = \sum_i V_i$, while the capacity remains unchanged. Conversely, to increase the overall capacity, cells with individual capacity C_i can be connected in parallel, leading to a total capacity $C_p = \sum_i C_i$, with the voltage remaining equal to that of each branch. Through these configurations, the inherently limited voltage and current levels of individual cells can be adapted to values suitable for practical use.

The combination of cells through series-parallel arrangements gives rise to the intermediate level of the hierarchy, referred to as the module. Modules are therefore assemblies of cells electrically interconnected according to a predefined configuration. This level provides a compromise between assembly complexity, maintainability, and electrical performance, while also enabling greater flexibility in system design. Additionally, modularization facilitates manufacturing processes and allows for easier replacement or servicing of specific sections of the battery system.

At the highest hierarchical level is the battery pack, which results from the interconnection of multiple modules and constitutes the complete functional unit

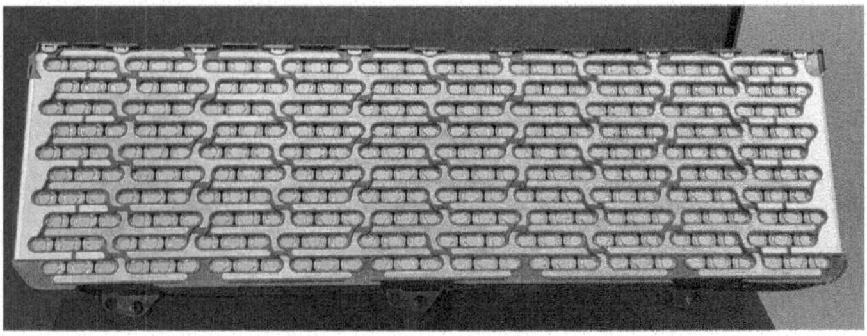

Figure 7.4 Battery pack of Lithium-ion cells. (Image courtesy of Wikimedia Commons.)

installed in the vehicle. At this level, not only are the necessary electrical connections implemented, but additional components related to structural support, monitoring, and control are typically integrated. These may include the BMS, further explained in Section 7.2.4, as well as thermal management subsystems (see Section 7.2.5).

As illustrated in Fig. 7.4, battery packs are realized as highly compact systems in which a large number of cells are arranged in optimized geometric patterns. These layouts are designed to maximize energy density, efficiently utilize the available space, and facilitate effective thermal management. From a mechanical perspective, the battery pack must ensure sufficient structural robustness to withstand vibrations, shocks, and other operational stresses, thereby preventing premature component degradation.

In modern EVs, the battery pack is commonly located at the bottom of the vehicle chassis. This configuration not only enables efficient packaging but also contributes to lowering the center of gravity, which improves vehicle stability and overall dynamic performance. Consequently, the battery pack structure represents a key element that directly influences system performance, safety, and integration within the electric powertrain.

7.2.4 Battery management system (BMS)

Within the battery pack, the BMS acts as the central control unit responsible for its management, ensuring that appropriate operating conditions are maintained in order to guarantee both efficient and safe operation of the set of cells composing the energy storage system. To achieve this objective, the BMS continuously monitors a set of key parameters such as voltage, current, and temperature, at different hierarchical levels including cell, module, and pack. Based on these measurements, the BMS estimates two fundamental parameters that serve as the basis for control and decision-making within the battery system:

- **SoC**: Refers to the current energy stored in the cell. It is typically defined as the ratio between the stored energy and the total available capacity, and is commonly expressed either in per unit or as a percentage. For example, if a cell with a total capacity of 1 Wh stores 0.5 Wh, its SoC is equal to 50% (or 0.5 per unit). Although defining SoC in terms of energy provides an intuitive interpretation, it can also be defined in terms of charge (Ah). This formulation is often more robust in practical applications, as it does not depend on the cell voltage, which varies during charge and discharge processes (see Section 7.1.2).

- **SoH**: Is associated with the degradation of the cell and represents its current condition relative to its initial state. It is commonly quantified based on the reduction in usable capacity, although it may also reflect performance degradation such as increased internal resistance. Similarly to SoC, it is usually expressed as a percentage relative to the initial capacity of the cell. For instance, if a cell initially has a usable capacity of 1 Wh and this value decreases to 0.7 Wh after a certain period of operation, the corresponding SoH is 70% (or 0.7 per unit).

Based on the estimated values of SoC and SoH, the BMS performs several essential functions to ensure proper battery operation. One of the most relevant is cell balancing. Through this function, the BMS redistributes energy among cells within the same pack in order to maintain similar SoC levels across all cells. This process improves the utilization of the total available capacity and prevents situations in which certain cells reach their operational limits earlier than others.

In addition to balancing, the BMS carries out critical protection functions aimed at preventing adverse operating conditions. These include protection against overcharge, overcurrent, overdischarge, and unfavorable thermal conditions. Such conditions could accelerate cell degradation and reduce battery lifetime (see Section 7.5.2), as well as introduce safety risks.

7.2.5 Thermal management systems

Thermal control and management of battery packs is a fundamental aspect in improving their performance and lifetime, while simultaneously ensuring safe operating conditions. Electrochemical cells are particularly sensitive to temperature, both at high and low extremes. Elevated temperatures can accelerate degradation mechanisms within the cells, leading to premature aging and reduced lifetime. Conversely, low temperatures can decrease the available capacity and limit the power output due to reduced electrochemical kinetics and increased internal resistance.

To maintain the battery within an appropriate operating temperature range, thermal management systems are employed. These systems can be broadly classified into passive and active solutions. Passive systems rely on natural heat dissipation mechanisms such as conduction and convection. They are typically simple, low-cost, and do not require additional energy input. However, their ability to regulate temperature is limited, especially under high power demand or harsh environmental conditions.

In contrast, active thermal management systems incorporate forced heat transfer mechanisms, such as air cooling through ventilation systems or liquid cooling circuits. These systems enable a more precise and efficient control of the temperature distribution within the battery pack. Although they involve higher complexity, cost, and energy consumption, active systems are widely adopted in modern EVs due to their superior performance.

7.3 Advantages of batteries vs other storage technologies

7.3.1 Efficiency and energy conversion

One of the main advantages of batteries compared to other energy storage technologies is their high efficiency (performance). Most battery technologies exhibit remarkably high efficiencies, often exceeding 90% [2]. This high performance is primarily a consequence of the limited number of energy conversion stages involved during battery charge and discharge processes, which are largely governed by reversible electrochemical reactions.

In contrast, alternative energy storage solutions such as hydrogen typically involve more complex energy conversion chains. In hydrogen-based systems, electrical energy must first be converted into hydrogen through electrolysis. The hydrogen is then stored and subsequently reconverted into electrical energy using FCs. This multi-stage generation–storage–conversion process introduces additional losses at each step, leading to significantly lower overall efficiencies, typically in the range of 25–40% [3].

High efficiency values are particularly relevant in the context of electric mobility, where maximizing the utilization of stored energy is a critical requirement. Efficient energy storage systems directly translate into improved vehicle performance, extended driving range, and reduced energy demand during charging processes. In this regard, battery systems enable a more direct and effective use of electrical energy, minimizing intermediate losses and simplifying the overall system architecture.

Furthermore, the relatively straightforward energy conversion pathway of batteries contributes to their robustness and ease of integration within electric powertrains. As a result, batteries are well suited for a wide range of mobility applications, where efficiency, reliability, and system simplicity are essential design criteria. This combination of high efficiency and practical implementation advantages largely explains the widespread adoption of battery technologies across modern EVs.

7.3.2 Technological maturity and scalability

Another key aspect that positions batteries as the dominant energy storage technology in the electric mobility sector is their high level of technological maturity. In recent years, battery technologies, particularly those based on lithium, have undergone sustained development, enabling significant improvements in both performance and manufacturing processes. This continuous progress has resulted in a very high level of technological readiness compared to other energy storage alternatives, for which large-scale deployment is still limited or emerging.

The growing demand for batteries in sectors such as consumer electronics and electric mobility has further accelerated their industrial development. As a consequence, battery production has reached large-scale manufacturing levels, supported by highly optimized and increasingly automated processes. This industrial scaling has contributed to the consolidation of global supply chains and has enabled a steady reduction in production costs, largely driven by economies of scale and process standardization.

Scalability is another defining advantage of battery systems. In contrast to many other storage technologies, batteries are inherently modular (see Section 7.2.3), allowing straightforward adaptation to different power and energy requirements. Individual cells can be combined into modules and packs of varying sizes, which facilitates their integration across a wide range of applications, from small portable devices to full-scale EVs. This flexibility significantly enhances their applicability in diverse sectors.

Finally, the degree of standardization achieved in battery technologies is notably higher than that of most competing storage solutions. Standardized cell formats, interfaces, and management systems contribute to improved interoperability and simplify system integration. Altogether, these factors reinforce the position of batteries as a mature, scalable, and industrially viable solution for modern electric mobility applications.

7.3.3 Limitations and trade-offs

Although the advantages of batteries as an energy storage technology are well established, they also present several limitations that may constrain their applicability in certain scenarios. One of the most significant limiting factors is aging. Batteries are subject to two primary degradation mechanisms (see Section 7.5.2). On the one hand, cyclic aging is directly related to usage, i.e., the number of completed charge and discharge cycles. On the other hand, calendar aging occurs independently of usage and is primarily driven by time-dependent physicochemical processes. Both mechanisms contribute to a progressive loss of energy capacity and power capability over the battery lifetime. This degradation is often non-linear and depends on operating conditions such as temperature, DoD, and current rates. As a result, battery systems often need to be oversized to meet long-term performance requirements, or cells must be replaced once degradation reaches levels that compromise their suitability for specific applications [1].

Another important limitation is the relatively low energy density of batteries. Despite significant technological progress in recent years, which has enabled substantial improvements compared to earlier generations, batteries still exhibit much lower energy densities than conventional fossil fuels such as gasoline, as well as lower values than alternative energy vectors such as hydrogen [4] (see Chapter 8). This limitation is particularly critical in electric mobility applications, where it translates into heavier and bulkier storage systems in order to achieve adequate driving range, thereby impacting overall vehicle efficiency, packaging constraints, and structural design considerations.

In addition, battery manufacturing is highly dependent on critical raw materials (see Section 7.8.1). Current battery technologies rely heavily on elements such as lithium, cobalt, and nickel, whose production is geographically concentrated. This concentration introduces vulnerabilities in the supply chain, increasing exposure to geopolitical risks, market fluctuations, and potential supply constraints [5]. Furthermore, extraction and processing of these materials raise environmental and ethical concerns, which are increasingly influencing regulatory frameworks and industrial strategies.

Finally, safety represents a key concern in battery operation. Batteries are susceptible to failure modes such as overcharge, short circuits, or mechanical damage. If not properly managed, these conditions may trigger thermal runaway events, characterized by an uncontrolled increase in temperature that can lead to fires or explosions. Furthermore, exposure to specific environmental conditions, such as high-salinity environments, can accelerate corrosion processes in electrical connections and terminals, potentially compromising system integrity and increasing the risk of electrical faults and safety incidents [6]. Although modern BMSs and advanced design strategies significantly mitigate these risks, they cannot be entirely eliminated.

7.4 Main battery technologies

7.4.1 Lithium-ion batteries

7.4.1.1 General operating principles

Lithium-ion batteries are based on the reversible intercalation of lithium ions (Li^+). One of the main advantages of this technology compared to other electrochemical storage systems is that the insertion and extraction of Li^+ ions does not involve significant structural changes in the active materials. Instead, lithium ions are accommodated within the crystalline structure of the anode and cathode host materials, preserving their structural integrity and enabling a high degree of reversibility over multiple charge–discharge cycles.

Considering a typical graphite anode and a metal oxide cathode, the electrochemical reactions during discharge can be expressed as follows:

Anode (oxidation):

$$LiC_6 \rightarrow C_6 + Li^+ + e^-$$

Cathode (reduction):

$$Li_{1-x}MO_2 + xLi^+ + xe^- \rightarrow LiMO_2$$

Overall reaction:

$$LiC_6 + Li_{1-x}MO_2 \rightarrow C_6 + LiMO_2$$

During discharge, Li^+ ions migrate from the anode to the cathode through the electrolyte, while electrons flow through the external circuit, generating an exploitable electric current. This coupled ionic and electronic transport mechanism is fundamental to the operation of lithium-ion batteries and determines their efficiency and power delivery capability.

During charging, an external energy source is applied across the battery terminals, forcing the electrochemical reactions to proceed in the opposite direction. As a result, lithium ions migrate back from the cathode to the anode, where they are re-intercalated into the graphite structure. The corresponding reactions are:

Anode (reduction):

$$C_6 + Li^+ + e^- \rightarrow LiC_6$$

Cathode (oxidation):

$$LiMO_2 \rightarrow Li_{1-x}MO_2 + xLi^+ + xe^-$$

Although this process is highly reversible, it is not ideal in practice. The performance and long-term stability of lithium-ion batteries are influenced by several factors, including electrochemical kinetics, lithium-ion diffusion within the electrode materials, and the stability of the electrode–electrolyte interfaces. Additionally, parasitic reactions and side effects at the interfaces can gradually degrade the materials. These aspects introduce inefficiencies and contribute to degradation over repeated charge–discharge cycles, ultimately affecting both capacity retention and power capability.

7.4.1.2 Cathode chemistries

Within the family of lithium-ion batteries, different technologies can be distinguished based on the cathode chemistry. This classification is particularly relevant from an operational perspective, as the cathode material largely determines key parameters such as energy density, specific power, safety, and overall cost. In the context of electric mobility, several cathode chemistries have consolidated their industrial adoption in recent years:

- **Nickel-Manganese-Cobalt (NMC):** This chemistry has become one of the most widely adopted solutions in electric mobility applications. Its main advantage lies in its well-balanced trade-off between energy density, cycle life, and cost. Furthermore, NMC is a highly versatile chemistry, as different performance profiles can be achieved by adjusting the relative proportions of nickel, manganese, and cobalt. For instance, increasing the nickel content typically enhances specific capacity and energy density, while manganese contributes to improved structural stability and thermal robustness. However, one of its main drawbacks is the presence of cobalt, a critical material whose supply is highly concentrated in the Democratic Republic of the Congo (see Section 7.8.1). This supply concentration introduces economic and geopolitical concerns, which have driven the development of low-cobalt or cobalt-reduced variants [7].
- **Nickel-Cobalt-Aluminum (NCA):** This chemistry is characterized by its high energy density compared to other lithium-ion technologies. This feature makes NCA particularly suitable for electric mobility applications and, more generally, for use cases where high energy storage capacity is required. As a result, NCA-based batteries enable a reduction in the overall weight of the vehicle while maintaining competitive driving range. However, this benefit comes at the expense of reduced thermal stability compared to other chemistries. Consequently,

Table 7.2 Simplified comparison of main lithium-ion cathode chemistries.

Chemistry	Energy Density	Power	Safety	Cost
NMC	High	Medium	Medium	Medium
NCA	Very High	Medium	Low	High
LFP	Medium-Low	Medium	High	Low
LMO	Medium	High	Medium	Medium

NCA systems typically require more advanced thermal management and control strategies to ensure safe operation and mitigate risks associated with elevated temperatures.

- **Lithium Iron Phosphate (LFP):** This chemistry addresses several limitations associated with other cathode materials, particularly in terms of supply chain sensitivity. LFP relies on abundant and widely available materials, which reduces both cost and geopolitical dependency. In addition, it exhibits excellent thermal and chemical stability, resulting in superior safety characteristics and a significantly longer cycle life compared to nickel-based chemistries. These features make LFP especially suitable for intensive-use applications. Nevertheless, its main limitation is its relatively low energy density, primarily due to the absence of nickel in its composition. This constraint represents a significant challenge for applications in electric mobility where maximizing driving range is a priority.
- **Lithium Manganese Oxide (LMO):** The primary advantage of this chemistry lies in its high power capability, enabled by a crystal structure that facilitates fast lithium-ion diffusion. This property makes LMO particularly attractive for applications requiring rapid charge and discharge dynamics, as well as strong transient response. It also allows for more effective utilization of fast-charging systems (see Chapter 6). However, this high power performance is associated with accelerated degradation of the active materials, leading to a shorter expected lifetime. For this reason, LMO is often used in combination with other cathode chemistries, forming hybrid systems that balance power capability and durability [8].

As a summary, Table 7.2 provides a synthesized comparison of the main strengths and weaknesses of each cathode chemistry in terms of key performance indicators such as energy density, power capability, safety, and cost. As can be observed, lithium-ion battery chemistries exhibit complementary characteristics, and no single solution currently satisfies all performance requirements across different applications.

In particular, nickel-rich chemistries such as NMC and NCA prioritize high energy density and are therefore well suited for maximizing vehicle range. In contrast, LFP emphasizes safety, durability, and cost-effectiveness, making it an attractive option for applications where these factors are critical. Meanwhile, LMO is primarily oriented toward high-power applications, where fast response and charge acceptance are required.

Consequently, the selection of a specific cathode chemistry typically involves a detailed analysis of the target application, taking into account technical, economic, and operational constraints. In many cases, this leads to the adoption of compromise solutions or even hybrid configurations that combine different chemistries in order to achieve a balanced overall performance.

7.4.1.3 Performance characteristics

The suitability of lithium-ion batteries for electric mobility applications is primarily determined by four key metrics: energy density, lifetime, safety, and cost. The performance of these batteries across these dimensions ultimately defines their applicability within the electric mobility sector.

- **Energy density:** Lithium-ion batteries provide high energy density values in both gravimetric and volumetric terms, although still significantly lower than alternative energy storage technologies such as hydrogen. In electric mobility applications, typical energy density values range between 150 and 300 Wh/kg at the cell level [9]. However, the effective energy density at the battery pack level is lower due to the presence of structural components, thermal management systems, and associated electronics. This reduction from cell to pack level highlights the importance of system-level design, where packaging efficiency and auxiliary components play a critical role in determining the overall energy storage capability of the vehicle. Among the different lithium-ion chemistries, nickel-rich compositions such as NMC and NCA generally exhibit higher energy density values.
- **Lifetime:** In electric mobility applications, battery lifetime is commonly evaluated in terms of the number of charge and discharge cycles that can be completed before the available capacity decreases to approximately 80% of its initial value. In this context, lithium-ion batteries typically achieve lifetimes ranging from several hundred to several thousand cycles [2,9]. However, this parameter strongly depends on multiple factors, including the DoD, operating temperature, and charging strategy (see Section 7.5.2). Within the lithium-ion family, LFP batteries generally exhibit extended cycle life compared to other chemistries. In addition to cycle life, calendar aging also plays a relevant role, as degradation mechanisms may occur even when the battery is not actively cycled.
- **Safety:** Lithium-ion batteries require careful management to ensure safe operating conditions. This is primarily due to two factors: the reactive nature of their constituent materials, which implies that under abnormal conditions the system may release energy in an uncontrolled manner, and the widespread use of flammable organic electrolytes. Under such conditions, the development of advanced systems such as BMSs, together with improvements in materials and cell design, has become a major area of research aimed at enhancing battery safety. Furthermore, thermal management systems are essential to maintain the battery within a safe operating temperature range and to prevent the onset of critical failure modes such as thermal runaway.

- **Cost:** Although lithium-ion batteries have experienced a sustained cost reduction throughout the 21st century (see Chapter 1), mainly driven by economies of scale, manufacturing process optimization, and supply chain improvements, their cost remains closely linked to the dependence on certain materials with significant geographical concentration, such as cobalt and nickel. This has encouraged the development and deployment of alternative chemistries such as LFP, which exhibit a significantly lower reliance on critical materials. Additionally, ongoing research efforts are focused on reducing material intensity and improving recyclability, with the aim of further decreasing costs and enhancing long-term sustainability.

7.4.1.4 Advantages and limitations

The widespread adoption of lithium-ion batteries in electric mobility applications is justified by the combination of several key advantages, which have positioned this technology as the dominant energy storage solution in the sector.

Among these advantages, their relatively high energy density stands out. Lithium-ion batteries achieve high gravimetric and volumetric energy density values compared to other battery technologies, making them particularly suitable for mobility applications. This is especially relevant in systems where mass sensitivity is a critical constraint, such as UAVs (see Chapter 4). Nevertheless, the achievable energy density remains significantly lower than that of alternative energy carriers such as hydrogen.

Another important advantage is their high efficiency and technological maturity. The high efficiency of lithium-ion batteries enables more effective energy utilization, reducing losses during charge and discharge processes and improving overall system performance. At the same time, their advanced level of technological maturity has enabled significant optimization of manufacturing processes, leading to cost reductions, large-scale production capabilities, and improved integration into modular systems adaptable to a wide range of applications.

One of the most distinctive features of lithium-ion batteries is their versatility. The existence of multiple technological solutions (see Section 7.4.1.2) allows for the tailoring of active material chemistry in order to optimize the trade-off between energy performance, power capability, lifetime, and safety. This flexibility has been a key enabler for the widespread deployment of lithium-ion batteries across different segments of electric mobility, ranging from light-duty vehicles to more demanding applications with higher energy requirements.

Despite these advantages, lithium-ion batteries also present several limitations. One of the most relevant is their progressive degradation over time. As a result of usage (cycle ageing) and time-dependent mechanisms (calendar ageing), the effective capacity of lithium-ion batteries decreases over time. If operating conditions and usage profiles are not properly managed, this degradation can significantly reduce the useful lifetime of the battery system.

Finally, material dependency and safety concerns must also be considered. On the one hand, the manufacturing of lithium-ion batteries relies heavily on the availability of critical materials such as nickel and cobalt, whose supply is geographically concentrated. This introduces pressure on the supply chain and makes the technology sensitive to geopolitical factors. On the other hand, the presence of reactive materials and flammable electrolytes may lead to adverse events under abnormal conditions if appropriate monitoring, management, and protection strategies are not implemented. In addition, although costs have decreased significantly over recent decades, further reductions remain constrained by material availability and processing complexity. Consequently, ongoing research efforts are focused on alternative chemistries and improved system designs aimed at mitigating these limitations while preserving the core advantages of lithium-ion technology.

7.4.2 Nickel-based batteries

7.4.2.1 Nickel-metal hydride (NiMH) technology

In contrast to lithium-ion batteries discussed in Section 7.4.1, whose widespread adoption in electric mobility emerged during the 21st century in parallel with significant cost reductions (see Chapter 1), nickel-based batteries have been utilized since the early stages of transport electrification. Within this family, Nickel-Metal Hydride (NiMH) technology has historically represented the most relevant solution in electric mobility applications.

NiMH batteries consist of a positive electrode typically based on nickel hydroxide (H_2NiO_2), a negative electrode formed by a metal alloy capable of reversibly absorbing hydrogen, and an alkaline aqueous electrolyte. The operation of these batteries relies on the transport of hydroxyl ions (OH^-), which close the internal ionic circuit. During discharge, oxidation occurs at the anode through the release of hydrogen from the metal hydride, while reduction takes place at the cathode involving nickel oxyhydroxide. These processes can be described by the following electrochemical reactions:

Anode (oxidation):

$$MH + OH^- \rightarrow M + H_2O + e^-$$

Cathode (reduction):

$$NiOOH + H_2O + e^- \rightarrow Ni(OH)_2 + OH^-$$

Overall reaction:

$$MH + NiOOH \rightarrow M + Ni(OH)_2$$

During charging, the direction of ion transport is reversed, forcing OH^- ions to migrate in the opposite direction. As a result, reduction takes place at the anode and oxidation at the cathode, restoring the active materials to their original chemical states and enabling reversible operation.

Within the electric mobility domain, NiMH batteries have been extensively deployed in HEVs. This is largely due to their ability to withstand demanding operating conditions, particularly frequent partial charge and discharge cycles.

In HEVs, the onboard battery is typically recharged through an internal energy generation system based on ICEs, which operate during braking, deceleration, or cruising phases (see Chapter 2). This operational profile requires high cycling robustness rather than maximized energy storage capacity, a requirement well matched by NiMH technology. Additionally, the use of an aqueous electrolyte contributes to enhanced thermal stability and robust performance under varying operating conditions.

Despite these advantages, NiMH batteries exhibit several limitations that constrain their applicability in certain electric mobility scenarios. One of the most notable is the so-called memory effect. This phenomenon refers to an apparent reduction in usable capacity when the battery is repeatedly subjected to partial charge and discharge cycles. Under such conditions, the battery may effectively "remember" a reduced charge level, limiting its operational capacity window. Furthermore, NiMH batteries are characterized by relatively high self-discharge rates, leading to a gradual loss of stored energy during idle periods. These factors must be considered when evaluating their suitability for specific applications within electric mobility.

7.4.2.2 Comparison with lithium-ion

It is relevant to compare NiMH batteries with lithium-ion technologies, as the latter currently dominate the field of electric mobility. Such a comparison provides a comprehensive perspective on the technological evolution of energy storage systems and clarifies the factors that have driven the preferential adoption of each technology depending on application requirements.

In terms of energy density, NiMH batteries exhibit typical values in the range of 60–120 Wh/kg and 140–300 Wh/L [10], which places them significantly below the values reported for lithium-ion batteries (see Section 7.4.1.3). This limitation makes NiMH systems less suitable for applications where driving range is a critical parameter, as is the case in most electric mobility solutions, particularly in vehicles where mass sensitivity plays a key role in overall performance and efficiency.

Regarding energy efficiency, NiMH batteries show a relatively wide variation depending on operating conditions and usage profiles, typically ranging between 60% and 92% [11]. This positions NiMH technology below lithium-ion systems in terms of round-trip efficiency, as lithium-ion batteries generally achieve values in the range of 90–95% [2]. In addition, NiMH batteries are characterized by higher self-discharge rates, which further reduces the effective utilization of stored energy, especially during idle periods or intermittent operation.

Despite these drawbacks, NiMH batteries offer advantages in specific operational aspects. In particular, they exhibit high thermal robustness and stable performance across a wide range of operating conditions. These characteristics make them well suited for applications involving demanding and dynamic duty cycles, where the battery is subjected to frequent partial charge and discharge events. Such conditions are typical in vehicles combining multiple traction modes, including both thermal and electric propulsion systems.

As a consequence of these advantages and limitations, the current application domain of NiMH batteries is largely restricted to HEVs. In these systems, operational robustness and cycling capability are more critical than maximizing energy storage capacity, since vehicle range is not solely dependent on the battery. Conversely, lithium-ion technology has progressively displaced NiMH in other segments, such as BEVs and plug-in HEVs, where higher energy density and reduced mass are essential requirements. This transition reflects the increasing importance of energy-related performance metrics in modern electric mobility systems.

7.4.3 Lead-acid batteries

7.4.3.1 Operating principles

Lead-acid batteries represent one of the oldest and most widely deployed technologies within the field of electrochemical energy storage. Their operation is based on the interaction of three species, involving metallic lead and lead dioxide as active materials, while an acidic solution acts as the electrolyte, typically sulfuric acid (H_2SO_4). This technology is characterized by a high level of technological maturity, combined with structural simplicity and a well-established operating principle.

During the discharge process, the anode composed of metallic lead (Pb) and the cathode formed by lead dioxide (PbO_2) react with the acidic electrolyte, producing lead sulfate ($PbSO_4$) and releasing electrons to the external circuit. These electrochemical processes can be described by the following reactions:

Anode (oxidation):

$$Pb + SO_4^{2-} \rightarrow PbSO_4 + 2e^-$$

Cathode (reduction):

$$PbO_2 + SO_4^{2-} + 4H^+ + 2e^- \rightarrow PbSO_4 + 2H_2O$$

Overall reaction:

$$Pb + PbO_2 + 2H_2SO_4 \rightarrow 2PbSO_4 + 2H_2O$$

During charging, these reactions are reversed through the application of an external energy source, regenerating the original active materials at both electrodes and restoring the electrolyte composition.

A distinctive feature of lead-acid batteries is the direct participation of the electrolyte in the electrochemical reactions. As a result, unlike lithium-ion and NiMH systems, the electrolyte concentration varies throughout the charge and discharge processes. This variation has a direct impact on cell parameters such as voltage and SoC. Consequently, the electrolyte density can be used as a practical indicator to estimate these parameters, providing a simple and effective method for monitoring battery state.

Overall, the operating principle of lead-acid batteries is defined by its clarity and robustness, which has enabled their extensive use across a wide range of applications over more than a century.

7.4.3.2 Applications and limitations

Despite their high level of technological maturity, lead-acid batteries have been almost completely displaced as traction batteries in electric mobility applications. In this context, their use has been progressively restricted to scenarios with relatively low operational demands, where range is not a critical requirement and where cost-effective and simple solutions are prioritized.

Indeed, their low cost remains the primary advantage of this technology when compared to alternatives such as lithium-ion or NiMH batteries. This cost competitiveness is the result of a fully mature technology, supported by highly optimized manufacturing processes and a well-established global supply chain. In addition, lead-acid batteries exhibit good dynamic behavior, allowing them to deliver relatively high power rates over short periods. This characteristic has historically made them suitable for supporting engine starting processes in ICE vehicles. Furthermore, these batteries present a very high recyclability rate, enabling material recovery levels that are significantly higher than those of most other electrochemical storage technologies.

However, lead-acid batteries also exhibit several critical limitations that explain their reduced role in modern electric mobility. The most significant drawback is their low energy density, both gravimetric and volumetric, typically in the range of 30–50 Wh/kg [12]. This limitation results in heavy and bulky battery packs for a given energy capacity, which directly impacts vehicle efficiency and driving range. Consequently, this constraint has been the main factor leading to the replacement of lead-acid technology in traction and propulsion applications.

Additional disadvantages include lower energy efficiency compared to more advanced battery technologies, as well as a relatively limited cycle life, particularly under deep discharge conditions. Moreover, lead-acid batteries are prone to degradation mechanisms such as sulfation, which can significantly reduce their usable capacity if operating conditions are not properly controlled. Environmental concerns must also be considered, as the presence of lead introduces toxicity issues that complicate waste management and limit the feasibility of second-life applications.

As a result, within the field of electric mobility, lead-acid batteries are currently almost exclusively used for auxiliary low-voltage ($\sim$ 12 V) systems, including control electronics, monitoring units, starting systems, and climate control subsystems. This architecture is common across BEVs, HEVs, and even conventional ICE vehicles. Therefore, despite their limited role in propulsion, the use of lead-acid batteries in mobility applications remains significant, largely driven by their low cost and reliability.

As a final summary, Table 7.3 provides a qualitative comparison of the main characteristics of different battery technologies in the context of electric mobility. As shown, there is a clear complementarity in the advantages and limitations of

Table 7.3 Comparative overview of major battery technologies used in electric mobility.

Parameter	Lithium-ion	NiMH	Lead-acid
Energy Density	✓✓✓	✓✓	✓
Cost	✓	✓✓	✓✓✓
Efficiency	✓✓✓	✓✓	✓
Robustness	✓✓	✓✓✓	✓✓
Main Application	BEVs	HEVs	Auxiliary systems

each technology, which explains their specific domains of application. This landscape suggests a scenario of technological coexistence, at least in the short to medium term.

7.5 Performance metrics and degradation mechanisms

7.5.1 Key performance indicators

The evaluation of battery performance in electric mobility applications relies on a set of indicators that synthesize its behavior under typical operating conditions. Although a wide range of metrics can be considered, four of them are particularly relevant, namely Cycle life, calendar life, coulombic efficiency, and the evolution of the available capacity over time.

- **Cycle life:** The capacity of a battery is not constant over time and is strongly influenced by aging and usage conditions. In particular, the cycle life is governed by internal physico-chemical phenomena that affect both the active materials and the structural integrity of the cell, leading to a progressive degradation of the usable capacity. In the context of electric mobility, the cycle life is typically defined as the number of charge and discharge cycles that can be completed before the usable capacity falls below a given threshold, commonly around 80% of the initial capacity. It is worth noting that this parameter can be defined based on full cycles, equivalent partial cycles, or a combination of both, depending on the application and operating profile.
- **Calendar life:** In contrast to the cycle life, the calendar life is independent of the usage of the battery and represents the loss of effective capacity over time. This type of aging is the result of degradation mechanisms that occur even under storage conditions. Although it does not depend directly on cycling, it is strongly influenced by operating conditions such as temperature and the SoC. As a result, batteries stored at high temperatures or high SoC levels tend to exhibit accelerated degradation, even in the absence of active use.
- **Coulombic efficiency:** Unlike the round-trip efficiency, which is defined in terms of energy, the coulombic efficiency relates the charge extracted during discharge processes to the total charge supplied during charging. This parameter is therefore directly associated with internal losses caused by parasitic reactions

and irreversible processes occurring within the cell. In high-quality batteries, the coulombic efficiency typically exhibits values very close to 1. However, even small deviations from unity, when accumulated over a large number of cycles, can result in a noticeable loss of usable capacity, making this parameter particularly relevant for long-term degradation analysis.

- **Capacity evolution:** The usable capacity of a battery, understood as the amount of energy it can store, does not remain constant throughout its lifetime. Instead, it progressively decreases due to different degradation mechanisms (see Section 7.5.2). In this context, capacity fade is one of the main indicators of the SoH of a cell (see Section 7.2.4) and has a direct impact on both vehicle range and performance. A reduction in capacity not only limits the available energy but is often accompanied by a decrease in power capability, thereby affecting the dynamic response of the system.

Taken together, these indicators provide a comprehensive framework for assessing battery performance and durability in electric mobility applications. Their combined analysis is essential for the proper design, sizing, and management of battery systems, as well as for predicting their lifetime under different operating conditions.

7.5.2 Degradation mechanisms

The progressive loss of capacity and performance in batteries is driven by a variety of degradation mechanisms, typically arising from complex physico-chemical processes. In practice, these mechanisms do not operate independently but interact with each other, leading to cumulative effects that depend both on the underlying battery technology and on the operating conditions. Focusing on lithium-ion batteries, which are the most widely used in electric mobility, three primary degradation mechanisms can be identified as dominant contributors to capacity fade: the growth of the Solid Electrolyte Membrane (SEI), the lithium plating phenomenon, and the structural degradation of the cathode. These mechanisms are described in detail below.

- **SEI growth:** This is one of the most critical degradation processes in lithium-ion batteries. The SEI refers to a passivation layer formed on the surface of the anode, typically graphite, composed of various compounds resulting from electrolyte reduction, such as Li_2CO_3, LiF, and Li_2O. The SEI formed during the initial cycles plays a beneficial role, as it stabilizes the electrode–electrolyte interface. As illustrated in Fig. 7.5, this layer is selectively conductive: it allows the transport of Li^+ ions while preventing electron conduction. This property is essential to avoid continuous electrolyte decomposition. Despite its initial protective function, the SEI tends to grow progressively over time due to ongoing parasitic reactions, especially under elevated temperatures, high SoC conditions, or prolonged cycling. This growth leads to a continuous consumption of cyclable lithium, effectively trapping it within the SEI and reducing the amount of active lithium available for electrochemical reactions. As a consequence, the battery experiences a gradual loss of capacity. Furthermore, the thickening of the SEI increases the internal resistance of the cell, negatively impacting both energy efficiency and power capability.

Therefore, SEI growth represents a key trade-off between interfacial stability and long-term degradation.

- **Lithium plating:** This phenomenon constitutes another critical degradation mechanism in lithium-ion batteries. During the charging process, metallic lithium may deposit on the surface of the anode instead of being intercalated into its structure. This occurs when the intercalation kinetics are insufficient to accommodate the incoming flux of Li^+ ions. Lithium plating is particularly exacerbated under high charging currents, low temperatures, or high SoC conditions, where diffusion limitations and increased overpotentials hinder proper intercalation. The main issue associated with lithium plating is that the deposited lithium is not always reversible. A fraction of this metallic lithium becomes electrochemically inactive, leading to a loss of active lithium inventory and, consequently, capacity fade. Additionally, as illustrated in Fig. 7.6, the accumulation of plated lithium can lead to the formation of dendritic structures (whiskers). These dendrites may grow across the separator and establish an electrical connection between the anode and cathode, resulting in internal short circuits. Such events pose severe safety risks, including thermal runaway.
- **Cathode structural degradation:** During repeated charge and discharge cycles, variations in lithium concentration and changes in the oxidation state of transition metals occur, particularly in cathodes based on layered metal oxides. These processes induce mechanical stresses within the crystalline structure of the material, potentially leading to particle cracking, loss of electrical contact, and irreversible phase transformations. Over time, these structural changes degrade the ability of the cathode to reversibly store and release lithium ions. Additionally, transition metal dissolution may occur, with dissolved species migrating through the electrolyte and depositing on the anode surface. This phenomenon further destabilizes the SEI and contributes to additional degradation pathways. Overall, cathode degradation results in a reduction of the available active material and a consequent loss of capacity and performance.

Importantly, these degradation mechanisms are strongly interdependent and often act synergistically. Battery degradation should therefore be understood as a cumulative and multifactorial process rather than as the result of isolated phenomena. For instance, lithium plating can promote additional SEI growth due to the high reactivity of metallic lithium with the electrolyte. Conversely, the thickening of the SEI can increase internal resistance, leading to higher overpotentials that favor lithium plating. Similarly, transition metal dissolution from the cathode can catalyze parasitic reactions at the anode, further accelerating SEI growth.

This complex interplay highlights the importance of controlling operating conditions and material properties in order to mitigate degradation and extend battery lifetime. Parameters such as temperature, current rate, and SoC window play a critical role in determining the dominant degradation pathways (see Section 7.5.3). Consequently, advanced battery management strategies are essential to minimize the impact of these mechanisms, ensuring both performance and safety over the battery lifetime.

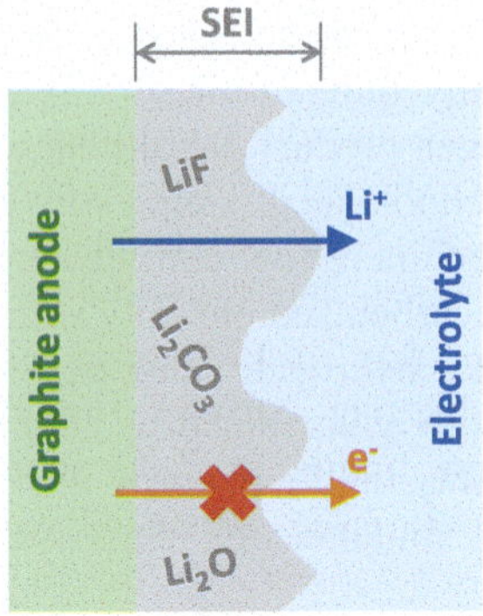

Figure 7.5 Schematic representation of the SEI formed at the graphite anode. The SEI acts as an electronically insulating but ionically conductive layer, allowing Li^+ transport while blocking electron transfer, thereby preventing continuous electrolyte decomposition.

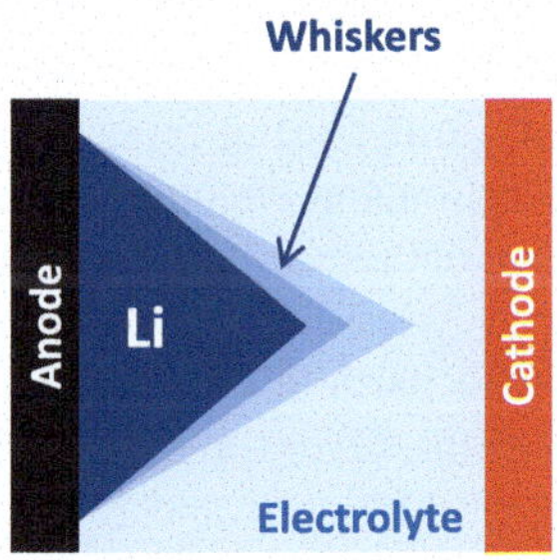

Figure 7.6 Schematic representation of lithium plating and dendritic growth (whiskers) at the anode surface. Under unfavorable operating conditions, metallic lithium deposits on the anode instead of intercalating, potentially leading to dendrite formation, internal short circuits, and safety hazards.

7.5.3 Factors affecting battery lifetime

Understanding the capacity fade process in lithium-ion batteries requires not only a detailed knowledge of the degradation mechanisms described in Section 7.5.2, but also a careful analysis of the factors that influence the severity and rate at which these mechanisms evolve. In the context of lithium-ion batteries, particularly in electric mobility applications, four key factors play a critical role in determining battery degradation:

- **Temperature:** Lithium-ion batteries must operate within an optimal temperature range, avoiding both excessively high and low temperatures, as extreme conditions significantly accelerate degradation. Elevated temperatures enhance parasitic chemical reactions, promoting phenomena such as SEI growth and electrolyte decomposition, which lead to capacity loss and increased internal resistance. Conversely, low temperatures reduce electrochemical kinetics, limiting performance and increasing the likelihood of lithium plating during charging. In this context, the implementation of thermal management systems is essential to maintain the operating temperature of electrochemical cells within a suitable range, thereby mitigating degradation and preserving battery lifetime (see Section 7.2.5).

- **DoD:** Defined as the fraction of total capacity utilized during each cycle, DoD is a critical parameter that must be carefully controlled, as it directly affects both battery utilization and degradation rate. Higher DoD values correspond to deeper charge/discharge cycles, which induce larger variations in the structure of active materials, lithium concentration in the electrodes, and associated mechanical and chemical stresses. As a result, repeated deep cycling accelerates degradation. However, operating at low DoD values reduces the effective utilization of the available capacity. Therefore, the selection of DoD typically involves a trade-off between minimizing degradation and maximizing energy usage.
- **Charge and discharge rate:** Commonly referred to as C-rate, this parameter describes the rate at which a battery is charged or discharged relative to its nominal capacity, effectively representing a normalized measure of current. High C-rate values imply large currents, which can lead to increased internal heating, higher polarization, and, in extreme cases, lithium metal deposition. These effects are particularly critical during fast charging conditions. In contrast, moderate C-rate operation reduces internal stress and contributes to improved battery longevity.
- **Usage profile:** In electric mobility applications, batteries are subjected to highly variable operating conditions, including acceleration events, regenerative braking, and rest periods. These usage patterns result in the interaction of multiple degradation mechanisms, driven by transient high-power demands, fast charging cycles, and varying environmental conditions. Consequently, the resulting aging processes are complex and difficult to predict using simplified models, highlighting the importance of considering realistic operating scenarios when evaluating battery lifetime.

7.6 Applications of batteries in electric mobility

7.6.1 Role of batteries in electric powertrains

In electric propulsion systems, the battery fulfills a dual role. Indeed, it constitutes the primary energy storage and supply system. Unlike ICE vehicles, where propulsion energy is generated through the direct conversion of fuel, EVs require a prior energy storage stage before its utilization. Although different storage technologies can be employed, electrochemical batteries are by far the most widely used solution in EV applications. As a result, the role of the battery directly conditions both the architecture of the powertrain and its dynamic behavior.

While the main function of the battery is to store and supply electrical energy to the propulsion system, this role varies depending on the type of EV. In BEVs, the battery is the sole energy source for propulsion, which imposes stringent requirements in terms of energy capacity, power delivery, and overall efficiency. In contrast, in HEVs, the battery typically assumes a secondary role, acting as a support element to the ICE, which remains the dominant source of propulsion (see Chapter 2). In such architectures, the battery operates as an intermediate energy buffer, storing energy recovered during regenerative braking and supplying additional power during high-demand events such as acceleration. At an intermediate level, Plug-in HEVs increase the relevance of the battery, as it combines both functions: supporting the thermal powertrain and acting as the primary energy source during purely electric driving modes.

Beyond its energy-related role, the battery is a central element in EV design. The architecture, size, and weight of the battery pack significantly influence key aspects such as mass distribution, structural integration, and thermal management. In particular, its placement within the vehicle affects the center of gravity, with direct implications for vehicle stability and dynamic performance. Furthermore, the requirement to operate within specific temperature ranges (see Section 7.5.3) often necessitates the integration of advanced thermal management systems. These systems introduce additional constraints, both in terms of structural integration and auxiliary energy consumption.

In summary, the battery not only serves as the primary energy storage system in EVs but also plays a fundamental role in shaping vehicle design and performance. Consequently, battery pack design becomes a key factor in defining the overall efficiency, dynamic behavior, and architectural configuration of electric powertrains.

7.6.2 Battery requirements for EV applications

In the context of electric mobility, batteries employed in EVs must satisfy a set of stringent technical requirements in order to constitute a viable energy storage solution. These requirements arise from both the operational conditions of the vehicle and the expectations of end users. Moreover, it should be emphasized that the battery does not merely act as an energy storage system, but rather as an integral component of the EV, with a significant impact on its overall design, performance, and operational constraints.

Energy density represents one of the most critical parameters. This metric directly determines both the mass and volume of the storage system and, consequently, has a direct influence on the driving range of the vehicle. In this regard, a high energy density (both gravimetric and volumetric) is essential to achieve acceptable driving ranges while minimizing weight and space requirements. This aspect becomes particularly critical in vehicles with high sensitivity to mass, such as UAVs (see Chapter 4).

In addition to energy density, a high power density is also required. This parameter defines the rate at which the battery can deliver or absorb power and plays a key role in the dynamic behavior of the vehicle. High power capability is especially important under demanding operating conditions, such as rapid accelerations or take-off phases. Furthermore, a high power density enables more effective energy recovery during regenerative braking or charging events, thereby improving the overall energy efficiency of the system.

Batteries must also meet strict safety and reliability requirements. In electric mobility applications, they are expected to operate safely and stably over a wide range of conditions, including variations in temperature, mechanical stresses, and load profiles. Appropriate monitoring, control, and protection systems must be implemented to prevent or mitigate undesirable phenomena such as thermal runaway (see Section 7.3.3), which may compromise both system integrity and user safety.

Finally, long service life constitutes a key requirement from both technical and economic perspectives. Batteries in EVs are expected to withstand several thousand charge and discharge cycles (cyclic life) over multiple years of operation (calendar life). In addition, capacity fade and performance degradation should remain within acceptable limits to ensure that the battery retains a significant portion of its initial capacity over time. This requirement not only affects the functional performance of the vehicle, but also plays a decisive role in its overall economic viability.

7.6.3 Design constraints in automotive batteries

A set of physical, structural, and safety constraints strongly conditions the design of batteries for automotive applications, thereby limiting the range of feasible technological solutions. Since the battery constitutes an integral part of the vehicle, these constraints extend beyond the selection of materials and electrochemical configurations, and must also account for how the battery pack is integrated within the overall vehicle architecture.

Available space represents one of the most critical constraints. Unlike stationary applications, where volume is often not a limiting factor, the space within a vehicle is typically restricted and highly constrained by the overall design. In this context, achieving high volumetric energy density becomes essential in order to maximize driving range within a limited footprint. Consequently, compact battery pack designs are required, often involving optimized packaging strategies and efficient use of available geometric volumes.

Weight constitutes another major limiting factor. The total mass of the vehicle has a direct impact on energy consumption, dynamic performance, and driving range. As a result, the mass allocated to the energy storage system must be carefully managed. This necessitates a trade-off between energy capacity and overall weight, favoring the use of lightweight active materials and optimized structural components. In addition, minimizing inactive mass within the battery pack, such as casing and support structures, is a key design objective.

The structural integration of the battery pack also imposes significant constraints. In modern EV designs, the battery is not merely an energy storage unit but often contributes to the mechanical integrity of the vehicle. Therefore, battery packs must be engineered to withstand mechanical loads, vibrations, and potential impact events without compromising their functionality. The location of the battery pack within the vehicle further influences both its design and dynamic behavior, including aspects such as center of gravity and weight distribution.

Finally, safety requirements introduce additional constraints that must be considered when selecting and designing battery systems. Battery packs must ensure stable and reliable operation under a wide range of operating conditions. These requirements influence not only the choice of materials, but also the structural design and the implementation of associated safety systems. As a result, safety considerations play a central role in defining the final architecture of automotive battery systems.

7.6.4 Operational profiles in electric mobility

In electric mobility applications, batteries operate under highly dynamic conditions that impose strongly variable operational profiles. These conditions arise from real driving cycles, in which continuous changes in operating states occur, including acceleration, idle periods, take-off, or cruising phases, among others. As a result, the battery rarely operates under steady-state conditions, which introduces significant complexity in both its control and the prediction of its behavior.

One of the defining characteristics of battery operation in automotive applications is the presence of frequent partial charge and discharge cycles, largely associated with processes such as acceleration or regenerative braking (see Chapter 2). These microcycles, typically observed in urban driving conditions, induce continuous fluctuations in the SoC. Such variations can have a pronounced impact on battery aging, as degradation mechanisms are strongly dependent not only on the number of cycles, but also on their depth and distribution.

In addition, batteries in EVs are often subjected to high power charge and discharge events. These occur not only during demanding operating conditions such as rapid acceleration or take-off, but also during fast charging processes (see Chapter 6). High power transients can promote the onset of degradation mechanisms such as lithium plating (see Section 7.5.2), particularly under unfavorable conditions, thereby accelerating performance deterioration.

Furthermore, operational profiles are significantly influenced by external factors such as ambient temperature, driving style, and route characteristics. Harsh conditions, including elevated temperatures or aggressive driving patterns with frequent acceleration and braking, increase both thermal and electrochemical stress within the battery. Consequently, these factors can substantially accelerate degradation processes and reduce the overall lifetime of the system.

7.7 Emerging battery technologies

7.7.1 Solid-state batteries

Solid-state batteries represent one of the most promising development pathways for next-generation energy storage systems, particularly in the field of electric mobility [13]. The main distinguishing feature of this technology is the replacement of the liquid or gel electrolyte (commonly used in conventional batteries such as lithium-ion systems) with a solid electrolyte. This fundamental change has significant implications for both the internal operation and overall performance of the cell.

The operating principle remains broadly similar to that of lithium-ion batteries and is based on the transport of lithium ions (Li^+) between the anode and the cathode during charge and discharge processes. However, in solid-state batteries, ion transport occurs through a solid electrolyte, typically composed of ceramic, polymeric, or hybrid materials. In many configurations, the solid electrolyte can also fulfill the role of the separator, acting as both an ionic conductor and a physical barrier between electrodes, thereby simplifying the internal structure of the cell.

Another key advantage lies in the potential for significantly higher energy densities. In the long term, values exceeding 800 Wh/kg have been projected [14]. This improvement is largely attributed to the possibility of employing lithium metal anodes instead of conventional graphite-based materials, which substantially increases the specific capacity of the system. Such an increase in energy density could have major implications for electric mobility, enabling extended driving ranges and potentially addressing one of the main limitations for the electrification of transport sectors with stringent energy requirements, such as aviation (see Chapter 5).

Safety is another important advantage associated with solid-state batteries. The absence of flammable liquid electrolytes reduces the risks related to leakage, thermal instability, and combustion. Furthermore, solid electrolytes can, in principle, suppress or mitigate dendrite formation, thereby reducing the likelihood of internal short circuits caused by electrode contact (see Section 7.5.2). These characteristics make solid-state systems particularly attractive for applications with strict safety requirements, such as automotive systems.

Despite these potential benefits, solid-state batteries still face several critical challenges that must be addressed to ensure their practical viability. One of the most significant limitations is their relatively low power capability, primarily due to the lower ionic conductivity of many solid electrolytes compared to liquid counterparts. This limitation may restrict their applicability in scenarios requiring high dynamic response, which is a key requirement in many EV applications.

Another important challenge is the instability of the solid–solid interface between the electrodes and the electrolyte. Poor interfacial contact can lead to increased internal resistance, reduced efficiency, and accelerated degradation over time. In addition, manufacturing processes for solid-state batteries remain complex and costly. The production and processing of suitable materials for both electrodes and electrolytes currently rely on techniques that are not yet fully optimized for large-scale industrial deployment [13]. As a result, further technological advancements are required before solid-state batteries can achieve widespread commercialization.

7.7.2 Lithium-sulfur batteries

Lithium-sulfur batteries represent one of the most promising emerging technologies for energy storage in electric mobility, primarily due to their high theoretical energy density [15]. These batteries typically employ sulfur as the active material in the cathode and lithium metal as the anode, resulting in distinctive operating principles and performance characteristics compared to conventional lithium-ion systems.

The operating mechanism of lithium-sulfur batteries is based on electrochemical reactions between lithium and sulfur, leading to the formation of intermediate compounds known as lithium polysulfides. During discharge, sulfur is progressively reduced at the cathode, generating a sequence of soluble polysulfide species within the electrolyte. During the charging process, this reaction is reversed. This mechanism differs fundamentally from the intercalation-based processes observed in lithium-ion batteries, and it introduces specific challenges related to material transport and stability.

Lithium-sulfur batteries exhibit a very high theoretical energy density, with reported values exceeding 500 Wh/kg in certain studies [15]. This high energy density is primarily attributed to the high specific capacity of sulfur and the use of lithium metal anodes. In addition, sulfur is an abundant, low-cost, and relatively non-toxic material, which provides advantages from economic, logistical, and sustainability perspectives when compared to other battery chemistries.

However, the large-scale deployment of lithium-sulfur batteries is currently hindered by several significant challenges. One of the most critical issues is the so-called shuttle effect, associated with the dissolution and migration of lithium polysulfides within the electrolyte during cycling. This phenomenon leads to the loss of active material, reduced coulombic efficiency, and rapid capacity fading over time. Furthermore, the cathode undergoes substantial volumetric expansion and contraction during charge and discharge cycles, which can induce mechanical stresses and compromise the structural integrity of the battery.

Additional limitations include the relatively short cycle life and lower energy efficiency compared to more mature technologies, which may impact their practical viability in electric mobility applications. These factors can reduce the effective contribution of the battery to vehicle range over its operational lifetime. Finally, the use of lithium metal anodes introduces safety concerns related to dendrite formation (see Section 7.5.2), which may lead to internal short circuits and potential failure under certain operating conditions.

7.7.3 Sodium-ion batteries

Sodium-ion batteries are emerging as a viable alternative to lithium-ion batteries, particularly in applications where cost and raw material availability are more critical than energy density. Their operating principle is analogous to that of lithium-ion technology, relying on the reversible transfer of sodium ions (Na^+) between electrodes during charge and discharge processes. However, the key differences between these technologies arise from the distinct physicochemical properties of sodium.

The primary limitation of sodium-ion batteries is their relatively low energy density, typically in the range of 100–160 Wh/kg [16], which is significantly lower than that of lithium-ion systems (see Section 7.4.1). This limitation is mainly attributed to the larger ionic radius and higher atomic mass of Na^+ compared to Li^+, resulting in lower specific capacities and reduced cell voltages. From an automotive perspective, this constraint represents a major challenge, as achieving competitive

Table 7.4 Summary of advantages and limitations of emerging battery technologies.

Technology	Advantages	Limitations
Solid-State	High energy density potential Improved safety (non-flammable electrolyte)	Limited power capacity High cost and manufacturing challenges
Lithium-Sulfur	Very high theoretical energy density Low-cost and abundant sulfur	Shuttle effect Poor cycle life Low coulombic efficiency
Sodium-Ion	Abundant and low-cost materials Good low-temperature performance	Lower energy density

driving ranges would require larger and heavier battery packs, ultimately penalizing vehicle efficiency.

Despite this limitation, sodium-ion batteries exhibit several operational advantages. Notably, they tend to demonstrate improved performance at low temperatures and reduced sensitivity to demanding operating conditions. In addition, under certain operating regimes, these systems can exhibit relatively moderate degradation rates, which may translate into acceptable cycle life for specific applications.

The most significant advantage of sodium-ion batteries lies in the abundance and wide geographical distribution of sodium. This characteristic reduces reliance on critical materials such as lithium, cobalt, and nickel (see Section 7.8.1), thereby improving supply chain resilience. Furthermore, these batteries can employ hard carbon as anode material and cathode chemistries based on Prussian Blue analogues or layered oxides, avoiding the need for highly processed graphite. This feature contributes to lower production costs and mitigates geopolitical risks associated with constrained supply chains.

Given these characteristics, the current application scope of sodium-ion batteries in electric mobility is likely to be limited to segments where cost, safety, and robustness are prioritized over energy density. Representative use cases include urban vehicles, micromobility platforms, and certain hybrid configurations. As a summary, Table 7.4 provides a comparative overview of the emerging battery technologies discussed in Section 7.7.

7.7.4 Other emerging concepts

Given their potential role as a critical element in the decarbonization of multiple energy sectors, research on novel battery technologies remains a highly active field, with diverse development pathways targeting different applications. Within the broad landscape of emerging electrochemical energy storage systems, several concepts are still at a significantly earlier stage of maturity compared to the technologies reviewed in Table 7.4.

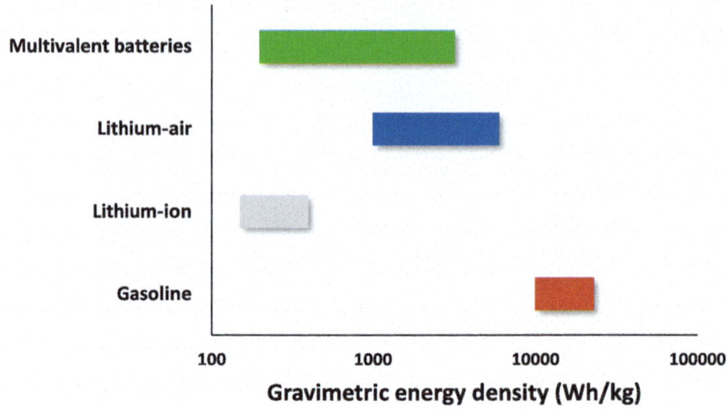

Figure 7.7 Comparison of gravimetric energy density ranges for gasoline, lithium-ion batteries, lithium-air batteries, and multivalent battery systems. While gasoline and lithium-ion values correspond to typical practical ranges, the values reported for lithium-air and multivalent batteries are based on theoretical or projected estimates and do not represent performance achieved in commercial devices or validated prototypes.

Among these proposals, which can largely be classified as experimental or theoretical, lithium-air batteries stand out. These systems could, in principle, achieve extremely high energy densities by using oxygen as the cathode reactant. However, their development is currently hindered by major challenges, including poor chemical stability, low round-trip efficiency, and limited reversibility of the electrochemical reactions [17].

Another relevant research direction is that of multivalent batteries. These systems employ multivalent metals such as magnesium or aluminum, which are more abundant and offer the potential for higher volumetric capacities [18]. Nevertheless, they face substantial limitations associated with sluggish ionic diffusion kinetics and unresolved issues in electrolyte and electrode material compatibility.

Fig. 7.7 illustrates typical ranges of gravimetric energy density for conventional fossil fuels, lithium-ion batteries, and the emerging technologies discussed in this section. As shown, a substantial gap exists between the specific energy of conventional fuels and that of lithium-ion batteries. While some emerging concepts, could partially bridge this gap, their projected performance remains largely theoretical. Consequently, the values reported in Fig. 7.7 should be interpreted with caution, as their translation into practical, commercial systems is still subject to significant uncertainty.

7.8 Industrial, economic and geopolitical context

7.8.1 Critical raw materials

The development and manufacturing of lithium-ion batteries rely on a set of critical raw materials, namely lithium, cobalt, and nickel. The availability and geographic distribution of these elements strongly influence battery production and, consequently, the feasibility, deployment, and scalability of a wide range of electric mobility

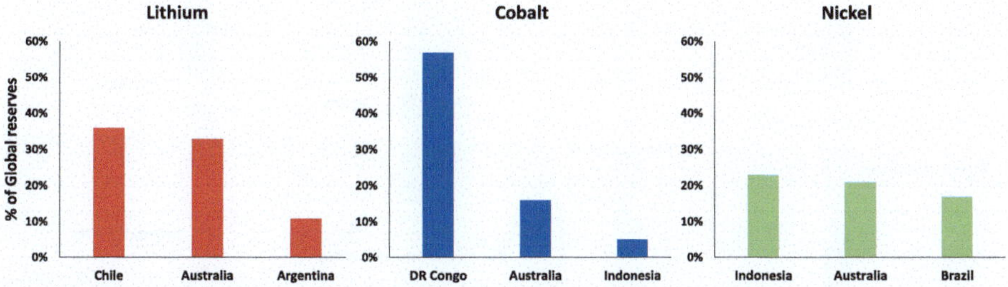

Figure 7.8 Share of global reserves for key battery raw materials (lithium, cobalt, and nickel), highlighting the geographic concentration among leading countries.

solutions that predominantly rely on this technology. To assess the geopolitical dimension of resource availability, Fig. 7.8 presents the global share of reserves for lithium, cobalt, and nickel across the three countries with the highest concentration, while Chapter 3 provides a complementary analysis focused on production.

Lithium constitutes the fundamental element of lithium-ion batteries, owing to its favorable electrochemical properties, including low atomic weight and high electrochemical potential. Global lithium reserves are largely concentrated in South America and Australia, reflecting a relatively limited geographic distribution. This concentration is even more pronounced in the case of cobalt, with nearly 60% of global reserves located in the Democratic Republic of the Congo. In contrast, nickel reserves are more widely distributed, with significant shares in countries such as Indonesia, Australia, and Brazil, although high-purity nickel suitable for battery applications introduces additional supply constraints.

In this context, the reliance on critical raw materials exposes the battery industry to geopolitical risks that directly impact cost, supply security, and scalability. Fig. 7.8 highlights the extent of this concentration, illustrating the potential vulnerability of supply chains to geopolitical instability, trade restrictions, or disruptions in extraction and processing capacities.

These challenges are driving the implementation of mitigation strategies aimed at reducing dependency on constrained resources. Such strategies include the diversification of supply sources, the development of alternative battery technologies with lower reliance on critical materials (see Sections 7.7.2 and 7.7.3), and the expansion of recycling pathways to recover valuable materials at the end of battery life [1]. Together, these approaches are essential to ensure the long-term sustainability and resilience of the electric mobility ecosystem.

7.8.2 Global supply chains

The global lithium-ion battery supply chain must be understood as a multi-layered system involving a sequence of highly specialized processes, including material extraction and processing, battery manufacturing, vehicle manufacturing, and final markets. Fig. 7.9 synthesizes this structure based on the data reported in [19], illustrating the relative weight of key strategic regions across the entire value chain,

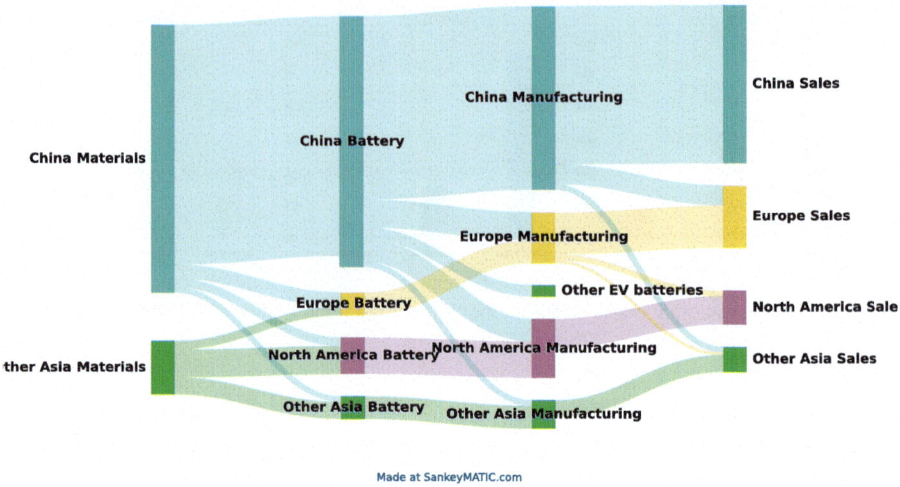

Figure 7.9 Simplified representation of the global supply chain associated with lithium-ion batteries, illustrating the main stages from raw materials to final vehicle sales. The diagram focuses exclusively on battery ECs and does not include other electric mobility segments. For clarity, nodes with marginal contributions have been omitted, and the structure has been simplified to highlight the dominant industrial flows. Node sizes and flow magnitudes are indicative and intended to convey relative importance rather than exact quantitative values.

from raw material processing to final vehicle sales. The figure focuses on the market for ECs, as this segment is representative of the electric mobility sector; as shown in Fig. 7.9, the contribution of batteries for other EVs remains comparatively limited.

At the upstream level, corresponding to material extraction and processing, a strong geographical concentration can be observed, with Asian countries (particularly China) playing a dominant role. This concentration constitutes a first structural bottleneck in the supply chain, as the availability of processed materials suitable for battery production depends heavily on a limited number of regions, and most notably on a single country. Such a configuration increases exposure to supply disruptions and amplifies geopolitical and economic risks.

At the intermediate level, the battery stage includes the manufacturing of cells and associated components. In this phase, Fig. 7.9 highlights a clear dominance of China, which acts as the central hub for battery production. However, Europe and the United States are progressively increasing their relative share, reflecting ongoing efforts to expand domestic manufacturing capacity. While this trend contributes to a gradual diversification of the supply chain, the current structure still reveals a significant dependence on a single geographic region.

The subsequent stages correspond to manufacturing and sales of ECs. These levels exhibit a more balanced geographical distribution compared to upstream activities, although China continues to maintain a leading position. In particular, the sales stage shows a notable contribution from Europe, as illustrated in Fig. 7.9, indicating strong demand in this region. Additionally, the flow associated with other EV batteries

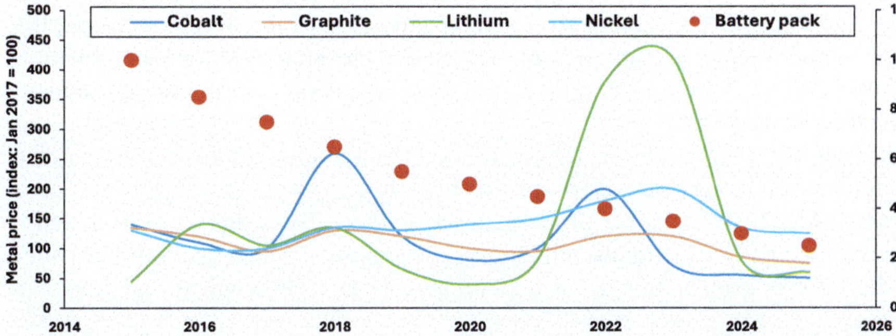

Figure 7.10 Evolution of selected battery metal prices (cobalt, graphite, lithium, and nickel) and lithium-ion battery pack prices over the period 2015–2025. Metal prices are expressed as an index (January 2017 = 100), while battery pack prices are given as an index (2015 = 100).

remains marginal when compared to lithium-ion batteries used in ECs, reinforcing the idea that ECs constitute the primary application driving the battery value chain.

Overall, Fig. 7.9 reveals a structurally imbalanced supply chain, in which Asian regions dominate the upstream and intermediate stages, while downstream activities such as vehicle manufacturing and sales are more geographically distributed. This misalignment underscores the dependency of regions such as Europe and the United States on external suppliers, highlighting the importance of diversification strategies aimed at strengthening local capabilities and enhancing resilience across the value chain.

7.8.3 Market dynamics and costs

The market dynamics of lithium ion batteries have been governed by both technological factors and the intrinsic behavior of raw material markets. In this context, Fig. 7.10 presents the evolution of the cost of key materials used in lithium ion battery manufacturing (namely graphite, cobalt, lithium, and nickel) together with the evolution of battery pack costs over the period 2015 to 2025, based on data reported in [19].

A key observation from Fig. 7.10 is the markedly different trends followed by raw material prices and battery pack costs. While the prices of critical metals exhibit significant volatility throughout the analyzed period, the cost of battery packs shows a clear and sustained downward trend. This divergence reflects the influence of distinct driving mechanisms. On the one hand, raw material markets are primarily governed by resource availability, supply and demand balance, and the capacity for extraction and refining. On the other hand, the battery industry benefits from economies of scale, process optimization, and cumulative industrial learning. These factors have enabled battery manufacturers to partially absorb fluctuations in raw material costs, maintaining a long term reduction in overall system costs.

A particularly relevant feature highlighted in Fig. 7.10 is the sharp increase in lithium prices between 2021 and 2023, reaching a peak around 2023. This surge can be attributed to a temporal mismatch between rapidly increasing demand and constrained supply. The demand for lithium rose significantly due to the accelerated adoption of EVs, while supply expansion was limited by the long development timelines associated with new mining and refining projects. As a result, lithium became a major contributor to cost pressures within the battery supply chain during this period.

Although this price spike was relatively short lived, it underscores a critical aspect of the lithium ion battery market: despite its historical trend towards cost reduction, it remains exposed to the volatility of raw material markets. These markets are inherently influenced by supply constraints and demand fluctuations, which can introduce short term disruptions even within a context of long term technological and industrial progress.

7.9 Current status and future outlook

7.9.1 Current technological readiness

Lithium ion technology can be considered to have reached a high level of technological maturity within the family of electrochemical energy storage systems. In particular, in the field of electric mobility, this technology has achieved a dominant position over alternative solutions, especially in BEVs. Its cost has benefited from a sustained reduction, largely driven by economies of scale and continuous process optimization (see Section 7.8.3).

Its widespread adoption in electric mobility applications has been enabled by progressive technological development, which has improved key parameters such as energy density and lifetime. These advancements have acted as a catalyst for the deployment of electric mobility solutions, particularly ECs, by helping to overcome both technological and economic barriers that previously limited their large scale adoption.

Within the lithium ion battery family, chemistries such as NMC, NCA, and LFP (see Section 7.4.1.2) have reached a high level of maturity and are widely used in EVs, being tailored to specific application requirements. This technological diversification enhances the robustness of the overall system and enables its adaptation across different segments of electric mobility.

Overall, lithium ion batteries can be regarded as a mature technology from an industrial perspective, although further improvements remain possible in areas such as energy density, charging performance, and material sustainability.

7.9.2 Trends in electric mobility

The current development of batteries for electric mobility is characterized by a combination of performance improvements in established technologies and the gradual introduction of new solutions. In this context, although lithium ion batteries can be considered a mature technology, their development remains active, with ongoing efforts focused on cell design and manufacturing processes, enabling further improvements in energy density, cost reduction, and safety.

A key trend within lithium ion batteries is the diversification of chemistries. In particular, a significant growth of LFP technologies has been observed, driven by their lower cost and higher thermal stability, especially in applications where energy density is not the primary constraint. In parallel, nickel based chemistries continue to evolve with a strong focus on increasing energy density, which is critical for applications requiring extended driving range.

In addition to material level developments, several research and industrial efforts are focused on design optimization. These include the development of larger cell formats, improved integration within the vehicle, and advanced architectures such as cell-to-pack [20], which enhance volumetric efficiency and reduce costs by simplifying structural integration.

Finally, there is a growing trend towards the progressive introduction of new battery technologies. Among these, solid state batteries have attracted significant attention in recent years, as they offer the potential for substantial improvements in safety and energy density. However, these technologies still face important technical challenges that limit their large scale commercial deployment (see Section 7.7.1).

7.9.3 Future challenges and opportunities

The future development of lithium ion batteries will be largely determined by the ability to overcome a set of limitations across three key domains: technological, industrial, and sustainability related. From a technological perspective, future developments are expected to focus on increasing the energy density of lithium ion technologies, which remains a critical parameter directly influencing the driving range and overall efficiency of EVs. In addition, improving power capability is essential, as it would enable faster charging times and, consequently, enhance user experience.

From an industrial and resource perspective, reducing the dependence on critical materials such as nickel, lithium, and cobalt is a major challenge. The geographical concentration of these materials introduces uncertainty and stress in the supply chain (see Section 7.8.2). In this context, significant efforts are being directed towards the development of efficient recycling strategies [21], as well as the exploration of alternative chemistries that can reduce or diversify material requirements.

At the same time, important opportunities arise from the development of emerging battery technologies (see Section 7.7). Some of these technologies promise substantial improvements in energy density and safety, potentially overcoming key limitations of current lithium ion systems. However, many of these solutions are still at an early stage of development, and their transition to large scale industrial deployment remains uncertain.

References

[1] M. Gómez-González, M. Tostado-Véliz, M. Valverde, F. Jurado, Optimal design of photovoltaic domestic installations considering second-purpose batteries, IEEE Access 11 (2023) 140126–140137, https://doi.org/10.1109/ACCESS.2023.3340017.

[2] I. Alsaidan, A. Khodaei, W. Gao, A comprehensive battery energy storage optimal sizing

model for microgrid applications, IEEE Transactions on Power Systems 33 (4) (2018) 3968–3980, https://doi.org/10.1109/TPWRS.2017.2769639.
[3] M. Tostado-Véliz, A. Rezaee Jordehi, S.A. Mansouri, Y. Zhou, F. Jurado, A local electricity-hydrogen market model for industrial parks, Applied Energy 360 (2024) 122760, https://doi.org/10.1016/j.apenergy.2024.122760.
[4] B. Zakeri, S. Syri, Electrical energy storage systems: a comparative life cycle cost analysis, Renewable and Sustainable Energy Reviews 42 (2015) 569–596, https://doi.org/10.1016/j.rser.2014.10.011.
[5] M. Laha, V. Weber, M.P. Riera, A. Adachi, The geopolitics of batteries, Tech. Rep., German Council on Foreign Relations, 2025, https://dgap.org/en/research/publications/geopolitics-batteries. (Accessed 27 March 2026).
[6] L. Song, Z. Zhou, X. Ju, B. Wang, U. Rojas-Alva, X. Zhou, L. Yang, Experimental study on thermal runaway characteristics and fire hazards of lithium-ion batteries in semi-confined space of transportation, Journal of Energy Storage 128 (2025) 117220, https://doi.org/10.1016/j.est.2025.117220.
[7] G. Azimi, K.H. Chan, A review of contemporary and emerging recycling methods for lithium-ion batteries with a focus on nmc cathodes, Resources, Conservation and Recycling 209 (2024) 107825, https://doi.org/10.1016/j.resconrec.2024.107825.
[8] C. Tian, K. Qin, T. Xu, L. Suo, Hybrid li-rich cathodes for anode-free lithium metal batteries, Next Nanotechnology 7 (2025) 100114, https://doi.org/10.1016/j.nxnano.2024.100114.
[9] M. Hasan, R. Haque, M. Jahirul, M. Rasul, I. Fattah, N. Hassan, M. Mofijur, Advancing energy storage: the future trajectory of lithium-ion battery technologies, Journal of Energy Storage 120 (2025) 116511, https://doi.org/10.1016/j.est.2025.116511.
[10] D.F. Quintero Pulido, C.F. Covrig, M. Bruchhausen, On the performance of portable nimh batteries of general use, Batteries 11 (1) (2025), https://doi.org/10.3390/batteries11010030, https://www.mdpi.com/2313-0105/11/1/30.
[11] S. Arya, S. Verma, Nickel-Metal Hydride (Ni-MH) Batteries, John Wiley & Sons, Ltd, 2020, pp. 131–175, Ch. 8, https://onlinelibrary.wiley.com/doi/abs/10.1002/9781119714774.ch8.
[12] D.R. Battlebury, A high performance lead–acid battery for ev applications, Journal of Power Sources 80 (1) (1999) 7–11, https://doi.org/10.1016/S0378-7753(98)00247-X.
[13] M. Chen, J. Guan, Optimization strategies of electrolytes in solid-state batteries: from laboratory to industrialization, Materials Science & Engineering. R, Reports 169 (2026) 101212, https://doi.org/10.1016/j.mser.2026.101212.
[14] BSS, Welion's solid state battery achieves 824 wh/kg energy density in lab, aims 1000 wh/kg, https://www.battery-energy-storage-system.com/news/welion-824-wh-kg-solid-state-battery-1000.html, Dec. 2025. (Accessed 24 April 2026).
[15] N. Xu, X. Song, et al., Over 500 wh kg^{-1} solid-state lithium metal batteries with long cycling stability using in situ polymerized electrolyte, Journal of the American Chemical Society (2026), https://doi.org/10.1021/jacs.5c17294.
[16] J.-Y. Hwang, S.-T. Myung, Y.-K. Sun, Sodium-ion batteries: present and future, Chemical Society Reviews 46 (12) (2017) 3529–3614, https://doi.org/10.1039/C6CS00776G.
[17] P.G. Bruce, S.A. Freunberger, L.J. Hardwick, J.-M. Tarascon, Li–o_2 and li–s batteries with high energy storage, Nature Materials 11 (1) (2012) 19–29, https://doi.org/10.1038/nmat3191.
[18] A. Ponrouch, C. Frontera, F. Bardé, M.R. Palacín, Towards a calcium-based rechargeable battery, Nature Materials 15 (2) (2016) 169–172, https://doi.org/10.1038/nmat4462.
[19] I.E. Agency, Global ev outlook 2025, Tech. Rep., IEA, Paris, 2023, https://www.iea.org/reports/global-ev-outlook-2025. (Accessed 27 March 2026).

[20] H. Wang, Q. Wang, Z. Zhao, C. Jin, C. Xu, W. Huang, Z. Yuan, S. Wang, Y. Li, Y. Zhao, J. Sun, X. Feng, Thermal runaway propagation behavior of the cell-to-pack battery system, Journal of Energy Chemistry 84 (2023) 162–172, https://doi.org/10.1016/j.jechem.2023.05.015.

[21] M. Rezaei, A. Nekahi, A. Kumar, M. R, A. Nizami, X. Li, S. Deng, J. Nanda, K. Zaghib, A review of lithium-ion battery recycling for enabling a circular economy, Journal of Power Sources 630 (2025) 236157, https://doi.org/10.1016/j.jpowsour.2024.236157.

Hydrogen storage, refuelling, and conversion

8

8.1 Fundamentals of hydrogen as an energy vector

8.1.1 Physical and chemical properties

Hydrogen exhibits a set of distinctive physical and chemical properties that simultaneously explain its relevance as an energy vector for electric mobility and the major technical challenges associated with storage, transport, and final utilization.

In particular, hydrogen possesses exceptionally high gravimetric energy density, meaning that it can store large amounts of energy per unit mass. Specifically, in terms of lower heating value (LHV), hydrogen reaches approximately 120 MJ/kg. This value places it well above conventional energy carriers such as gasoline, diesel, methane, or methanol, as shown in Table 8.1. Consequently, on a mass basis, hydrogen can store more chemical energy than most commonly used fuels. This characteristic makes it particularly attractive for mobility applications in which system mass represents a critical design constraint, including heavy-duty transport and other sectors where payload efficiency remains highly relevant.

However, the major gravimetric advantage of hydrogen is counterbalanced by very low volumetric energy density. As illustrated in Table 8.1, hydrogen stores substantially less energy per unit volume than most conventional fuels, even under compressed conditions such as 200 bar. This limitation implies that large storage

Table 8.1 Comparison of LHV, gravimetric energy density, and approximate volumetric energy density of hydrogen and selected energy carriers under representative storage conditions. Values are indicative and may vary depending on purity, exact pressure, and temperature conditions.

Energy Carrier	Storage Condition	LHV (MJ/kg)	Volumetric Energy Density (MJ/L)
Hydrogen	Gas, 200 bar	~120	~1.9
Hydrogen	Gas, 700 bar	~120	~4.7
Hydrogen	Liquid (−253°C)	~120	~8.5
Methane	Gas, 200 bar	~50	~7.2
Methane	Liquefied	~50	~22.2
Methanol	Liquid (ambient)	~20	~15.8
Gasoline	Liquid (ambient)	~43–44	~32
Diesel	Liquid (ambient)	~42–43	~36
Lithium-ion	Pack level	~0.5–0.9	~1.0–2.5

Electric Mobility Systems. https://doi.org/10.1016/B978-0-44-349397-3.00016-1

volumes are required to contain practical amounts of usable energy. For this reason, current hydrogen mobility solutions rely primarily on high-pressure storage systems (typically 350–700 bar) or cryogenic liquefaction at approximately −253 °C. These approaches significantly increase volumetric energy density, but they do so at the expense of higher system cost, greater technological complexity, additional energy consumption, and, in certain cases, more demanding safety requirements (see Section 8.3).

Hydrogen may be stored in either gaseous or liquid form, with each option presenting distinct operational implications. In gaseous form, extremely low molecular weight and small molecular size result in very high diffusivity, meaning that hydrogen disperses rapidly and can more easily permeate seals or structural materials. This property can increase leakage risk and may also contribute to hydrogen embrittlement in certain metals. In liquid form, hydrogen achieves much higher volumetric density, but maintaining this state requires extreme cryogenic conditions. Liquefaction itself is energy intensive, while long-term storage introduces additional challenges such as boil-off losses caused by gradual evaporation.

From a chemical perspective, hydrogen also presents significant safety considerations. It has a wide flammability range in air (approximately 4% to 75% by volume) and particularly low minimum ignition energy. These properties require stringent safety, monitoring, and control systems to mitigate operational risk. Although rapid diffusion may reduce certain hazards in open-air leak scenarios compared with heavier fuels, nearly invisible flame and high reactivity make hydrogen potentially dangerous in confined or flammable environments, with elevated fire or explosion risk.

Overall, the role of hydrogen in electric mobility is defined by strong duality. On one hand, exceptional gravimetric energy density makes hydrogen an attractive solution for applications where storage mass represents a limiting factor. On the other hand, low volumetric efficiency, high diffusivity, and demanding storage conditions require non-standard pressure or temperature architectures, introducing technological, energetic, and safety penalties. This duality fundamentally shapes the engineering strategies and vehicle architectures employed in hydrogen-based EVs.

8.1.2 Hydrogen as an energy carrier

Unlike primary energy sources such as solar radiation, wind, coal, or natural gas, hydrogen cannot be directly exploited as an intrinsic energy source. Instead, it must first be produced from other primary energy sources, then transported, stored, and ultimately converted for final use. In this sense, hydrogen does not constitute a primary energy source, but rather an energy vector capable of acting as an intermediary carrier between different sectors, processes, and stages of the energy chain.

To better understand this concept, hydrogen may be compared to electricity, another well-established energy vector. Electricity is also generated from primary sources such as solar energy, wind, nuclear power, or fossil fuels, and is subsequently transmitted, stored, and converted for end-use applications. However, electricity is primarily characterized by highly efficient and immediate transmission through electrical

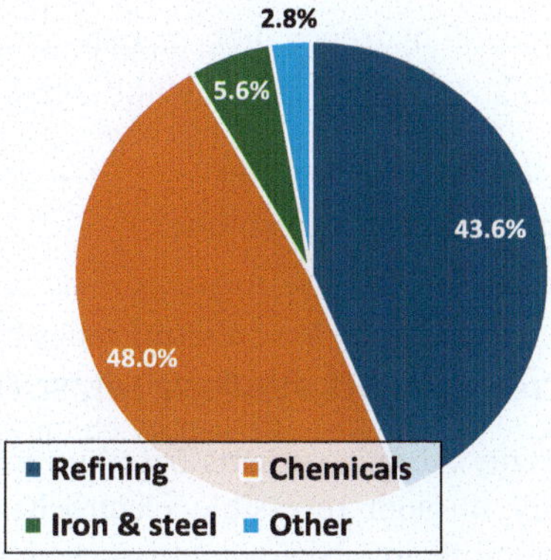

Figure 8.1 Global hydrogen demand by end-use sector in 2023.

networks. In contrast, hydrogen offers distinct advantages associated with long-duration chemical storage, large-scale physical transport, and the ability to temporally or spatially decouple energy generation from final consumption. These characteristics make hydrogen particularly relevant in applications where direct electrification may be technically challenging, inefficient, or impractical.

This functional distinction explains why hydrogen is increasingly considered a complementary decarbonization tool rather than a universal substitute for electricity. In sectors requiring seasonal energy storage, long-range transport, or high operational flexibility, hydrogen may provide capabilities that conventional electrical systems alone cannot easily deliver.

Despite growing interest in hydrogen-based mobility, current global hydrogen demand remains overwhelmingly concentrated in traditional industrial sectors. According to [1], global hydrogen demand reached approximately 100 Mt in 2024. As illustrated in Fig. 8.1, most hydrogen consumption remains concentrated in three major industrial applications: oil refining, chemical synthesis (methanol and ammonia), as well as iron and steel production. These sectors have historically dominated hydrogen utilization at industrial scale, primarily due to their dependence on hydrogen as a chemical feedstock rather than as an energy carrier. Although emerging applications such as electric mobility are beginning to expand, their contribution to total hydrogen demand remains marginal when compared with these established industrial uses.

In this context, the future penetration of hydrogen as an energy vector within electric mobility depends not only on technological progress in storage systems, refuelling infrastructure, or FC development, but also on a broader structural transformation of end-use demand. Such transformation is inherently complex, as it requires substantial adaptation of supply chains, infrastructure, and industrial frameworks that have historically evolved around conventional industrial consumption patterns.

Within this broader framework, hydrogen should therefore be understood primarily as a complementary energy vector rather than a direct replacement for electricity.

The greatest strategic value of hydrogen lies in potential sector coupling, where electricity generation, energy storage, industry, and transport may be integrated into a more flexible and interconnected energy architecture. Through this role, hydrogen can contribute to systemic optimization by linking sectors that are otherwise difficult to decarbonize through direct electrification alone.

8.1.3 Energy efficiency and system perspective

Unlike electrochemical energy storage systems based on batteries, where electricity is stored relatively directly through reversible electrochemical reactions, the use of hydrogen in electric mobility involves a multi-stage chain comprising multiple conversion, storage, and reconversion processes. This sequence, commonly described as electricity → hydrogen → electricity, inherently introduces cumulative energy losses at each stage. Consequently, hydrogen in electric mobility should not be evaluated solely on the basis of its physicochemical properties (see Section 8.1.1), but rather through a broader system-level perspective that considers the overall efficiency of the complete pathway from primary electricity input to final traction or propulsion output.

In particular, the conversion of electricity into hydrogen through water electrolysis (see Section 8.2.2) typically achieves overall efficiencies in the range of approximately 60–80%, depending on electrolyzer technology, operating conditions, and system integration factors [2]. Following production, hydrogen must be stored and transported, often requiring compression or liquefaction, both of which impose additional energy penalties due to the low volumetric energy density of hydrogen. Finally, hydrogen must be reconverted into electricity onboard the vehicle using FCs (see Section 8.5), whose mature mobile applications can reach efficiencies of up to approximately 60%. When considering the entire chain, total system efficiency may fall within the range of roughly 25–40% [3], substantially below the roundtrip efficiency typically associated with lithium-ion battery systems, which can approach 90% under favorable conditions (see Chapter 7).

Nevertheless, the relevance of hydrogen in electric mobility should be understood primarily from a technological and systemic perspective rather than from efficiency alone. Hydrogen offers competitive advantages in parameters such as driving range, operational flexibility, refuelling speed, long-duration storage capability, and suitability for applications where battery mass or charging constraints become limiting factors. These characteristics position hydrogen as a potentially viable solution for sectors where direct electrification faces significant technical or economic barriers, including long-haul heavy-duty transport, selected non-electrified rail applications, and certain maritime or aviation segments.

From a broader energy systems perspective, hydrogen also enables temporal and geographical decoupling between electricity generation and final energy use. Unlike electricity, which often requires near-immediate consumption or dedicated storage infrastructure, hydrogen can be produced during periods of surplus generation, stored over extended timescales, transported across regions, and later reconverted when demand arises. This capability may be particularly relevant in renewable-dominated systems, where excess wind or solar generation could be redirected toward hydrogen production instead of being curtailed.

Therefore, batteries and hydrogen should not necessarily be interpreted as strictly competing solutions, but rather as complementary energy vectors within a diversified electric mobility ecosystem. While batteries maximize efficiency in many light-duty and short-range applications, hydrogen may provide strategic value where energy density, operational speed, or sector integration are more critical. This complementary framework suggests a future mobility landscape in which battery-electric and hydrogen-based technologies coexist, each addressing distinct operational requirements.

8.2 Hydrogen production pathways

8.2.1 Conventional production (grey/blue)

At present, conventional hydrogen production from fossil fuels remains the dominant pathway for large-scale global hydrogen supply. Among these methods, Steam Methane Reforming (SMR) is by far the most widely deployed technology. In this process, methane reacts with steam at high temperatures to produce hydrogen together with carbon-rich gaseous compounds, primarily carbon monoxide and carbon dioxide. In a subsequent stage, hydrogen yield is further increased through the water-gas shift reaction, where carbon monoxide reacts with steam to generate additional hydrogen and CO_2. SMR is a highly mature industrial technology supported by well-established infrastructure, optimized processes, and comparatively low production costs. These advantages have consolidated SMR as the prevailing solution for industrial hydrogen generation.

Hydrogen produced through SMR without carbon mitigation is commonly referred to as grey hydrogen and is associated with a substantial carbon footprint due to the significant CO_2 emissions generated throughout the process. Consequently, although grey hydrogen may later be used in applications with zero direct tailpipe emissions, such as FCECs, its full life-cycle environmental impact remains strongly conditioned by upstream emissions. This significantly limits its compatibility with deep decarbonization objectives. In addition to SMR, hydrogen can also be conventionally produced through coal gasification. While this pathway has historically maintained relevance in certain coal-dependent regions, its even higher carbon intensity has increasingly constrained its role in future low-emission energy systems.

To partially address these emissions, conventional SMR can be integrated with carbon capture, utilization and storage (CCUS) technologies, giving rise to so-called blue hydrogen. Through this approach, a significant share of the CO_2 generated during production can be captured before atmospheric release, thereby reducing net carbon emissions while preserving the industrial maturity of fossil-based hydrogen production. Blue hydrogen has gained increasing strategic relevance as a transitional pathway because it leverages existing infrastructure without requiring immediate large-scale deployment of less mature alternatives. However, its decarbonization

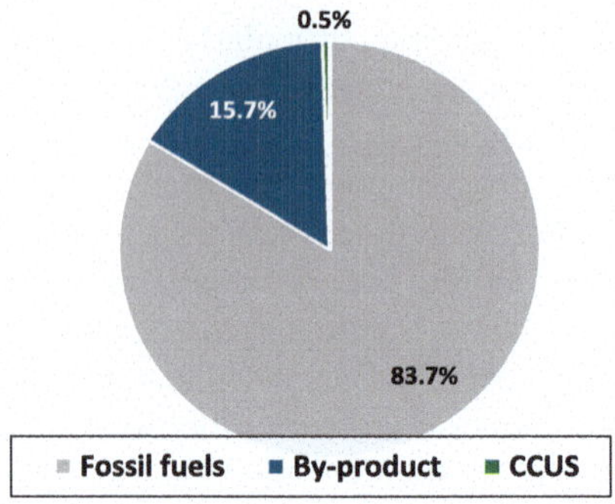

Figure 8.2 Global hydrogen production pathways in 2023.

potential remains limited by imperfect capture efficiencies, methane leakage across natural gas supply chains, and the economic and logistical challenges associated with CCUS deployment.

As illustrated in Fig. 8.2, global hydrogen production remains overwhelmingly dependent on fossil-fuel-based technologies, which account for approximately 83.7% of total output [1]. Industrial by-products represent a secondary contribution, whereas CCUS-based pathways and other low-emission technologies still constitute only a marginal fraction. This distribution highlights the considerable technological, economic, and infrastructural challenges that must be overcome to transition toward truly low-carbon hydrogen production systems.

8.2.2 Renewable hydrogen (green)

Renewable hydrogen, commonly referred to as green hydrogen, is primarily produced through water electrolysis using a device known as an electrolyzer. In this process, an electrical current drives the decomposition of water (H_2O) into hydrogen and oxygen. Unlike SMR, electrolysis eliminates the direct dependence on fossil fuels by requiring only electrical energy as its primary input. Consequently, when this electricity is supplied by renewable sources such as solar, wind, or hydropower, the resulting hydrogen can be considered a very low-emission energy carrier. This characteristic positions green hydrogen as a strategically important pathway for the decarbonization of multiple sectors, including electric mobility.

Several electrolyzer technologies currently exist, each with distinct maturity levels, operational characteristics, and economic implications. Among the most widely adopted are alkaline electrolyzers, Proton Exchange Membrane (PEM) electrolyzers, and, to a lesser extent, solid oxide electrolyzers [4]. Alkaline systems remain the most technologically mature and economically competitive solution due to their established industrial deployment. In contrast, PEM electrolyzers offer superior operational flexibility, allowing them to rapidly adapt hydrogen production to the variable generation profiles characteristic of renewable energy systems. This dynamic capability makes PEM particularly attractive for renewable integration, although it is generally associated with higher costs and greater dependence on critical materials [5]. As a result, while electrolysis itself is inherently free from direct carbon emissions, the actual environmental performance of green hydrogen depends entirely on the carbon intensity of the electricity used.

From a system-level perspective, large-scale electrolyzer deployment offers significant advantages for renewable energy integration. Electrolysis can absorb surplus renewable electricity during periods of overgeneration, converting otherwise curtailed power into hydrogen that can be stored for extended periods and used later across multiple sectors. Nevertheless, despite its environmental promise, renewable hydrogen production remains constrained by substantial structural and economic barriers, including electrolyzer capital costs, renewable electricity availability, and infrastructure scalability, all of which currently limit the pace of widespread green hydrogen deployment [6].

8.2.3 Implications for mobility

The different hydrogen production pathways fundamentally condition the penetration of hydrogen-based technologies within the electric mobility sector through three major determining factors: cost, carbon footprint, and future availability. The evolution of these three variables will play a decisive role in the large-scale adoption of hydrogen-based EVs.

From an economic perspective, conventional hydrogen production technologies, particularly SMR, currently remain the most cost-competitive solution in terms of hydrogen cost per kilogram. This advantage is primarily driven by technological maturity, established industrial infrastructure, and existing fossil fuel supply chains. In contrast, green hydrogen production still faces substantial economic disadvantages relative to conventional pathways. These disadvantages are largely associated with the high capital cost of electrolyzers, the cost and availability of low-carbon electricity, and the comparatively limited economies of scale currently achieved by renewable hydrogen systems. As a result, despite its environmental benefits, green hydrogen remains less economically competitive in many present-day mobility applications.

From an emissions reduction standpoint, the widespread use of grey hydrogen can significantly undermine climate policy objectives due to the high carbon intensity associated with SMR. This substantially weakens one of hydrogen principal intrinsic advantages: its potential role as a low-emission energy carrier. The integration of CCUS technologies can partially mitigate these emissions by reducing the carbon footprint of fossil-based hydrogen production, although only incompletely and often at the expense of additional technological complexity and cost. Consequently, green hydrogen remains the only genuinely low-emission hydrogen production pathway fully aligned with long-term decarbonized mobility strategies.

Finally, the development of a hydrogen value chain specifically oriented toward electric mobility remains strongly constrained by future availability considerations. This availability depends not only on the maturity of hydrogen vehicle technologies themselves, but also on the simultaneous deployment of production capacity, transportation networks, and refuelling infrastructure. Structural bottlenecks such as limited renewable electricity generation in many regions, insufficient hydrogen distribution systems, and the scarcity of refuelling stations continue to restrict widespread deployment. For these reasons, the present role of hydrogen in electric mobility appears more realistically focused on specific applications where long driving range, rapid refuelling, or operational intensity represent critical limitations for alternative technologies, rather than as a universal replacement for all forms of electrified transport.

8.3 Hydrogen storage technologies

8.3.1 Compressed hydrogen storage

Compressed gaseous hydrogen storage currently represents the most mature and widely deployed solution for hydrogen storage in electric mobility applications, particularly in EVs based on FCs. Its technological predominance is primarily explained by the fact that it enables hydrogen to be stored in gaseous form through comparatively mature and industrially established mechanical compression technologies. This approach avoids the substantial technical complexity associated with hydrogen liquefaction and cryogenic handling, while also circumventing the lower technological maturity of alternative storage pathways such as metal hydrides or liquid organic hydrogen carriers.

Despite this maturity, the effective integration of hydrogen into electric mobility remains constrained by one of hydrogen most fundamental physical limitations: its extremely low volumetric energy density under ambient conditions. As illustrated in Fig. 8.3, gaseous hydrogen at atmospheric pressure exhibits a volumetric energy density that is far too low for practical mobile applications. Increasing storage pressure significantly improves hydrogen density, thereby increasing the amount of usable energy stored per unit volume. However, even under very high-pressure conditions, hydrogen remains substantially below the volumetric energy density of conventional liquid fuels and still presents limitations relative to other chemical energy carriers (see Table 8.1).

For this reason, industrial hydrogen mobility systems rely on very high-pressure storage architectures. In practice, the dominant standards are 350 bar and 700 bar systems. Historically, 350 bar storage has been widely adopted in heavy-duty road vehicles such as buses, trucks, and fleet transport, where larger available chassis volumes partially mitigate the volumetric penalty of compressed hydrogen. By contrast, 700 bar systems have become the dominant standard for lighter vehicles such as FCECs, where packaging constraints are considerably more severe. At 700 bar, hydrogen can achieve nearly twice the volumetric density obtainable at 350 bar, enabling smaller onboard tanks and improved driving range within passenger vehicle geometries.

Temperature also plays a critical role in compressed hydrogen storage performance. As shown in Fig. 8.3, hydrogen pre-cooling to low temperatures increases achievable storage density for a given pressure by reducing thermal expansion effects. This

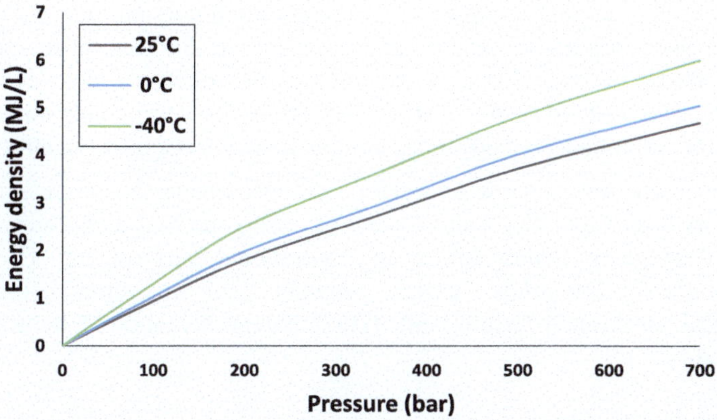

Figure 8.3 Volumetric energy density of compressed hydrogen as a function of storage pressure under different temperature conditions.

improves both storage effectiveness and refuelling efficiency. Consequently, modern hydrogen refuelling protocols such as SAE J2601 integrate strict thermal management requirements to control gas temperature during fast-filling operations (see Section 8.4.2). Thermal conditioning is therefore not merely a safety feature but also a performance-enabling parameter.

Nevertheless, hydrogen compression introduces additional energy penalties. Mechanical compression to 700 bar typically requires approximately 10–15% of the hydrogen own energy content, thereby reducing overall well-to-tank efficiency. Moreover, storing gas at such elevated pressures introduces significant engineering and safety challenges, including material fatigue, hydrogen embrittlement, leakage risks, thermal stresses, and impact resistance requirements. These factors necessitate sophisticated pressure regulation systems, leak detection technologies, thermal relief devices, and stringent certification protocols.

As a critical structural component, hydrogen tanks must simultaneously withstand extreme internal pressures while minimizing system mass and resisting corrosion. Accordingly, compressed hydrogen storage vessels are generally classified into four principal categories:

- **Type I:** Fully metallic tanks, typically manufactured from steel or aluminum. These systems provide high mechanical robustness and relatively low manufacturing costs but suffer from excessive mass, making them poorly suited for most EV applications.
- **Type II:** Metallic tanks reinforced with partial composite wrapping in high-stress regions. These systems improve the weight-to-strength ratio relative to Type I designs but remain comparatively heavy for advanced mobility applications.
- **Type III:** Tanks incorporating a metallic inner liner fully wrapped in composite materials such as carbon or glass fiber. These designs significantly reduce structural mass while maintaining strong pressure resistance, though they involve higher production costs.

- **Type IV:** Advanced tanks employing a non-structural polymer liner fully wrapped in carbon fiber composites. These currently represent the state-of-the-art solution for EVs based on FCs due to their superior gravimetric efficiency, corrosion resistance, and suitability for 700 bar operation. However, their manufacturing complexity and reliance on advanced composite materials result in elevated costs.

Overall, compressed gaseous hydrogen storage remains the dominant practical solution for hydrogen mobility today because it offers the best compromise between technological maturity, deployability, and operational performance. However, its intrinsic volumetric limitations, energy penalties, and structural cost challenges remain central constraints shaping the future competitiveness of hydrogen-based electric mobility.

8.3.2 Liquid hydrogen storage

Liquid hydrogen storage represents an alternative to compressed hydrogen storage aimed at overcoming some of the principal limitations associated with gaseous high-pressure systems. This storage pathway is based on hydrogen liquefaction, a process in which gaseous hydrogen is cooled to cryogenic temperatures near −253 °C (20 K), below its boiling point at atmospheric pressure, thereby enabling the phase transition from gas to liquid. In this liquid state, hydrogen reaches a density of approximately 70–71 kg/m^3, substantially exceeding that of compressed gaseous hydrogen even at 700 bar. As a result, volumetric energy density of hydrogen increases significantly (see Table 8.1), allowing considerably smaller storage volumes for an equivalent onboard energy capacity.

This improvement makes liquid hydrogen an especially attractive option in mobility sectors where volumetric constraints are particularly severe, such as long-haul heavy-duty transport, aviation, and certain maritime applications. In these cases, the substantial reduction in storage volume may partially offset the additional technological and energetic burdens associated with cryogenic storage systems.

At present, however, the energy demand of the liquefaction process constitutes the principal limiting factor of this technology. Hydrogen liquefaction requires a complex sequence of compression, expansion, and extreme refrigeration stages, all of which consume substantial energy. In practical terms, the liquefaction process may require approximately 30–40% of the hydrogen own energy content, compared to roughly 10–15% for compressed storage. This severe energy penalty significantly reduces overall system efficiency and represents one of the highest well-to-tank losses among currently available hydrogen storage technologies.

A further challenge lies in the complexity of the storage vessels themselves. Liquid hydrogen storage requires highly specialized cryogenic tanks, typically based on double-wall architectures with high-vacuum insulation layers designed to minimize thermal transfer from the external environment. Despite these advanced thermal protection strategies, some degree of external heat ingress remains unavoidable. This heat leakage gradually causes hydrogen evaporation, a phenomenon known as boil-off. Boil-off leads to progressive fuel losses and can become a major operational challenge. In stationary or industrial systems, boil-off may be partially mitigated through reliquefaction or auxiliary hydrogen use, but in electric mobility applications these strategies are often difficult to implement efficiently. Furthermore, evaporated hydrogen can generate internal overpressure, requiring dedicated venting, pressure relief, and monitoring systems.

From a safety perspective, liquid hydrogen introduces additional risk dimensions beyond those already associated with gaseous hydrogen. In addition to the inherent flammability of hydrogen, cryogenic storage systems must address hazards such as material embrittlement, cryogenic burns, insulation degradation, and thermal overpressure management. These factors substantially increase engineering complexity and operational requirements. Consequently, while liquid hydrogen storage offers clear advantages in volumetric performance, its widespread adoption in EVs remains constrained by high liquefaction energy demand, cryogenic infrastructure complexity, boil-off losses, and elevated system costs.

8.3.3 Solid-state storage

The limitations associated with compressed and liquid hydrogen storage, particularly regarding volumetric energy density and storage-related energy losses, respectively, have driven the search for alternative solutions capable of overcoming these constraints. In this context, solid-state hydrogen storage is based on the reversible incorporation of hydrogen into solid or liquid materials. This approach enables hydrogen storage through chemical bonds or physical interactions that are ultimately more stable than conventional gaseous-phase storage.

Among the various technological approaches to solid-state hydrogen storage, metal hydrides represent one of the most extensively studied options. This technology exploits the capacity of certain metals and alloys, such as lanthanum-nickel, magnesium, or titanium-based materials, to absorb substantial quantities of hydrogen per unit volume. During the charging process, hydrogen molecules dissociate at the surface of the absorbing material and are incorporated into its crystalline lattice. During discharge, this process is reversed through thermal input, allowing controlled hydrogen release. Metal hydride systems offer two major competitive advantages. First, they can achieve volumetric energy densities comparable to those of compressed hydrogen storage at 700 bar. Second, by avoiding both high-pressure compression stages and extreme cryogenic liquefaction, these systems are intrinsically safer than compressed or liquid hydrogen storage alternatives.

Despite these advantages, metal hydride hydrogen storage faces substantial challenges for mobility applications. First, hydrogen desorption typically requires additional thermal energy, reducing overall system efficiency. Second, although these systems are competitive from a volumetric perspective, their gravimetric energy density is generally poor due to the substantial mass of the storage medium itself. This limitation significantly constrains their applicability in transportation sectors where weight is a critical parameter, particularly passenger vehicles and aviation.

An alternative to metal hydrides is represented by Liquid Organic Hydrogen Carriers (LOHCs). This storage pathway employs organic liquids, such as toluene/methylcyclohexane or dibenzyltoluene, which can absorb and release large amounts of hydrogen through reversible hydrogenation and dehydrogenation reactions. One of the principal advantages of LOHC systems is that both charging and discharging processes can occur under near-ambient temperature and pressure conditions. This characteristic substantially simplifies storage logistics and enables the reuse of existing infrastructure developed for liquid fuel transport and distribution. Consequently, LOHCs present an important deployment advantage by facilitating integration into established fuel supply chains and broader industrial hydrogen ecosystems.

However, despite these logistical benefits, LOHCs currently face major barriers for direct electric mobility applications. Their primary limitation lies in the energetically intensive catalytic processes required for both hydrogen loading and unloading, which significantly penalize system-level efficiency. Additionally, the onboard management of the carrier liquid introduces further complexity in portable or vehicle-integrated platforms, making their practical implementation in light-duty mobility particularly challenging.

Finally, advanced materials such as Metal-Organic Frameworks (MOFs), carbon-based nanomaterials, and cryogenic adsorbents aim to combine high specific surface area, elevated adsorption capacity, and improved operating conditions. Among these, MOFs have emerged as a particularly promising research field within hydrogen storage, and laboratory-scale studies have demonstrated encouraging results [7]. Nevertheless, these technologies remain at very early stages of development, and their future commercialization remains subject to substantial uncertainty due to unresolved issues related to cost, cyclic durability, material stability, and operating requirements.

Although solid-state hydrogen storage presents highly attractive theoretical advantages for electric mobility applications, most of its associated technologies remain far from technological maturity. As a result, they currently represent less commercially deployed solutions than compressed or liquid hydrogen systems. Their long-term viability will largely depend on overcoming present limitations, particularly in relation to mass, cost, kinetics, thermal efficiency, and scalable industrial manufacturability.

8.3.4 Comparative analysis

Throughout Section 8.3, multiple technological solutions for hydrogen storage have been examined, employing different physical states and storage mechanisms for hydrogen. Each of these alternatives presents distinct competitive advantages and disadvantages, as summarized in Table 8.2, thereby outlining a complementary technological landscape in which different storage solutions may coexist by targeting specific applications where their respective strengths provide a differential competitive advantage.

In terms of energy density, compressed hydrogen currently represents the most technologically mature solution for electric mobility and the one most widely implemented in FC-based EVs. This storage pathway offers a significant competitive advantage in gravimetric energy density, particularly at high storage pressures such as 700 bar. However, its relatively low volumetric energy density constitutes an important limitation in mobility segments where onboard storage space is constrained.

Table 8.2 Comparative analysis of the main hydrogen storage technologies for mobility applications.

Technology	Advantages	Disadvantages
Compressed	Technologically mature High gravimetric energy density	Low volumetric energy density High-pressure safety requirements
Liquid	High volumetric energy density	High energy penalty Boil-off losses Complex thermal insulation Cryogenic safety challenges
Metal Hydrides	High volumetric density Improved intrinsic safety Stable storage conditions	Poor gravimetric density Thermal management complexity
LOHC	Existing liquid fuel infrastructure compatibility Safe transport and handling	Energy-intensive catalytic processes Complex carrier management Limited onboard suitability
MOF	Potentially high storage capacity Lower pressure operation	Low technological maturity High material cost Uncertain large-scale commercialization

By contrast, liquid hydrogen storage provides substantially higher volumetric energy density, significantly reducing the physical volume required for hydrogen storage. This characteristic offers a strategic advantage in certain sectors, including long-distance heavy-duty transport, maritime applications, or aviation. Nevertheless, liquid hydrogen also presents important limitations due to the high energy demand of liquefaction, the complexity of cryogenic storage systems, and hydrogen losses associated with boil-off.

Both compressed and liquid hydrogen storage pathways also raise significant safety challenges because of their reliance on either extreme pressures or cryogenic temperatures. In this context, solid-state storage technologies, particularly metal hydrides and certain LOHC-based systems, offer meaningful safety advantages. Metal hydrides enable comparatively stable storage conditions, while some LOHC pathways allow hydrogen handling under operational conditions similar to those of conventional liquid fuels, thereby simplifying logistics and infrastructure compatibility. However, both approaches also present major disadvantages, especially from an energy efficiency standpoint, due to the presence of thermally or catalytically intensive processes. Furthermore, metal hydrides suffer from poor gravimetric energy density, whereas LOHC systems introduce additional complexity related to carrier liquid management.

When system cost and technological maturity are considered, compressed hydrogen remains the most economically attractive solution at present, benefiting from

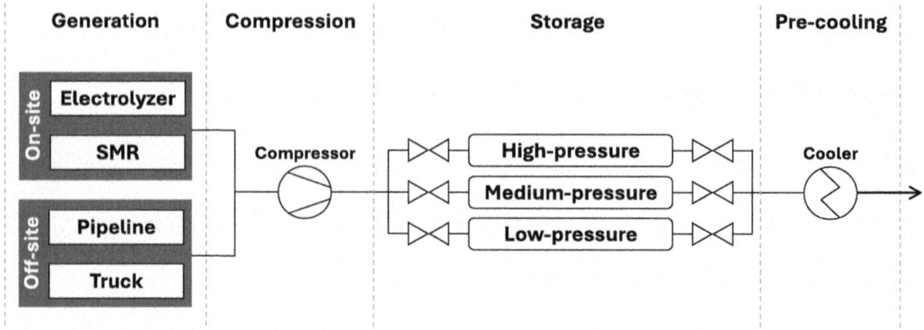

Figure 8.4 Generalized architecture of a HRS, illustrating the main functional stages of hydrogen supply, compression, cascade intermediate storage pre-cooling, and final vehicle dispensing.

comparatively mature technologies, established industrial platforms, and a relatively well-developed regulatory framework. In this hierarchy, liquid hydrogen occupies an intermediate position, with partial deployment focused on specialized niches, while emerging technologies such as MOFs remain at very early developmental stages and continue to face substantial uncertainty regarding their future viability for electric mobility applications.

In practical commercial electric mobility applications, the selection of a specific hydrogen storage technology ultimately depends on balancing multiple factors, including range, mass, volume, safety, and system-level efficiency. At present, compressed hydrogen remains the dominant solution for commercial FC-based vehicles. Conversely, solid-state storage technologies remain primarily prospective solutions, with potential future relevance in stationary storage, heavy-duty transport, or highly specialized mobility sectors.

8.4 Hydrogen refuelling technologies

8.4.1 Refuelling station architecture

The viability of hydrogen-based electric mobility solutions depends not only on efficient onboard hydrogen storage and electricity conversion technologies, but also on the availability of infrastructure capable of supplying hydrogen to vehicles in a safe, rapid, and efficient manner. In this context, Hydrogen Refuelling Stations (HRSs) represent a fundamental component of the hydrogen supply chain for EVs.

The architecture of an HRS can vary primarily depending on the hydrogen supply pathway used by the station itself, whether through onsite generation via electrolysis or SMR, or through external logistical delivery such as pipelines or road transport. Nevertheless, regardless of the hydrogen generation or supply method, most HRSs typically incorporate four common functional blocks: hydrogen generation/supply, compression, intermediate storage, and pre-cooling, as illustrated in Fig. 8.4.

The first block corresponds to hydrogen generation or supply. If hydrogen is produced onsite through electrolysis or SMR, it is generally generated at moderate pressures insufficient for direct mobility applications, thus requiring subsequent compression. If the hydrogen is externally supplied, it may arrive already compressed, potentially reducing compression requirements depending on delivery conditions, although additional conditioning may still be necessary. Hydrogen may also be delivered in liquid form, in which case a vaporization stage with thermal input is required to convert the liquid hydrogen into gaseous form prior to further processing.

The compression subsystem constitutes one of the most critical stages within an HRS. Its primary objective is to increase gaseous hydrogen pressure to levels suitable for vehicle refuelling, typically 350 bar for heavy-duty vehicles and 700 bar for passenger vehicles. This stage generally requires substantial energy input and has major implications for both the technical and economic viability of the infrastructure, given its impact on operational costs, maintenance requirements, and system complexity.

Following compression, HRSs typically incorporate an intermediate storage stage. This storage is often configured using a cascade system composed of multiple storage vessels operating at different pressure levels, commonly low, medium, and high pressure. During refuelling, hydrogen is initially supplied from the lowest-pressure reservoir compatible with the vehicle tank pressure, which is generally assumed to be substantially depleted. As tank pressure rises during filling, hydrogen is progressively drawn from higher-pressure storage vessels. This strategy optimizes energy efficiency, reduces instantaneous compressor workload, and accelerates the refuelling process.

Finally, a pre-cooling stage may be incorporated immediately prior to final dispensing. This subsystem is particularly common in high-pressure (700 bar) stations and aims to reduce hydrogen temperature to approximately −40 °C. This pre-cooling compensates for the thermal increase associated with rapid filling, improves the final mass of hydrogen transferred (see Section 8.3.1), and ensures compliance with thermal safety limits and operational refuelling protocols.

8.4.2 Refuelling protocols and operation

Unlike conventional liquid fuels such as gasoline or diesel, hydrogen refuelling in EVs is typically performed under substantial variations in pressure, temperature, and gas density, particularly in high-pressure applications at 350 and 700 bar, which currently represent the dominant standards for heavy-duty and light-duty vehicles, respectively. These operating conditions make hydrogen refuelling considerably more complex than conventional fuel dispensing, requiring significantly more advanced monitoring and process control to prevent overpressurization, excessive thermal loads, or inefficient tank filling. Consequently, refuelling protocols play a critical role in ensuring that hydrogen delivery to EVs is conducted under safe, efficient, and technically optimized conditions.

The SAE J2601 standard [8] currently constitutes the principal regulatory framework governing gaseous hydrogen refuelling processes for surface vehicles. One of its most important contributions is the definition of standardized filling profiles. Rather than relying on a constant mass flow rate, SAE J2601 specifies dynamic refuelling conditions based on predefined algorithms combined with real-time sensor feedback, thereby enabling continuous adjustment of operating parameters throughout the filling event. This dynamic approach significantly improves both safety and operational efficiency. In addition, SAE J2601 establishes pressure thresholds, thermal compensation strategies, and filling parameters according to key variables such as nominal tank pressure, ambient temperature, and the initial state of charge of the onboard storage system.

To effectively implement these dynamic filling profiles, dispenser-to-vehicle interfaces in HRSs typically incorporate dedicated communication systems, commonly based on infrared or digital communication technologies. These systems enable the real-time exchange of critical information, including tank capacity, maximum allowable pressure, and thermal conditions. Through this continuous data transfer, the station can apply customized filling protocols adapted to the specific operational state and technical characteristics of each vehicle, thereby optimizing both refuelling speed and system protection.

To further ensure efficient and safe refuelling, modern HRSs generally incorporate a pre-cooling stage immediately downstream of intermediate storage (see Fig. 8.4). The objective of this stage is to reduce hydrogen gas temperature to approximately $-40\,^{\circ}C$ before dispensing it into the vehicle. This thermal conditioning is essential because, during rapid refuelling, hydrogen undergoes significant temperature increases due to fast compression as it enters the onboard tank. Excessive temperature rise can compromise the structural integrity of the storage vessel while simultaneously reducing the effective transferred hydrogen mass, since hydrogen density decreases with increasing temperature (see Fig. 8.3). By mitigating this thermodynamic effect, pre-cooling systems help maintain safe operating conditions, comply with protocol limitations, and maximize the final storage capacity of the EV.

8.4.3 Refuelling performance

The refuelling performance of compressed hydrogen represents one of the most significant operational advantages of FC-based EVs over BEVs [9]. Unlike battery charging systems, hydrogen refuelling systems are capable of transferring large amounts of energy within very short timeframes, thereby offering a user experience more comparable to conventional liquid fuel refuelling than to electrochemical charging processes.

Under standard operating conditions, a vehicle equipped with a 700 bar storage tank can typically complete refuelling in approximately 3 to 5 minutes [10], whereas this duration may increase for EVs operating with 350 bar storage systems depending on tank capacity and HRS characteristics. In particular, refuelling performance is strongly influenced by factors such as the effective integration of intermediate storage, compression systems, dynamic filling protocols, and pre-cooling technologies. Nevertheless, actual refuelling duration is not solely determined by nominal system design, as external and operational variables including ambient temperature, initial tank thermal state, and station utilization rate can significantly affect overall performance. Under intensive operating conditions, thermal or supply-related limitations may temporarily reduce refuelling efficiency if the infrastructure is not adequately dimensioned.

Consequently, FC vehicles maintain a clear advantage over BEVs in terms of refuelling time, substantially outperforming slow charging systems and, in many scenarios, even fast-charging technologies. This advantage becomes particularly relevant in applications where vehicle downtime is a critical operational parameter, such as commercial fleets, long-distance transport, or mobility systems with demanding availability requirements. In such contexts, minimizing refuelling time can directly improve operational productivity and logistical efficiency.

However, the speed advantage of hydrogen refuelling is intrinsically dependent on technologically sophisticated and highly specialized infrastructure. In contrast, BEVs can leverage a significantly more widespread electrical grid and broader opportunities for decentralized charging. Therefore, although rapid hydrogen refuelling constitutes a major competitive advantage for FC-based EVs, this benefit must be evaluated from a broader systemic perspective. Its practical superiority may be reduced or challenged in scenarios where infrastructure availability, economic considerations, or distributed charging flexibility becomes more decisive than refuelling speed alone.

8.4.4 Safety considerations

Hydrogen refuelling presents specific technical and safety characteristics that distinguish it from other conventional refuelling or charging processes. These particularities derive primarily from the physicochemical properties of hydrogen itself, including its low minimum ignition energy, wide flammability range, and high diffusivity. In addition, the demanding operational conditions associated with hydrogen dispensing, especially the elevated pressures involved, introduce further engineering and safety challenges. Collectively, these factors require the deployment of dedicated safety strategies that differ substantially from those applied to conventional liquid fuels or electrical charging infrastructures.

To establish effective safety protocols, it is essential to understand the principal risks associated with the refuelling process. One of the most significant is the possibility of hydrogen leakage. Due to its extremely small molecular size, hydrogen is more prone than many other gases to escaping through microcracks, defective seals, or improperly designed connections. Such leaks may create combustible atmospheres because hydrogen exhibits a broad flammability range in air. Although its very low density promotes rapid vertical dispersion in open environments, thereby potentially reducing prolonged accumulation, HRSs must nevertheless integrate leak detection sensors, active ventilation systems, automatic shutoff valves, and emergency shutdown protocols specifically designed to minimize operational hazards.

A second major source of risk arises from the high gas pressures handled during refuelling. Dispensing hydrogen at 350 or 700 bar requires the use of tanks, piping, valves, dispensers, and connectors capable of withstanding extreme mechanical stresses under repeated operational cycles. Consequently, the structural design of these systems is highly demanding, requiring advanced materials, rigorous resistance testing, engineered redundancies, and periodic safety inspections to minimize the probability of mechanical failure, structural fatigue, or accidental depressurization.

Several international regulatory frameworks, including SAE J2601, ISO 19880-1, and multiple national regulations, provide the normative foundation supporting these safety strategies. These frameworks establish strict requirements for infrastructure design, operational procedures, maintenance protocols, and emergency response measures. Through the implementation of such standardized criteria, hydrogen refuelling infrastructure can achieve consistent operational safety while supporting broader technological interoperability across different vehicles, stations, and regional deployment contexts.

8.5 Hydrogen-to-electricity conversion

8.5.1 FC fundamentals

In hydrogen-powered EVs, the FC constitutes the core component responsible for converting the chemical energy stored in hydrogen into electrical energy usable for vehicle traction or propulsion. In FCs, hydrogen chemical energy is converted directly into electricity without intermediate combustion, which fundamentally distinguishes them from ICEs, where fuel energy is first transformed into heat through combustion and subsequently into mechanical energy through thermodynamic cycles. This direct electrochemical conversion allows FCs to avoid many of the intrinsic limitations associated with conventional thermal cycles and, in general terms, enables higher operational efficiencies, near-zero local emissions (with water vapor as the principal byproduct), and substantially reduced mechanical noise and vibration.

From a structural perspective, an FC is conceptually similar to a battery (see Chapter 7), consisting of two electrodes (anode and cathode) separated by an electrolyte, as illustrated in Fig. 8.5. Similar to batteries, FCs can be associated in series of parallel forming stacks. However, unlike batteries, the primary function of an FC is not energy storage but continuous energy conversion. Its operating principle is based on the controlled separation of oxidation and reduction reactions. The process begins by supplying hydrogen to the anode, where it undergoes oxidation, releasing electrons and protons (H^+). At this stage, the electrolyte acts as an electronic barrier while selectively allowing ionic transport, thereby forcing electrons to circulate through an external electrical circuit. This electron flow generates usable electric current that can be employed for vehicle propulsion or auxiliary systems.

Simultaneously, at the cathode, oxygen is typically supplied directly from ambient air. There, oxygen reacts with the protons that have crossed the electrolyte and the electrons returning from the external circuit, producing water as the principal reaction product. The global reaction of a hydrogen FC is described as follows:

$$2\mathrm{H_2} + \mathrm{O_2} \rightarrow 2\mathrm{H_2O} + \text{electricity} + \text{heat}$$

It is important to note that this global reaction also generates residual heat. Effective thermal management is therefore essential within FC-powered vehicles,

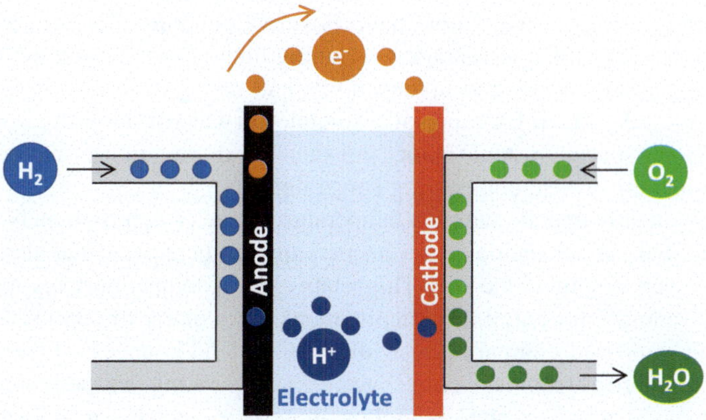

Figure 8.5 Schematic operating principle of a hydrogen FC.

often requiring dedicated cooling systems to prevent material degradation, maintain stable operating conditions, and improve overall system efficiency.

A fundamental distinction relative to batteries is that FCs do not possess intrinsic energy storage capability. While batteries store electrochemical energy internally within active materials and deliver it until depletion, FCs function as continuous onboard electricity generators as long as hydrogen fuel and oxygen are supplied. Consequently, in FC-based EVs, energy storage is typically associated with the hydrogen tanks rather than the FC itself. In this sense, the FC should be interpreted primarily as an onboard power generation system dependent on external refuelling infrastructure.

Because of this lack of direct storage functionality, FC systems are commonly integrated into hybrid architectures alongside batteries. This configuration enables regenerative braking energy recovery (see Chapter 2), improves transient power response, and reduces operational fluctuations imposed on the FC, thereby enhancing durability and dynamic performance.

Currently, multiple FC technologies exist, differentiated primarily by electrolyte type and operating conditions. Among these, alkaline and Proton Exchange Membrane FCs (PEMFCs) are among the most technologically mature, while Solid Oxide FCs and other emerging approaches remain at different stages of development [11]. For electric mobility applications, parameters such as operating temperature, power density, start-up time, and dynamic response are particularly critical, as these characteristics largely determine the practical suitability of each FC technology for mobile platforms.

8.5.2 PEMFCs

Currently, PEMFCs (also known as polymer electrolyte membrane FCs) represent the most widely adopted FC technology in FC-based vehicles, particularly in light-duty platforms, although their presence in heavy-duty applications is also increasing progressively. This dominant position is primarily the result of a combination of competitive advantages, including high power density, relatively low operating temperature, rapid start-up capability, and strong dynamic response. These characteristics make PEMFCs especially suitable for electric mobility applications, where available space, system mass, transient load variations, and operational flexibility constitute critical engineering constraints.

PEMFCs typically operate within a temperature range of approximately 60–80 °C, which places them at substantially lower operating temperatures than alternative FC technologies such as solid oxide FCs. This relatively low-temperature regime provides important advantages for automotive deployment, particularly through reduced start-up times and improved compatibility with frequent load changes. However, it also introduces significant technical challenges. Low operating temperatures increase sensitivity to contaminants such as carbon monoxide, which can poison catalytic sites and reduce electrochemical performance. In parallel, PEMFCs require highly controlled humidity and thermal management. Proper membrane hydration is essential because excessive drying can significantly reduce proton conductivity, whereas excessive water accumulation may cause flooding, thereby obstructing reactant transport and reducing cell efficiency.

The defining structural component of a PEMFC is the proton-conducting membrane itself, typically manufactured from polymeric materials, with perfluorosulfonic acid polymers such as Nafion being among the most widely used due to their high proton conductivity under properly hydrated conditions. This membrane simultaneously functions as electrolyte and physical separator, enabling proton transport between anode and cathode while preventing direct electron transfer and limiting undesired gas crossover.

The electrodes in PEMFCs are porous multifunctional structures designed not only to conduct electrons but also to support catalytic materials and facilitate gas diffusion. Carbon-based materials are commonly employed due to their conductivity, structural properties, and compatibility with catalyst deposition. At the anode, hydrogen oxidation is promoted through catalyst layers commonly based on platinum or platinum-alloy materials, which accelerate hydrogen dissociation into protons and electrons. At the cathode, oxygen reduction occurs as protons crossing the membrane combine with electrons returning through the external circuit and oxygen supplied from ambient air. Because the oxygen reduction reaction is kinetically slower, cathode catalyst design is particularly critical for overall system efficiency.

In addition to membrane and catalyst-coated electrodes, PEMFCs incorporate gas diffusion layers, generally carbon-based, which ensure homogeneous reactant distribution across catalytic surfaces, facilitate product water removal, and maintain electrical conductivity. The PEMFC architecture is further completed by bipolar plates, which provide gas channeling, thermal regulation, water management, and electrical interconnection among multiple individual cells assembled into a stack capable of delivering practical vehicle-scale power.

Despite their technological maturity, PEMFCs continue to face important limitations. Their dependence on critical materials, particularly platinum-based catalysts, significantly affects cost and scalability. Platinum availability, price volatility, and supply chain concentration remain substantial industrial concerns. Furthermore, membrane degradation, catalyst aging, carbon support corrosion, and sensitivity to impurities continue to constrain long-term durability. These challenges directly influence lifecycle cost, maintenance requirements, and broader commercialization potential. Consequently, while PEMFCs currently represent the leading FC technology for electric mobility, continued advances in materials science, catalyst reduction, and durability enhancement remain essential for achieving wider economic competitiveness [12].

8.5.3 Performance characteristics

The practical deployment of PEMFCs in electric mobility depends not only on their electrochemical principles, but also on a series of performance characteristics that determine their suitability for specific vehicle applications. Among the most relevant parameters are efficiency, power density, dynamic response, and durability, as these directly influence vehicle range, acceleration capability, system packaging, and operational cost. Furthermore, several of these characteristics, particularly efficiency, are highly dependent on operating conditions, making system optimization a critical aspect of PEMFC implementation in transportation platforms.

PEMFCs commonly employed in electric mobility applications typically achieve overall efficiencies in the range of 40% to 60% [13]. This places them above ICEs in energy conversion performance, although generally below battery-based systems when considering the complete pathway from stored energy to final traction output. PEMFC efficiency also varies significantly with load conditions. As current demand increases, efficiency declines due to higher activation losses, ohmic resistance, and mass transport limitations. Bu this reason, FCs rarely operate as standalone power sources in vehicle architectures. Instead, they are frequently integrated into hybrid systems with complementary energy storage technologies, such as batteries or supercapacitors, which can absorb transient power peaks while allowing the FC to operate closer to its optimal efficiency range.

Power density is another critical parameter in vehicle integration, as it strongly affects FC size, mass, and packaging flexibility. Among commercially viable FC technologies, PEMFCs offer some of the most competitive power densities, typically reaching approximately 1 W/cm^2 at stack level [14]. High power density supports improved vehicle compactness and lower mass, both essential in mobility applications. However, achieving such performance generally requires advanced membrane materials, precise thermal regulation, and greater catalyst loading, increasing both technical complexity and cost.

Dynamic response is equally important in transportation scenarios characterized by rapid acceleration, regenerative braking, and variable power demands. Although PEMFCs outperform high-temperature FC technologies in transient response, they remain slower than batteries due to limitations associated with hydrogen and oxygen supply regulation, membrane hydration control, and compressor inertia. This comparatively slower response further reinforces the need for hybrid vehicle architectures, where auxiliary storage systems provide immediate power during transient events.

Finally, durability remains one of the principal challenges limiting broader PEMFC deployment in mobility. Vehicle applications demand extended operational lifetimes capable of sustaining thousands of cycles without significant degradation. In PEMFC systems, catalyst degradation, membrane deterioration and component wear can severely compromise long-term viability. Therefore, successful vehicle integration requires balancing efficiency, power density, transient capability, and durability to ensure both technical competitiveness and economic feasibility across different mobility segments.

8.5.4 System integration in vehicles

The integration of FC systems into EV architectures is considerably more complex than in BEVs due to the broader range of subsystems required for operation, control and energy management. Unlike battery-electric configurations, where the battery pack typically functions as both the primary energy source and dynamic power provider, FC-based vehicles generally rely on a hybridized architecture in which the FC operates in coordination with an auxiliary electrical energy storage system, most commonly batteries. This auxiliary system absorbs transient power peaks, supports dynamic load variations and stabilizes overall vehicle operation. In addition, proper FC functionality requires multiple auxiliary subsystems dedicated to monitoring, control, and protection, collectively grouped under the general concept of balance of plant (BoP).

The performance, efficiency and durability of PEMFCs are strongly dependent on operating parameters such as pressure, temperature, humidity and reactant supply conditions. Deviations from optimal ranges can significantly reduce efficiency, accelerate degradation or compromise system safety. Consequently, FC vehicles require an extensive BoP designed to continuously regulate these variables. In general, the BoP in hydrogen-powered EVs includes air compressors, humidifiers, thermal management systems, cooling circuits, electronic controllers, valves, sensors, DC/DC converters, and water management subsystems. These components ensure appropriate reactant flow, membrane hydration, and temperature stability while protecting the FC from harmful operating conditions. However, the BoP often represents one of the principal contributors to system complexity, mass, and packaging constraints in FC-powered mobility platforms.

The integration of auxiliary electrical storage, typically in the form of batteries, is equally essential. This storage system functions as an energy buffer, allowing the FC to operate closer to its optimal efficiency point while the battery manages short-term fluctuations in power demand. Such an arrangement is particularly beneficial during rapid acceleration, start-up phases and regenerative braking events. By reducing exposure of the FC to aggressive transient cycles, the battery not only improves vehicle responsiveness but also contributes to extending FC lifetime. Furthermore, regenerative braking energy recovery would be significantly less effective without an electrical storage subsystem capable of rapid energy absorption.

From an energy management perspective, FC vehicles share certain operational similarities with HEVs, as both require coordinated control between two complementary power systems. In FC vehicles, however, both propulsion subsystems are electrically coupled, with the FC and auxiliary battery relying on distinct energy sources. This creates a hybrid energy architecture in practical terms. Therefore, advanced energy management strategies are essential to coordinate subsystem interactions, maximize system-level efficiency and preserve long-term component durability. Successful FC vehicle integration ultimately depends on optimizing the interaction between the FC stack, BoP, and auxiliary storage system as a unified energy platform.

8.6 Applications in electric mobility

8.6.1 Passenger light-duty vehicles

Within the electric mobility sector, passenger vehicles have represented one of the most visible applications of FC-based propulsion technologies. In this segment, vehicles typically incorporate compressed hydrogen storage systems operating at 700 bar, coupled with a PEMFC that generates onboard electricity for electric propulsion, usually supported by an auxiliary battery system. This architecture enables a driving and refuelling experience relatively close to that of conventional ICE vehicles, particularly in terms of driving range and refuelling speed.

The perceived advantages of hydrogen-powered passenger vehicles in road mobility have motivated several major manufacturers to commercialize FC passenger models. Representative examples include the Toyota Mirai (see Fig. 8.6), one of the most internationally recognized, the Hyundai NEXO, primarily positioned within the SUV segment, and the Honda Clarity Fuel Cell, developed as a passenger sedan. These vehicles have played an important role in demonstrating the practical feasibility of hydrogen technologies for electric mobility while increasing public and industrial visibility of FC systems. Their deployment has been especially relevant in regions where dedicated hydrogen refuelling infrastructure exists, including Japan, South Korea, California, and selected European markets.

One of the principal strengths of FC passenger vehicles is their ability to offer operational characteristics comparable to ICE vehicles in terms of driving autonomy and rapid refuelling. This familiarity can reduce user resistance to adopting low-emission transportation technologies, particularly among consumers prioritizing convenience and minimal behavioral change relative to traditional vehicle usage patterns.

From a technological perspective, the deployment of FC passenger vehicles has also provided an essential real-world validation platform for PEMFC systems, high-pressure hydrogen storage technologies and advanced vehicle-level energy management strategies. Although their market penetration remains significantly below that of alternative electric mobility solutions such as BECs (see Chapter 3), passenger FC vehicles have served a critical role in validating, showcasing and refining hydrogen-based mobility technologies. Consequently, beyond their direct commercial relevance, these vehicles have contributed substantially to demonstrating

Figure 8.6 Toyota Mirai, an example of passenger FC vehicle. (Image courtesy of Wikimedia Commons.)

that hydrogen can function as a technically viable energy vector within the passenger mobility segment while supporting broader technological development across the FC ecosystem.

8.6.2 Heavy-duty transport

Within the mobility sector, heavy-duty transport represents one of the segments where hydrogen technologies have generated the greatest strategic interest. Unlike light-duty passenger vehicles, heavy-duty applications such as long-haul trucks, intercity buses, logistics machinery, and intensive commercial transport present particularly demanding operational requirements in terms of range, payload capacity, and refuelling time, where battery-based solutions may face substantial practical limitations.

In this context, heavy-duty transport benefits significantly from two of the principal strategic advantages of hydrogen technologies: high specific energy and rapid refuelling capability. First, the elevated gravimetric energy density of hydrogen allows compressed storage systems, commonly operating at 350 bar in many heavy-duty applications, to provide driving ranges suitable for long-distance transport without the prohibitive mass penalties associated with very large battery packs. In long-haul road transport, fully battery-based architectures may require extremely high-capacity storage systems, potentially reducing payload and operational efficiency due to increased vehicle mass. For this reason, FC-powered vehicle technologies have found some of their most commercially promising deployment opportunities in heavy-duty road transportation. Major manufacturers including Hyundai, Toyota, Nikola, Daimler Truck, and Volvo have promoted multiple FC-based projects targeting long-distance trucks and logistics platforms.

A second major strategic advantage is the relatively short refuelling time of hydrogen systems, which can approach the operational patterns of conventional ICE vehicles. This characteristic is particularly valuable in sectors where minimizing downtime is economically critical, such as public transportation fleets, freight corridors, and intensive logistics operations. In addition, many heavy-duty applications operate on fixed or predictable routes, which can facilitate more efficient deployment of hydrogen refuelling infrastructure despite its currently limited availability. Centralized fleet depots, logistics hubs or dedicated transport corridors can therefore partially mitigate one of the principal infrastructure barriers associated with hydrogen deployment. These operational advantages have motivated numerous pilot and commercial projects across multiple global regions, particularly in Europe, China, and South Korea [15]. Consequently, although economic and infrastructural challenges remain significant, heavy-duty transport is widely considered one of the most promising near- to medium-term applications for large-scale FC deployment within electric mobility.

8.6.3 Other applications

Beyond road transportation, hydrogen-based technologies have increasingly been explored across other electric mobility sectors, although with varying levels of technological maturity and deployment. In these applications, hydrogen-powered vehicles are primarily motivated by constraints related to mass, range, or insufficient charging infrastructure, where conventional battery-electric solutions may face substantial operational limitations.

Among these sectors, rail transport has arguably achieved one of the highest levels of practical hydrogen implementation. Projects such as the Coradia iLint and the FCH2RAIL initiative in Spain [16] have demonstrated the feasibility of FC-powered rail systems, particularly for regional or secondary rail lines where full catenary electrification may be economically impractical or technically challenging. In such contexts, hydrogen offers a zero direct-emission alternative capable of replacing diesel traction without requiring continuous electrified infrastructure. Similar concepts have also been explored in light rail systems, where reducing visual impact or minimizing infrastructure-related safety constraints may provide additional incentives.

In the maritime sector, hydrogen is considered a potential pathway for decarbonizing applications such as ferries, inland waterway transport and certain coastal operations. Although volumetric storage constraints, onboard safety considerations and long-range autonomy challenges remain substantial, hydrogen may be particularly suitable for vessels operating on predictable routes or from ports with centralized infrastructure. In addition to direct FC applications, hydrogen-derived fuels are also being investigated as broader maritime decarbonization options.

In aviation, FC-based propulsion currently faces major implementation barriers due to stringent requirements associated with volumetric energy density, certification and safety. Nevertheless, hydrogen technologies are actively being studied for UAVs, short-range aircraft and experimental regional aviation concepts. In these applications, the high gravimetric energy density of hydrogen may offer advantages in weight-sensitive scenarios, although storage complexity remains a major technological constraint.

8.7 Advantages and limitations

8.7.1 Advantages

As an energy vector, hydrogen offers several strategic advantages that have motivated its growing development and research within specific electric mobility applications. Among its most significant benefits are its high gravimetric energy density and rapid refuelling capability. These characteristics can provide competitive advantages over alternative energy storage systems, such as batteries, particularly in mobility sectors including heavy-duty road transport and certain rail applications.

High gravimetric specific energy is arguably one of the principal advantages of hydrogen in electric mobility. At a theoretical level, hydrogen possesses one of the highest energy-to-mass ratios among fuels commonly considered for transportation. This means that, for a given target driving range, a hydrogen-based storage system may require substantially less mass than an equivalent battery-based system. Such a property is especially valuable in applications where vehicle autonomy is critical and system mass strongly influences operational performance. Long-haul heavy-duty transport, rail systems and selected aviation applications can therefore particularly benefit from this characteristic, as excessive energy storage mass may otherwise impose significant technical or economic penalties.

A second major advantage is the high speed of hydrogen refuelling. Under appropriate infrastructure and operational conditions, conventional compressed hydrogen refuelling systems can complete full refuelling processes within timescales comparable to those of conventional ICE vehicles (see Section 8.4), significantly outperforming many conventional battery charging solutions in this regard. This reduced downtime can be especially advantageous in applications where vehicle availability is a critical operational parameter, such as mass public transportation, freight logistics or commercial fleet operations.

Finally, FC-based vehicles retain several important user-level advantages associated with electric mobility more broadly. These include quiet operation, zero direct CO_2 emissions at the point of use and high local conversion efficiency relative to combustion-based propulsion systems. As a result, hydrogen-powered FC vehicles can combine certain operational conveniences traditionally associated with liquid or gaseous fuels, such as rapid refuelling and long range, with environmental and functional benefits typically linked to electrified transportation.

8.7.2 Limitations

Despite its strategic advantages, hydrogen also presents major technical, economic and infrastructural limitations when considered as an energy vector for electric mobility. Among the most significant constraints specific to mobility applications are its low volumetric energy density, the requirement for specialized refuelling infrastructure, and its comparatively low overall system efficiency. Collectively, these limitations currently represent major barriers to the widespread deployment of FC-based mobility solutions across many transport segments when compared with alternative electrification pathways such as batteries.

First, under ambient conditions, hydrogen exhibits very low volumetric energy density. Although hydrogen offers exceptional gravimetric energy content, storing meaningful quantities of usable energy requires either very large storage volumes or the implementation of compression, cryogenic liquefaction or advanced storage materials (see Section 8.3). Each of these approaches introduces additional penalties related to system cost, engineering complexity, safety requirements, and energy consumption. High-pressure tanks increase structural and material demands, liquefaction requires substantial energy input, and alternative storage media often remain technologically immature for broad mobility deployment. Consequently, hydrogen storage remains one of the principal technical bottlenecks limiting practical implementation across many electric mobility applications.

A second major limitation concerns infrastructure. Unlike battery systems, which can partially leverage already widespread electrical grid networks, FC vehicles depend on dedicated hydrogen value chains involving production, purification, compression, transportation, storage and specialized hydrogen refuelling stations. Developing such infrastructure often requires substantial capital investment, geographic coordination and industrial standardization. This helps explain why HRS deployment remains limited in most global regions (see Section 8.8.1), thereby constraining accessibility and representing a substantial barrier to broad FC vehicle adoption.

Finally, the global efficiency of hydrogen-based propulsion systems remains a fundamental challenge. The complete electricity $\rightarrow H_2 \rightarrow$ electricity chain involves cumulative energy losses during hydrogen production (particularly electrolysis), compression or liquefaction, transport, storage and final electricity generation through FCs. As a result, the overall well-to-wheel efficiency of FC vehicles is generally significantly lower than that of direct battery-electric pathways. This reduced systemic efficiency has broad implications for energy demand, infrastructure utilization and total operational sustainability.

Together, these limitations explain why hydrogen mobility, despite its promising characteristics, remains primarily attractive only in specific applications where its operational benefits can outweigh its systemic disadvantages.

8.7.3 Comparison with battery systems

The comparison between hydrogen-based mobility systems and battery-based systems should not be approached primarily from the perspective of technological replacement, but rather from one of technological complementarity. Indeed, both technologies present distinct strengths that make them competitive within specific mobility niches, depending largely on factors such as application type, operational profile and associated technical constraints.

In this context, battery systems offer high overall efficiency, architectural simplicity, and direct integration with existing electrical grids. These characteristics make battery-electric systems particularly well suited for light-duty vehicles, urban mobility, and applications where charging infrastructure is widely accessible. By contrast, FC-based propulsion may be more appropriate in applications requiring extended range, rapid refuelling capability or reduced system mass in scenarios where volumetric constraints are manageable. For these reasons, hydrogen is increasingly being explored in sectors such as rail transport and heavy-duty long-distance road mobility.

Accordingly, while batteries currently dominate direct electrification across many transport segments, hydrogen may represent a medium- to long-term alternative for applications where battery limitations become operationally restrictive. Thus, rather than converging toward a single universal energy solution, the future evolution of electric mobility will likely benefit from the coexistence of multiple complementary technologies, each optimized for specific operational demands, infrastructure realities and economic conditions.

8.8 Industrial and infrastructure context

8.8.1 Infrastructure deployment

The deployment of an adequate network of HRSs is a critical prerequisite for the widespread adoption of FC vehicles, as it directly determines the practical accessibility, territorial coverage, and operational viability of this mobility solution. Hydrogen refuelling infrastructure faces substantial technical and economic barriers due to the highly specialized nature of the required facilities, which must integrate hydrogen production or supply, compression systems, intermediate storage, and dispensing technologies under stringent pressure and safety conditions. As a result, the expansion of hydrogen refuelling infrastructure requires coordinated interaction among public policy frameworks, industrial planning, and regional demand development.

Fig. 8.7 illustrates the deployment of hydrogen refuelling infrastructure for FC vehicles across different world regions according to data from [17]. A first notable observation is that the number of HRSs remains significantly lower than the number of charging points available for BEVs (see Chapter 6), while also exhibiting a far greater degree of geographical concentration. Nevertheless, infrastructure growth has remained continuous since 2020, with the majority of stations concentrated in three major Asian markets: China, Japan, and Korea. Among these, China has emerged as the primary driver of infrastructure expansion, showing the most pronounced growth during the analyzed period. Japan and Korea, although maintaining important leadership positions, have exhibited comparatively more moderate expansion rates. Europe has also demonstrated steady infrastructure growth, largely associated with strategic corridor development and decarbonization policies, whereas the United States remains comparatively underdeveloped in this area.

In this context, hydrogen refuelling infrastructure deployment is not determined solely by technological maturity, but is strongly influenced by public policy priorities, industrial strategies, and national-level economic planning. A persistent "chicken-and-egg" problem continues to constrain expansion: insufficient HRS coverage discourages FC vehicle adoption, while limited vehicle penetration reduces the economic incentive for large-scale infrastructure investment. This dynamic creates a self-reinforcing cycle that significantly restricts broader deployment. Consequently,

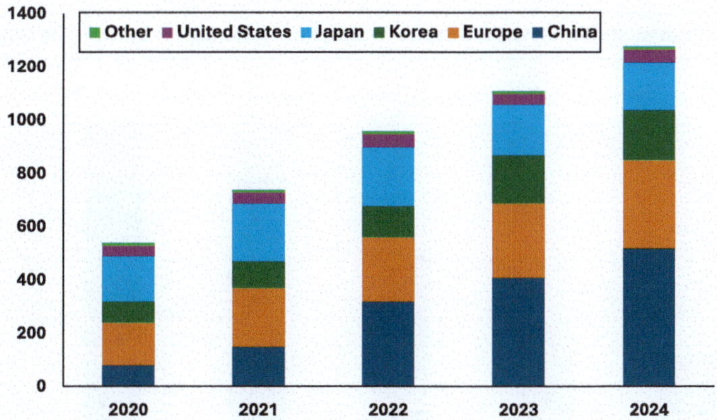

Figure 8.7 Global deployment of hydrogen refuelling stations by region between 2020 and 2024.

although the number of HRSs has increased steadily in recent years, infrastructure expansion has not yet reached the scale required to support substantial global growth in FC vehicle adoption.

8.8.2 Supply chain considerations

The viability of hydrogen as an energy vector for electric mobility depends not only on vehicle performance or HRS deployment, but also on the robustness and efficiency of its broader supply chain. Hydrogen is characterized by a highly complex and specialized value chain integrating production, conditioning, transport, and distribution. Many of these processes are themselves energy-intensive, adding significant cost, technical complexity, and infrastructure requirements to the overall system. Consequently, both the economic competitiveness and environmental sustainability of hydrogen are strongly determined by supply chain performance.

Hydrogen production represents the first critical stage of this value chain. At present, most global hydrogen production still relies on SMR, a pathway associated with substantial CO_2 emissions unless complemented by CCUS technologies (see Section 8.2.1). This dependence significantly compromises the environmental benefits of hydrogen as a clean mobility vector. In contrast, electrolysis powered by renewable electricity currently represents the most mature pathway for near-zero-emission hydrogen production. However, current electrolyzer technologies still face important scalability limitations linked to the availability of competitively priced renewable electricity, electrolyzer capital cost, and integration challenges within large-scale energy systems (see Section 8.2.2).

Although hydrogen production offers the strategic advantage of geographical decoupling between generation and final consumption, this flexibility requires an adequate transport and distribution network. Long-distance hydrogen transport can currently be achieved through dedicated pipelines, high-pressure trucks, cryogenic

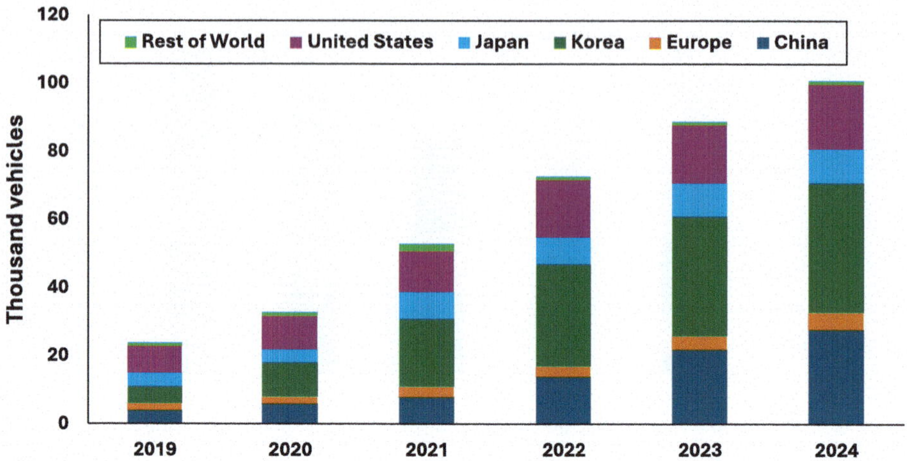

Figure 8.8 Global FC vehicle stock by region between 2019 and 2024.

transport, or chemical carriers such as LOHCs or ammonia. Each option presents distinct trade-offs in terms of cost, efficiency, safety, and technological maturity. Furthermore, hydrogen frequently requires additional conditioning prior to transport, whether through compression or liquefaction, introducing further energy penalties, technical complexity, and safety requirements (see Section 8.3).

Overall, the hydrogen value chain comprises a series of deeply interconnected processes requiring highly specialized infrastructure and coordinated industrial systems. Therefore, the future viability of hydrogen as an electric mobility energy vector will depend largely on the capacity to optimize this chain holistically, reducing systemic inefficiencies while improving economic and environmental performance across production, transport, and final use.

8.8.3 Market trends

Market trends for FC vehicles clearly indicate that this mobility segment remains in an early stage of development. Fig. 8.8 presents the global stock of FC vehicles according to data from [17]. As illustrated, the number of deployed FC vehicles remains significantly lower than that of other electric mobility alternatives. Market penetration is strongly shaped not only by technological maturity, but also by public policy frameworks, industrial strategies, and the availability of supporting infrastructure.

Fig. 8.8 shows a sustained increase in global FC vehicle stock over the analyzed period, although this growth remains substantially more modest than that observed for dominant electric mobility pathways, while also displaying strong geographical concentration. Korea remains the global leader in FC vehicle penetration, reflecting a long-term national strategy aimed at positioning hydrogen as a strategic industrial sector. China, however, represents the fastest-growing market, rapidly expanding its FC vehicle fleet as part of a broader industrial strategy focused particularly on heavy-duty transport, commercial applications, and domestic manufacturing leadership. Japan and the United States continue to occupy relevant positions, although with comparatively more moderate penetration levels, while Europe maintains a relatively limited market presence.

Future market development will largely depend on reductions in the cost of green hydrogen production, continued expansion of transport and refuelling infrastructure, and further technological progress in FC systems. Improvements in these interconnected domains will determine whether FC vehicles can expand beyond current niche applications. At present, hydrogen is expected to maintain a predominantly sector-specific role within electric mobility, focused on applications where prevailing technologies face substantial operational constraints, particularly in sectors requiring rapid refuelling, extended range, or high payload capacity. Consequently, rather than emerging as a universal mobility solution, hydrogen is more likely to develop as a complementary energy vector serving strategically selected transport segments.

References

[1] International Energy Agency, Global hydrogen review 2025, Technical report, International Energy Agency (IEA), Paris, France, published 12 September 2025. Licence: CC BY 4.0, https://www.iea.org/reports/global-hydrogen-review-2025, Sep. 2025.

[2] R. Aguado, M. Tostado-Véliz, U. Desideri, F. Jurado, Optimizing proton exchange membrane electrolyzer performance through dynamic pressure and temperature control: a mixed-integer linear programming approach, Energy Conversion and Management 345 (2025) 120402, https://doi.org/10.1016/j.enconman.2025.120402.

[3] M. Tostado-Véliz, A. Rezaee Jordehi, S.A. Mansouri, Y. Zhou, F. Jurado, A local electricity-hydrogen market model for industrial parks, Applied Energy 360 (2024) 122760, https://doi.org/10.1016/j.apenergy.2024.122760.

[4] A. M, P.K. G, Hydrogen towards sustainable transition: a review of production, economic, environmental impact and scaling factors, Results in Engineering 20 (2023) 101456, https://doi.org/10.1016/j.rineng.2023.101456.

[5] S. Shiva Kumar, V. Himabindu, Hydrogen production by pem water electrolysis – a review, Materials Science for Energy Technologies 2 (3) (2019) 442–454, https://doi.org/10.1016/j.mset.2019.03.002.

[6] A. Israr, M. Saadat, A. Israr, A. Rehman, M. Junaid, Green hydrogen energy, current trends, challenges and future developments, Sustainable Energy Technologies and Assessments 86 (2026) 104886, https://doi.org/10.1016/j.seta.2026.104886.

[7] B. Han, M.S. Ng, A. Chakraborty, Advancements in metal-organic frameworks (mofs) for high-capacity hydrogen storage: mechanisms, optimization strategies, and future perspectives, International Journal of Hydrogen Energy 220 (2026) 154063, https://doi.org/10.1016/j.ijhydene.2026.154063.

[8] SAE International, Fueling Protocols for Light Duty Gaseous Hydrogen Surface Vehicles, SAE Standard J2601_202005, SAE International, Warrendale, PA, USA, May 2020, https://doi.org/10.4271/J2601_202005, revised May 29, 2020, originally issued March 2010.

[9] G. Offer, D. Howey, M. Contestabile, R. Clague, N. Brandon, Comparative analysis of battery electric, hydrogen fuel cell and hybrid vehicles in a future sustainable road

transport system, Energy Policy 38 (1) (2010) 24–29, https://doi.org/10.1016/j.enpol.2009.08.040.

[10] M. Tostado-Véliz, A.A. Ghadimi, M.R. Miveh, M. Bayat, F. Jurado, Uncertainty-aware energy management strategies for pv-assisted refuelling stations with onsite hydrogen generation, Journal of Cleaner Production 365 (2022) 132869, https://doi.org/10.1016/j.jclepro.2022.132869.

[11] A. Kumar, R.K. Gupta, Fuel Cells Technology and Electrode Materials for a Sustainable Future, 1st edition, Elsevier, 2024, eBook ISBN: 9780443240393.

[12] E.F. Hidayat Juliandri, S.M. Zain, A.R. Noviyanti, Advances in proton exchange membrane fuel cell (pemfc) materials: a review of developments from 2021 to 2025, Journal of Power Sources 657 (2025) 238124, https://doi.org/10.1016/j.jpowsour.2025.238124.

[13] Y. Wang, K.S. Chen, J. Mishler, S.C. Cho, X.C. Adroher, A review of polymer electrolyte membrane fuel cells: technology, applications, and needs on fundamental research, Applied Energy 88 (4) (2011) 981–1007, https://doi.org/10.1016/j.apenergy.2010.09.030.

[14] N. Xie, J. He, W. Wei, J. Ba, T. Yang, Neural network-driven genetic algorithm for advanced fuel cell performance enhancement, Results in Engineering 29 (2026) 108426, https://doi.org/10.1016/j.rineng.2025.108426.

[15] Wikipedia contributors, Fuel cell bus — Wikipedia, the free encyclopedia, https://en.wikipedia.org/wiki/Fuel_cell_bus, 2026. (Accessed 28 April 2026).

[16] Adif, El tren demostrador del proyecto fch2rail, primer tren de hidrógeno que circula en pruebas en la red ferroviaria española, https://www.adif.es/w/circulacion-primer-tren-hidrogeno-linea-zaragoza-canfranc, Jun. 2023. (Accessed 28 April 2026).

[17] I.E. Agency, Global ev outlook 2025, Tech. Rep., IEA, Paris, 2023, https://www.iea.org/reports/global-ev-outlook-2025. (Accessed 27 March 2026).

Other storage technologies

9

9.1 Introduction to alternative energy storage in electric mobility

9.1.1 Positioning beyond batteries and hydrogen

In the electric mobility sector, both batteries (see Chapter 7) and hydrogen (see Chapter 8) clearly constitute the most widely deployed commercial energy storage solutions. However, neither technology represents a universal answer to all operational profiles within electric transportation.

Batteries, particularly lithium-ion systems, are characterized by high industrial maturity, relatively high efficiency, and straightforward integration into light-duty vehicles. Nevertheless, this technology faces important limitations associated with specific energy, both gravimetric and volumetric, comparatively long recharge times relative to conventional fuels, performance degradation under intensive cycling conditions, and significant dependence on critical raw materials. These constraints directly affect battery suitability in applications requiring extremely rapid power delivery, ultra-fast charging, or exceptional long-term durability under aggressive operational conditions.

Hydrogen storage, by contrast, offers major advantages in terms of potential driving range and high gravimetric energy density, making it particularly attractive for applications in which mass-to-range ratio is a determining factor, such as long-haul heavy-duty transportation. However, hydrogen systems also face substantial technical, economic, and safety challenges, especially those associated with compression or cryogenic liquefaction, low volumetric energy density, and dependence on complex, capital-intensive, and highly specialized infrastructure for production, distribution, and refuelling.

These limitations justify the continued exploration of alternative energy storage technologies. In this context, a fundamental distinction must be established between energy-oriented systems, whose primary objective is maximizing autonomy, and power-oriented systems, specifically designed to absorb or deliver large quantities of energy over very short time intervals. Depending on the intended operational role, different technological characteristics become desirable. Consequently, the storage technologies examined in this chapter are not primarily intended to replace batteries or hydrogen in electric mobility, but rather to complement them in applications where their specific performance attributes, particularly high power density, rapid transient response, or extreme cycle durability, provide strategic engineering advantages.

9.1.2 Classification of non-conventional storage technologies

The storage technologies reviewed in this chapter are classified according to the physical principle through which they store and release energy. Unlike conventional

Electric Mobility Systems. https://doi.org/10.1016/B978-0-44-349397-3.00017-3

storage systems such as batteries (electrochemical storage) and hydrogen (chemical energy storage and conversion), the systems analyzed here rely on distinct operating mechanisms. Broadly, they can be classified as follows:

- **Electrostatic storage:** Devices based on electrostatic storage accumulate energy through electrostatic charge separation. As a result, these systems typically provide extremely rapid charge and discharge times, very high power density, and exceptionally long cycle life. SCs are the most representative technology within this category, although hybrid capacitor architectures that partially combine electrostatic and electrochemical mechanisms also exist.
- **Mechanical storage:** These systems store energy in the form of mechanical energy. In electric mobility, FESSs are the most relevant example, storing energy as rotational kinetic energy through high-speed rotating masses. Their characteristics make them particularly suitable for transient high-power applications.
- **Emerging systems:** Additional storage technologies based on alternative physical principles remain less commercially consolidated in electric mobility. These include SMES, which stores energy within magnetic fields generated by superconducting coils, and structural storage materials, which seek to integrate storage capability directly into vehicle structural components.

Furthermore, it is important to distinguish between direct electrical discharge systems, such as SCs, and systems requiring intermediate mechanical or electromagnetic conversion, such as FESS. Storage systems may also be deployed either onboard vehicles or within stationary infrastructure. Finally, they can be classified according to their principal operational objective, including high-power delivery, rapid transient response, or broader energy stabilization applications.

9.1.3 Scope and relevance for electric mobility

Although the storage technologies classified in Section 9.1.2 receive substantially less industrial and commercial attention within the electric mobility sector, their importance remains significant in specific applications, particularly those requiring extremely high instantaneous power, rapid response, or exceptional cycle durability. In this context, alternative energy storage systems have found their most relevant applications primarily in urban and rail transportation, where operational profiles are frequently characterized by repeated acceleration, braking, and highly cyclical duty patterns.

Under these conditions, high-power storage systems such as SCs and FESSs have demonstrated particular relevance in rail systems, tramways, and electric buses. In these applications, such storage technologies are especially valuable for absorbing the large power peaks generated during regenerative braking, where the comparatively limited specific power of batteries and hydrogen systems could impose critical size, mass, or efficiency constraints, potentially leading to substantial recoverable energy losses. Similarly, these technologies can also enhance acceleration performance, where power demand is often highly transient, cyclical, and substantially elevated.

Beyond onboard vehicle applications, alternative storage systems also provide important functionality in stationary installations, including wayside configurations

[1]. In such cases, their role extends beyond individual vehicle support, enabling broader integration into larger energy ecosystems, where they can contribute to grid support, energy buffering, and infrastructure-level efficiency optimization.

9.2 Fundamental energy storage metrics and functional suitability

9.2.1 Specific energy, specific power, efficiency, and DoD

In order to evaluate and compare different energy storage technologies, a set of technical metrics must be considered which, although already introduced in previous chapters, acquire distinct implications when applied to the storage systems analyzed in this chapter due to their predominantly power-oriented operational nature. Thus, while batteries and hydrogen systems are typically assessed primarily through specific energy, whether gravimetric (energy-to-mass ratio) or volumetric (energy-to-volume ratio), alternative storage systems such as SCs and FESSs are more frequently characterized by their specific power, both gravimetric and volumetric.

Within this framework, specific energy remains an important parameter, but its role is largely comparative when benchmarked against conventional technologies such as batteries or hydrogen. In practical terms, for SCs and FESSs, limited specific energy generally constitutes a design constraint rather than a competitive advantage. These technologies typically exhibit substantially lower total storage capacities than conventional electrochemical or hydrogen-based systems. Consequently, their primary function is not to maximize onboard energy reserves or vehicle driving range, but rather to provide rapid energy absorption and delivery during short operational intervals.

Round-trip efficiency also remains a relevant metric, although the dominant loss mechanisms differ substantially from those of batteries. In SCs, efficiency is primarily influenced by resistive losses, self-discharge phenomena, and voltage balancing limitations, particularly in multi-cell architectures. In contrast, FESS efficiency is mainly governed by mechanical losses, residual friction, motor-generator conversion efficiency, and the effectiveness of vacuum enclosures or magnetic bearing systems. As a result, although efficiency remains important, it is generally less decisive than in long-duration storage applications where prolonged energy retention is critical.

Regarding DoD, this parameter is strongly associated with operational capacity and lifespan in batteries, where deep cycling can significantly accelerate degradation. However, in many alternative storage systems, DoD often exerts a more indirect influence, with operational limitations more closely linked to thermal stability, self-discharge, mechanical fatigue, or rotational safety margins.

9.2.2 Response time, safety, and operational constraints

Due to their power-oriented operational nature, response time constitutes one of the most distinctive characteristics of alternative energy storage systems. In this context, response time can be defined as the speed at which a storage system can capture, store, or release energy in response to variations in power demand. Accordingly, technologies such as SCs and SMES exhibit near-instantaneous response times, typically in the millisecond range, making them particularly suitable for transient power support, rapid energy recovery, and high-frequency load fluctuations. In the case of FESSs, although they are also characterized by very rapid response capabilities, this parameter is partially constrained by motor-generator dynamics as well as by factors associated with rotor acceleration and deceleration.

From a safety perspective, SCs introduce specific operational constraints primarily related to voltage management, module balancing, and overvoltage prevention, while generally maintaining high thermal stability due to the absence of significant bulk chemical transformations. In contrast, safety considerations in FESSs are closely linked to structural integrity, since high rotational speeds require advanced rotor materials and robust containment systems capable of preventing catastrophic mechanical failures. Meanwhile, certain supermagnetic systems introduce additional technical challenges associated with cryogenic operation, electromagnetic insulation, and magnetic field stability.

Other relevant operational constraints include self-discharge, environmental sensitivity, packaging requirements, and structural integration complexity. SCs typically present relatively high self-discharge coefficients, which limit their suitability for long-term energy storage, whereas FESSs may experience standby losses associated with residual friction, bearing systems, or vacuum maintenance. Furthermore, external factors such as temperature, vibration, mass, and volumetric constraints directly influence implementation feasibility in electric mobility applications.

9.2.3 Functional suitability framework for alternative storage systems

The suitability of an energy storage system for a specific application is determined by the advantages that such a system provides under the common operational profiles of that application. In this context, the alternative storage systems examined in this chapter offer competitive advantages primarily in three major areas: response time, power absorption capability, and the ability to operate efficiently under intensive cyclic profiles. These characteristics largely define the functional suitability of such storage technologies for specific electric mobility applications.

More specifically, SCs are particularly well adapted to applications dominated by short-duration transient events due to their excellent response under high-frequency cyclic operating conditions. Consequently, SCs have been widely employed in regenerative braking, instantaneous high-power pulse delivery, and peak shaving applications. By contrast, FESSs are also suitable for such applications, although their response time is generally somewhat lower than that of SCs due to motor-generator and rotor dynamic constraints. However, mechanical storage often offers superior performance for longer-duration energy retention because of lower self-discharge characteristics. These features make FESSs especially suitable for improving the efficiency of dynamic processes, particularly in repetitive energy recovery environments or load stabilization applications.

It is also important to evaluate the adaptability of each system to different deployment architectures. SCs can generally be integrated effectively in both onboard and wayside applications, whereas FESSs may present greater restrictions for onboard implementation, particularly in large-scale configurations, due to packaging, mass, and structural integration challenges. Finally, although SMES exhibits significant theoretical potential, its application in electric mobility remains at a very early and predominantly experimental stage of development.

9.3 SCs: operating principles and core technologies

9.3.1 SC fundamentals

An SC is an energy storage system based on electrostatic principles, storing energy in the form of an electric field through the physical accumulation and separation of electrical charges across conductive interfaces. Unlike electrochemical storage systems, its operation primarily relies on charge distribution at material interfaces rather than bulk chemical transformations. This predominantly physical storage mechanism enables extremely rapid charge-discharge rates, very high power density, and exceptionally long cycle life.

Structurally, an SC consists of two porous electrodes separated by an ionic electrolyte that permits ion transport while preventing direct electron flow between electrodes. Under idle conditions, with no stored energy, charges remain distributed without net polarization, and the system stays in electrical equilibrium, as illustrated in Fig. 9.1a. When the SC is connected to an external voltage source V, as shown in Fig. 9.1b, ions within the electrolyte migrate toward oppositely charged electrodes, accumulating on their surfaces and forming electric double-layers at each electrode-electrolyte interface. This molecular-scale physical charge separation creates an electrostatic field capable of storing energy. Once charged and disconnected from the source, as represented in Fig. 9.1c, the SC maintains a voltage difference equal to V across its electrodes, allowing stored energy to be later delivered to an external load.

The capacitance of an SC is fundamentally defined by the relationship between stored charge and applied voltage:

$$\mathcal{C} = \frac{q}{V} \tag{9.1}$$

where $\mathcal{C}$ is capacitance, q is electric charge, and V is voltage. From Eq. (9.1), the total stored energy can be expressed as:

$$E = \frac{1}{2}\mathcal{C}V^2 \tag{9.2}$$

where E is stored energy. Eq. (9.2) highlights the strong dependence of stored energy on operating voltage. However, during discharge, voltage decreases progressively, often requiring power electronic interfaces for output conditioning.

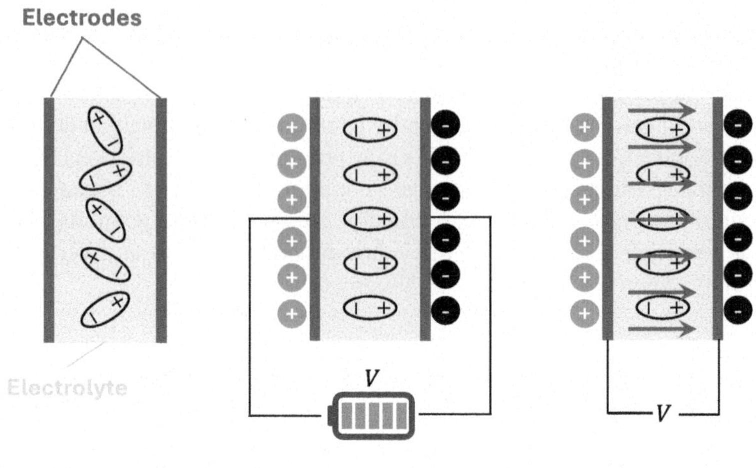

Figure 9.1 Schematic operating principle of a SC: (a) idle state, where ions in the electrolyte remain randomly distributed between porous electrodes; (b) charging process, during which an applied voltage drives ion migration toward oppositely polarized electrodes, forming electric double layers at the electrode–electrolyte interfaces; and (c) charged state, where electrostatic charge separation is established and energy is stored within the interfacial electric field.

9.3.2 EDLCs, pseudocapacitor,s and hybrid SCs

Commercial SCs do not constitute a homogeneous technological family, since multiple architectures can be distinguished according to their dominant storage mechanism, the materials employed and the internal configuration of the device. These technological differences directly determine their specific power capability, energy storage potential, operational stability, lifecycle durability, and industrial scalability. From a practical perspective, commercially relevant SCs can be broadly classified into three principal technological groups:

- **Electric Double-Layer Capacitor (EDLC):** this category represents the most mature and widely deployed SC technology. EDLCs primarily rely on electrostatic charge accumulation at the electrode-electrolyte interface, meaning their performance is strongly linked to maximizing accessible surface area. Consequently, porous electrode materials are central to their design, with activated carbon remaining the dominant industrial solution due to its relatively low cost, material abundance and stable long-term behavior. Beyond activated carbon, advanced carbonaceous materials such as carbon nanotubes, carbon aerogels, and graphene have attracted major research interest because of their potential to increase effective surface area, electrical conductivity, and overall device performance [2]. These materials may also improve pore accessibility and structural optimization, thereby enhancing charge storage efficiency. However, despite these advantages, their large-scale commercial deployment remains constrained by elevated production costs, manufacturing complexity, and scalability limitations.

- **Pseudocapacitors:** unlike EDLCs, pseudocapacitors aim to increase available capacitance through the incorporation of advanced functional materials capable of storing greater charge densities. In this context, certain metal oxides and conductive polymers have demonstrated significant potential for enhancing capacitance beyond purely electrostatic systems. This characteristic can provide strategic improvements in energy-related performance metrics, particularly where greater storage density is desirable without fully sacrificing rapid charge-discharge capability. Nevertheless, although pseudocapacitors may offer enhanced storage capability, they frequently exhibit disadvantages associated with reduced long-term stability, progressive material degradation, and higher manufacturing costs. These constraints have limited their widespread commercial adoption, particularly in applications requiring extreme durability, operational simplicity, and predictable lifecycle behavior.
- **Hybrid SCs:** hybrid SCs are designed to improve energy storage capability while preserving, as much as possible, the intrinsic high-power advantages of conventional SC architectures. To achieve this, they combine distinct storage approaches, often through asymmetric electrode configurations or mixed-material systems. This balanced-performance strategy is particularly attractive in electric mobility applications, where rapid power delivery, repetitive cycling, and improved usable energy capacity may all be simultaneously desirable. By partially overcoming the low specific energy limitations of traditional EDLCs, hybrid SCs represent an important technological pathway toward broader functional versatility, although this often introduces additional complexity in manufacturing, system balancing, and integration requirements.

9.3.3 Cell architecture, modules, and power electronics integration

SCs exhibit a modular architecture in which the cell represents the fundamental functional unit. A single SC cell consists of two electrodes, an electrolyte, and a separator. The electrodes are typically manufactured from activated carbon, carbon nanotubes, or graphene, and are engineered with extremely high specific surface area in order to maximize electric double-layer formation and thereby increase capacitance. The separator serves as a physical barrier that prevents electrical short circuits between electrodes while simultaneously allowing ionic transport, which is essential for charge storage and discharge processes.

The electrolyte is a critical component because it largely determines the operational voltage window of the cell. In general, aqueous electrolytes permit operating voltages of approximately 1 V, whereas organic electrolytes can extend this limit to nearly 3 V. Despite this improvement, such voltage levels remain insufficient for most electric mobility applications, which require substantially higher system voltages. Consequently, multiple cells are typically connected in series to increase total module voltage and form practical energy storage units. However, series stacking introduces voltage imbalances among cells due to manufacturing tolerances, leakage current variations, and differential aging. For this reason, passive or active balancing circuits are essential to prevent cell overvoltage, premature degradation, and reductions in module reliability.

Voltage output also varies significantly with SoC according to Eq. (9.2), making advanced control systems necessary to dynamically adapt module voltage to the requirements of traction systems or regenerative braking. These functions are commonly performed through bidirectional DC/DC converters, which regulate voltage and power flow while enabling applications such as regenerative braking, peak shaving, and power smoothing.

Although SCs generally demonstrate strong thermal stability relative to batteries, resistive heating during high-power cycling can negatively affect lifespan and efficiency if not properly managed. Therefore, effective thermal management strategies, optimized packaging, ventilation systems, and dedicated cooling solutions remain essential for reliable large-scale deployment.

9.3.4 Performance characteristics: strengths and weaknesses

SCs are particularly distinguished by their exceptional power performance. These devices exhibit extraordinary power absorption and delivery capabilities, making them highly suitable for applications involving intense transient power demands. Their response time is also exceptionally fast, typically in the millisecond range, which places them well beyond the transient response capabilities of conventional batteries. Such characteristics make SCs especially effective for applications including regenerative braking, peak shaving, and acceleration support, where rapid charge acceptance and discharge are essential.

Another major strength of SCs is their outstanding cycle durability. SCs can typically operate through hundreds of thousands to even millions of charge-discharge cycles while experiencing relatively limited degradation. As a result, they offer one of the most competitive cost-per-cycle profiles among energy storage technologies used in electric mobility. In addition, SCs generally provide high charge-discharge efficiency, low internal resistance, and strong thermal stability across broad operating conditions. These properties make them particularly robust for intensive cyclic applications such as urban transportation, rail systems, and other environments characterized by frequent power fluctuations.

However, SCs are also subject to significant limitations that constrain their broader deployment. The most notable is their low specific energy, which makes them poorly suited for prolonged energy storage or sustained propulsion applications. For this reason, SCs are rarely employed as primary energy storage systems in electric mobility platforms. Their limited energy density also results in relatively poor cost competitiveness on a per kilowatt-hour basis when compared with batteries or hydrogen-based systems.

Another important drawback is their relatively high self-discharge rate, which can lead to substantial charge loss during idle periods and reduce suitability for applications with long inactivity intervals. Furthermore, voltage varies significantly with SoC according to Eq. (9.2), as terminal voltage decreases nearly linearly during discharge, thereby requiring voltage stabilization and power conditioning systems.

9.3.5 Manufacturing trends and cost evolution

The mobility sector has served as one of the primary catalysts for the SC manufacturing industry during recent decades, largely driven by increasing demand for high-power applications in electric mobility. In this context, the market is dominated by specialized manufacturers producing technologies ranging from individual cells to fully integrated modules, with particular relevance among suppliers focused on urban transportation, rail systems, and hybrid power solutions. Although SC deployment in light-duty passenger vehicles remains relatively limited, their implementation in trams, electric buses, and regenerative energy recovery systems continues to expand [3].

The manufacturing cost of SCs is strongly influenced by the acquisition and processing costs of electrode materials. In most commercial applications, this cost is primarily associated with activated carbon, while more advanced systems may rely on graphene, carbon nanotubes, or selected metal oxides. Among these options, activated carbon maintains a major cost advantage due to its relative abundance, lower material cost, and industrial maturity. Consequently, although advanced nanomaterials may offer improved capacitance, conductivity, or energy density, their broader adoption remains constrained by scalability limitations, manufacturing complexity, and significantly higher production costs.

The carbon supply chain is generally less vulnerable to geopolitical constraints than the critical mineral supply chains associated with battery technologies, such as lithium, nickel, or cobalt (see Chapter 7). This characteristic provides SCs with an important strategic advantage in terms of material accessibility and supply stability. However, SC manufacturing often requires highly controlled production standards regarding porosity optimization, material purity, and precision assembly, all of which can restrict dramatic cost reductions.

From an economic perspective, SCs have shown progressive declines in cost per kilowatt over time [4], reinforcing their competitiveness in high-power applications. Nevertheless, cost per kilowatt-hour remains comparatively unfavorable due to inherently low energy density. Recycling considerations are generally favorable for carbon-based systems relative to many electrochemical technologies, although advanced materials may still present technical and economic recovery challenges.

9.4 SCs in electric mobility applications

9.4.1 Urban transport

In several urban transportation applications, braking events generate substantial amounts of recoverable energy characterized by high transient power peaks over very short time intervals. This operational profile is especially common in relatively heavy transport systems operating under frequent start-stop cycles, such as trams, light rail vehicles, metro systems, and certain electric bus fleets. In these scenarios, vehicle mass combined with repeated deceleration produces large quantities of kinetic energy that would otherwise be dissipated as heat through friction braking. As illustrated in [1], which examines a real tramway system in Cuenca (Ecuador), braking phases can generate power peaks exceeding 500 kW. Capturing and reutilizing this energy efficiently requires storage technologies with exceptional transient power response, which directly explains why regenerative braking constitutes one of the most important mobility applications for SCs.

During braking phases, SCs can absorb large bursts of electrical power almost instantaneously due to their low internal resistance, high specific power, and millisecond-scale response time. Once stored, this energy can be rapidly redeployed during subsequent acceleration phases, effectively reducing net grid energy demand. These repeated short-duration charge-discharge cycles are particularly well aligned with the operational strengths of SCs, whose exceptional cycle durability allows them to perform intensive energy exchanges with limited degradation. Under such conditions, the relatively high self-discharge rate of SCs becomes considerably less problematic because storage intervals are typically brief.

Another important advantage is the possibility of enabling catenary-free or partially autonomous operation along selected route segments where continuous overhead infrastructure may be undesirable. In these cases, SCs can operate independently or in hybrid configurations with batteries to extend functional flexibility (see Section 9.4.2). This strategy can reduce the visual impact, infrastructure cost, and urban integration challenges associated with overhead catenary systems, particularly in historical city centers.

Beyond onboard applications, SCs are also highly effective in wayside installations positioned at stations or strategic points along transport corridors. These systems can capture regenerative energy from multiple vehicles, redistribute it for subsequent vehicle acceleration, support station energy demands, or provide ancillary services such as voltage stabilization and peak demand reduction.

In urban electric buses, SCs are especially advantageous on high-frequency routes with predictable stop intervals, where ultra-fast charging at designated stations can complement or partially replace larger battery systems. Overall, SC implementation in regenerative braking demonstrates that power-oriented storage can significantly enhance urban mobility efficiency beyond purely energy-centric metrics.

9.4.2 Hybrid battery–SC systems

Hybrid energy storage systems aim to combine multiple energy storage technologies within a unified architecture in order to exploit the functional strengths of each while mitigating their individual limitations. Within electric mobility, battery-SC systems represent one of the most technically relevant hybrid configurations. In this arrangement, batteries provide the primary source of sustained energy due to their superior specific energy, while SCs supply high transient power for short-duration events such as rapid acceleration, regenerative braking, and sudden load fluctuations. This functional division allows the overall system to better match diverse mission profiles than either technology could independently.

One of the principal advantages of battery-SC hybridization is the reduction of peak power stress imposed on the battery pack. In this role, SCs operate as high-power buffers, absorbing and delivering short-duration power spikes that would otherwise subject batteries to aggressive charge and discharge rates. By reducing these transient loads, the hybrid architecture can extend battery cycle life, improve round-trip efficiency, and reduce thermal stress. Lower peak currents also mitigate resistive losses and slow degradation mechanisms associated with elevated C-rates, including internal heating and electrode wear. In parallel, SCs contribute to power smoothing by stabilizing fast load variations before they propagate through the drivetrain, thereby improving power quality, protecting sensitive power electronics, and potentially simplifying certain battery pack design constraints.

These advantages are particularly relevant in heavy-duty mobility, buses, delivery fleets, and performance-oriented EV platforms where frequent acceleration-deceleration cycles intensify transient power demands. In passenger EVs, hybridization may also improve regenerative braking capture efficiency by enabling more effective short-term energy absorption than batteries alone, especially under high-power braking events where battery acceptance rates may become restrictive.

However, integrating batteries and SCs into a single platform introduces additional complexity. Effective operation requires advanced energy management strategies capable of dynamically allocating power between both storage systems in real time, often through predictive or rule-based supervisory control [5]. This increases software, sensing, and converter requirements. Furthermore, the addition of SC modules, balancing electronics, and DC/DC converters increases system mass, packaging volume, and cost. These penalties can constrain deployment in space-sensitive or mass-sensitive platforms such as compact passenger vehicles or UAVs. Consequently, battery-SC systems are most attractive where transient power intensity, regenerative frequency, and durability gains clearly outweigh integration complexity.

9.4.3 High-performance automotive and motorsport

SCs have also found strategic applications in specialized mobility segments characterized by aggressive operating profiles with frequent and intense power transients. One of the most relevant examples is high-performance automotive and motorsport (see Fig. 9.2). In these applications, energy demand is dominated by elevated power peaks during acceleration, regenerative braking, and temporary dynamic assistance. Within this context, the role of the SC is not to increase vehicle autonomy, but rather to exploit its superior power response in order to optimize transient efficiency and dynamic performance under extreme operating conditions.

One of the most significant applications of SCs in motorsport appears in Kinetic Energy Recovery System (KERS) architectures. These systems aim to recover the substantial kinetic energy dissipated during intense braking events and temporarily

Figure 9.2 Toyota TS030 Hybrid LMP1 race car employing supercapacitor-based energy recovery for regenerative braking and acceleration boost during endurance competition. (Image courtesy of Wikimedia Commons.)

store it for subsequent acceleration assistance. In this application, the limited transient power absorption capability of many battery systems can constrain braking energy capture, making SCs particularly attractive due to their exceptional charge acceptance, rapid response, and high cycle durability. By enabling more efficient short-term storage of braking energy, SCs can improve launch acceleration, temporary power boost, and broader energy management strategies in race circuits characterized by repeated braking-acceleration sequences.

However, SC integration also introduces limitations related to mass, packaging volume, and increased control system complexity. These factors are particularly critical in motorsport, where vehicle weight, space optimization, and dynamic balance directly influence competitive performance. Additional DC/DC conversion and supervisory electronics may also increase system integration demands.

Consequently, the value of SCs in high-performance automotive applications lies primarily in their complementary role rather than as primary energy storage systems. Their principal contribution is to enhance regenerative energy capture, transient power delivery, and dynamic responsiveness in scenarios where instantaneous power performance is considerably more important than sustained energy capacity.

9.4.4 EV fast charging support

The expansion of EV charging infrastructure is rapidly increasing the deployment of fast and ultra-fast charging stations connected to electrical networks. These charging technologies can impose highly concentrated power demands reaching several hundreds of kilowatts over short time intervals (see Chapter 6). Such demand spikes may create substantial operational challenges for power systems, particularly when multiple vehicles charge simultaneously, potentially generating severe transient loads. In this context, SCs have emerged as a strategic stationary buffering solution due to their exceptional transient power response.

The primary function of SCs in charging-support applications is therefore not sustained energy provision, but peak power management. SCs can absorb energy from the grid during lower-demand periods and rapidly discharge it during charging peaks, thereby reducing the instantaneous power burden imposed on the electrical grid. This peak shaving capability can significantly mitigate stress on critical infrastructure components such as transformers, distribution lines, and substation equipment, potentially improving both reliability and asset longevity.

Another valuable role of SCs in EV charging systems is demand profile smoothing. Beyond managing charging peaks alone, SCs can compensate for short-term power fluctuations associated with intermittent renewable generation sources such as PV or wind systems. This capability is particularly relevant in public charging hubs, logistics centers, or heavy-duty transport corridors, where supply stability is essential and renewable integration may otherwise introduce operational volatility.

The economic implications of SC integration into charging infrastructure can also be significant. By reducing peak loads and stabilizing operational demand profiles, SCs may help defer, minimize, or avoid costly grid reinforcement investments. This includes reducing the need for oversized transformers, upgraded feeder lines, or expanded substation capacity. Furthermore, more stable power profiles can improve overall grid resilience and operational flexibility.

However, SC deployment is generally justified only under aggressive operating profiles characterized by frequent and pronounced transient power peaks. The economic viability of such systems therefore depends strongly on utilization intensity, charging frequency, and local grid constraints. Nevertheless, as high-power charging networks continue expanding, particularly for commercial fleets and heavy transport, SC-based buffering may become an increasingly important enabling technology for efficient and resilient electrified mobility infrastructure [6].

9.5 FESS: principles and architecture

9.5.1 Mechanical energy storage fundamentals

FESSs constitute an energy storage technology based on the accumulation of rotational kinetic energy. In a FESS, the stored energy is governed by the following equation:

$$E = \frac{1}{2}\mathcal{I}\omega^2 \tag{9.3}$$

where $\mathcal{I}$ is the moment of inertia and ω is the angular velocity. It is worth noting that using the International System of Units in (9.3) gives rise energy in Newton-meter dimension and not in its common kilowatt basis.

The fundamental operating principle of a FESS consists of absorbing electrical energy and converting it into rotational mechanical energy by accelerating a

Figure 9.3 Flywheel rotor architecture illustrating peripheral mass concentration, where a substantial portion of the rotor mass is distributed toward the outer radius to maximize the moment of inertia. (Image courtesy of Wikimedia Commons.)

rotor during charging. During discharge, this stored kinetic energy is reconverted into electrical energy through bidirectional motor-generator systems, following a functional principle analogous to regenerative braking (see Chapter 2).

As shown in (9.3), the energy stored in a FESS depends quadratically on angular velocity, making rotational speed the dominant parameter for maximizing specific energy. For this reason, FESS architectures typically prioritize operation at extremely high angular velocities, often involving rotors capable of very high revolutions per minute. Achieving such performance frequently requires advanced materials with high strength-to-weight ratios, precision balancing, and robust containment systems capable of withstanding severe centrifugal forces.

The moment of inertia also depends on rotor mass and, critically, on how that mass is distributed relative to the axis of rotation. Consequently, flywheel systems are commonly designed with a substantial proportion of rotor mass concentrated toward the outer radius, as illustrated in Fig. 9.3. This peripheral mass distribution increases $\mathcal{I}$, thereby enhancing energy storage capacity for a given rotational speed.

However, concentrating mass at larger radial distances also significantly increases internal mechanical stresses, particularly tensile hoop stresses generated during high-speed operation. As a result, structural design becomes a critical engineering constraint in FESS development, requiring careful optimization between energy capacity, material strength, rotor geometry, and operational safety. These factors strongly influence system feasibility, performance, and deployment potential in electric mobility applications.

9.5.2 Rotor materials, bearings, and vacuum enclosures

The structural design of a FESS constitutes one of the most critical engineering determinants of system feasibility, safety, and operational performance, often exerting a greater influence on practical deployment than in many other energy storage technologies. Unlike electrochemical systems, where material chemistry dominates, FESS performance is fundamentally constrained by mechanical architecture. This design depends primarily on three tightly integrated components: the rotor, the support and bearing system, and the operational environment intended to minimize parasitic energy losses. The interaction among these three subsystems directly defines maximum rotational speed, achievable specific energy, cycle durability, safety margins, and long-term reliability.

Regarding the rotor, material selection is among the most decisive variables in overall FESS capability. Historically, flywheels were predominantly manufactured from steel because of its industrial maturity, well-established fabrication methods, relatively low cost, and high structural strength. Steel remains viable in lower-speed or cost-sensitive systems, but its relatively high density introduces substantial centrifugal stresses as angular velocity increases. Since these stresses scale rapidly with rotational speed, steel rotors become progressively less suitable for ultra-high-speed applications, where structural failure risks increase significantly. As a result, contemporary advanced FESS increasingly employ composite materials such as carbon fiber, glass fiber, aramid fibers, or hybrid multilayer structures [7]. These materials offer substantially superior strength-to-weight ratios, enabling much higher rotational velocities while maintaining structural integrity. This directly enhances specific energy through the quadratic dependence on angular speed shown in Eq. (9.3). However, these advantages are accompanied by higher raw material costs, more complex fabrication and balancing procedures, anisotropic mechanical properties, and more difficult failure prediction compared with isotropic metallic rotors.

The bearing system represents the primary mechanical interface between the rotor and its support structure and therefore strongly affects frictional losses, maintenance requirements, and achievable speed. Traditional mechanical bearings remain common due to simplicity and technological maturity, but they generate friction, wear, lubrication demands, and thermal losses that progressively reduce efficiency. For this reason, many advanced FESS architectures incorporate magnetic bearings. These systems use actively or passively controlled electromagnetic fields to levitate and stabilize the rotor without direct physical contact, drastically reducing friction, minimizing wear, and enabling sustained operation at extremely high rotational speeds. This characteristic is especially valuable in high-speed FESS, where even small frictional losses can substantially reduce round-trip efficiency. Nevertheless, magnetic bearings require sophisticated sensors, control electronics, backup stabilization systems, and greater operational complexity.

Aerodynamic drag caused by rotor interaction with surrounding air constitutes a second major source of energy loss. To minimize these losses, many advanced FESS designs operate within vacuum chambers or low-pressure enclosures, substantially reducing air resistance and thermal dissipation. Finally, safety remains a fun-

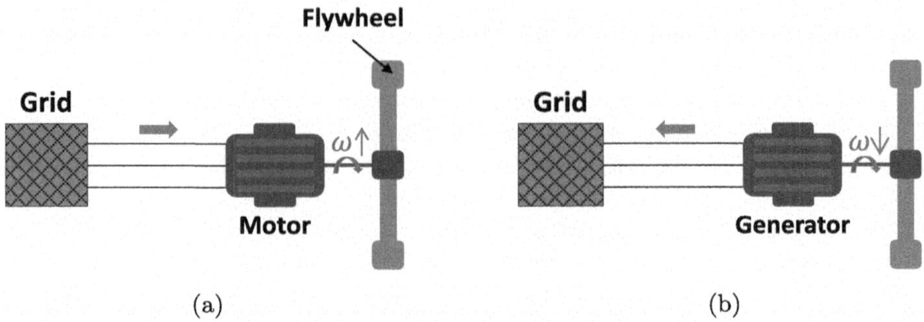

Figure 9.4 Bidirectional energy conversion in a FESS: (a) charging mode, where electrical energy from the grid drives the motor to accelerate the flywheel and store energy as rotational kinetic energy; (b) discharging mode, where the flywheel drives the generator to convert stored mechanical energy back into electrical power for grid supply.

damental design priority. Because catastrophic rotor failure may release extremely large quantities of stored kinetic energy almost instantaneously, FESS units typically employ highly reinforced containment structures capable of absorbing fragmentation events, limiting mechanical rupture, and protecting surrounding infrastructure.

9.5.3 Charging/discharging via motor-generator systems

FESSs employ bidirectional interfaces incorporating an integrated motor-generator assembly capable of transferring power both from the flywheel to external systems and in the reverse direction. The charging and discharging processes are summarized in Fig. 9.4. During charging (Fig. 9.4a), the system operates as a motor, absorbing electrical energy from the grid in order to accelerate the machine mechanically coupled to the flywheel. As a result, the angular velocity ω increases and, according to Eq. (9.3), the amount of stored kinetic energy rises accordingly through rotational acceleration.

During discharge, the process is reversed (Fig. 9.4b). In this operating mode, the interface functions as a generator, enabling the extraction of the kinetic energy accumulated in the flywheel and its transfer to external systems, such as secondary energy storage devices or electrical loads. This transition requires advanced control systems capable of regulating rotational speed, output frequency, and delivered power in order to maintain stable operational conditions throughout the energy conversion process.

The interface between the flywheel itself and the external electrical system frequently incorporates power electronics components, typically including AC/DC or DC/DC converters. These devices are responsible for controlling power flow between the flywheel and the connected electrical infrastructure. Such control enables the motor-generator interface to operate under different voltage and current profiles while optimizing conversion efficiency. In addition, these electronic systems improve synchronization, enhance operational flexibility, ensure voltage regulation, and allow the FESS to adapt effectively to varying grid or load requirements.

9.5.4 Performance advantages and limitations

FESSs offer a distinctive set of operational advantages directly derived from their mechanical storage principle, which fundamentally differentiates them from energy storage systems primarily based on electrochemical or electrostatic processes. Because FESSs do not rely on chemical reactions or significant material transformations during normal operation, they experience substantially lower degradation over repeated charge-discharge cycles. This characteristic provides exceptionally high cycle durability, allowing flywheel systems to achieve operational lifetimes that are generally superior to those of batteries and SCs, particularly in high-frequency cycling environments.

In addition, FESSs exhibit excellent dynamic response, since the conversion between rotational mechanical energy and electrical power can occur extremely rapidly. This fast response capability is especially advantageous in applications requiring high power absorption or delivery over very short time intervals, positioning FESSs in a functional domain partially comparable to SCs. Furthermore, flywheel systems demonstrate strong thermal resilience, as they are generally less vulnerable to extreme temperature conditions and inherently avoid failure mechanisms such as thermal runaway, which are commonly associated with certain electrochemical systems (see Chapter 7).

Despite these advantages, FESSs also face important technical limitations that constrain broader deployment. One of the principal disadvantages is their relatively high self-discharge rate, primarily caused by energy losses associated with residual bearing friction, aerodynamic drag, and auxiliary system consumption. These losses gradually reduce angular velocity, which directly decreases stored energy according to Eq. (9.3). Consequently, FESSs are generally unsuitable for long-duration energy storage applications.

Safety also represents a major engineering challenge. The presence of a rapidly rotating mass introduces significant risks in the event of structural failure, potentially releasing substantial mechanical energy. This necessitates robust containment systems and advanced safety engineering. Moreover, high-speed rotating assemblies can introduce additional complexities in vehicle integration, structural design, and dynamic stability. Overall, although FESSs provide substantial operational advantages in specific areas, their limitations make them highly specialized storage systems best suited for carefully defined niche applications.

9.5.5 Technological maturity and cost challenges

Mechanical energy storage using flywheels has been employed since the Industrial Revolution. However, FESSs have undergone substantial technological evolution in recent decades, meaning that their current maturity depends heavily on advances in materials, power electronics, magnetic bearings, and vacuum systems. Although FESSs have found application in certain electric mobility niches, particularly those requiring strong dynamic response, their adoption level remains significantly lower than that of batteries and even SCs, partly due to economic and integration constraints.

Maintenance represents an important source of cost for FESSs, especially because of friction associated with mechanical bearings, which not only increases wear but also directly affects system efficiency. Although these frictional losses can be reduced through the use of magnetic bearings, thus reducing maintenance requirements, such components simultaneously increase technical complexity, require specialized supervision, and raise both initial costs and those associated with manufacturing and installation. In addition, overall system cost may increase further due to the use of advanced high-strength materials, robust containment structures, or the placement of the flywheel within vacuum chambers designed to reduce aerodynamic drag.

These factors mean that, although FESSs may be considered technologically mature systems, they remain economically less competitive than other technologies such as lithium-ion batteries, particularly because of their lower specific energy and higher cost per unit of stored energy. Consequently, despite their technical strengths in specific power-oriented applications, their broader commercial deployment remains constrained by cost competitiveness, engineering complexity, infrastructure requirements, and application-specific integration challenges.

9.6 Flywheels in mobility applications

9.6.1 Rail and wayside energy recovery systems

FESSs have found one of their most established applications in the railway sector, particularly in tramway, metro, and light rail systems. In this domain, FESSs are primarily deployed as energy recovery devices during braking events. Railway operation is inherently characterized by repetitive duty cycles with frequent acceleration and deceleration phases, during which large amounts of energy are alternately consumed and generated.

During the braking of an electric train, traction machines operate in generator mode, converting the kinetic energy of the vehicle into electrical energy. However, in the absence of energy storage systems, the effective utilization of this regenerated energy depends on the simultaneous occurrence of an acceleration phase by another train within the same electrical section. Since this coincidence is relatively infrequent, a significant portion of the braking energy is typically dissipated through onboard or wayside resistive elements, reducing overall system efficiency.

FESSs, typically installed in wayside configurations due to onboard weight and packaging constraints, enable the temporary storage of this otherwise wasted energy. By introducing a temporal decoupling between energy generation and consumption, FESSs eliminate the need for synchronization between braking and acceleration events of different trains. This temporal decoupling significantly increases the effective rate of energy recovery within the system.

In addition to energy recovery, FESSs contribute to the stabilization of the railway electrical network. Their fast response time and high power capability make them suitable for mitigating voltage fluctuations, reducing peak power demand, and alleviating stress on traction substations. This can translate into reduced infrastructure requirements and lower operational costs associated with energy supply.

Another relevant advantage is the ability to operate with a certain degree of independence from the main electrical grid. This is particularly beneficial in urban environments where the installation of overhead catenary systems or third rail infrastructure may involve high costs or significant visual impact.

From an economic perspective, the feasibility of FESS deployment in railway systems is strongly dependent on network utilization intensity and the frequency of braking events. Dense urban networks with high train frequency and short inter-station distances represent optimal scenarios for their implementation. Nevertheless, factors such as initial investment cost, installation constraints, and safety considerations continue to limit widespread adoption to applications where the energy and operational benefits justify the investment.

9.6.2 Automotive and motorsport applications

FESSs have also found limited application in the automotive sector. In contrast to railway systems, where flywheels are typically deployed in wayside configurations, automotive applications generally require onboard integration. This introduces stringent constraints in terms of weight, volume, safety, and system integration, which strongly influence design choices and restrict their adoption to niche use cases.

One of the most established applications of FESSs in the automotive domain is their use in KERS. These systems recover energy during braking in a manner analogous to regenerative braking (see Chapter 2), providing support during subsequent acceleration phases. During braking, a high-speed rotor is accelerated, storing energy mechanically in the flywheel. In such applications, similarly to SCs (see Section 9.4.3), the high power capability of flywheels enables superior performance compared to batteries in KERS systems, allowing rapid energy absorption and delivery during acceleration. Furthermore, their minimal degradation under repetitive charge and discharge cycles makes them particularly suitable for high-performance vehicle applications.

In this context, the primary deployment of FESSs in automotive systems has been in motorsport. In high-performance racing environments, where intense braking events are followed by rapid acceleration, power performance becomes more critical than long-term energy storage. Under these conditions, the characteristics of flywheel systems make them well suited to enhance acceleration, optimize energy management, and improve overall vehicle performance under demanding operating conditions.

FESSs have also been explored in certain passenger transport applications, particularly in buses. Some implementations have investigated the feasibility of onboard flywheel storage systems, using charging points based on overhead lines at bus stops (see Fig. 9.5). However, weight-related constraints have limited their role

Figure 9.5 Gyrobus during charging. The gyrobus was an electric bus concept that used a high-speed flywheel as its primary energy storage system. Energy was stored mechanically in the rotating mass and periodically replenished at designated charging points. Although limited by energy density and operational constraints, the gyrobus represents one of the earliest real-world implementations of flywheel energy storage in electric mobility. (Image courtesy of Wikimedia Commons.)

primarily to energy recovery systems in largely experimental vehicles [8]. In such configurations, FESSs can operate as auxiliary storage systems, reducing battery stress under peak power demands and highly dynamic operating cycles (see Section 9.7.3).

9.6.3 Aerospace and UAV potential

FESSs have found highly specific applications in certain aerial systems and UAVs. In these platforms, flywheel systems are primarily employed to support maneuvers that require high power demand over very short time intervals, where the characteristics of FESSs are particularly well suited and can outperform other storage technologies such as batteries.

The high power density and fast dynamic response of flywheels make them well adapted to applications involving demanding transient loads. In this context, FESSs in aviation systems have been considered for supporting take-off phases, evasive maneuvers, and the activation of auxiliary systems with high energy demand. In such applications, the high cycle durability of FESSs makes them suitable for operational profiles characterized by frequent charge and discharge events, ensuring reliable performance over extended service lifetimes.

However, flywheels present significant constraints for aerospace applications. One of the main limiting factors is their relatively low specific energy, which restricts their use as primary energy storage systems responsible for providing vehicle autonomy, particularly in platforms with stringent mass limitations. In addition, the integration of high-speed rotating components introduces further challenges related to safety, vibrations, and dynamic control, especially in systems with strict stability requirements and sensitivity to mechanical disturbances.

9.6.4 Why flywheels remain limited in passenger EVs

Flywheels offer strong performance in terms of power delivery, durability, and transient response. However, their application in passenger EVs remains highly marginal compared to other storage technologies such as batteries. This is due to a combination of technical, economic, and integration constraints that hinder their adoption relative to more established and mature solutions.

One of the main limitations of FESSs in passenger vehicles is related to packaging requirements. Flywheel systems require robust enclosures capable of safely containing rotors operating at very high rotational speeds. These enclosures introduce additional volume and mass, which can be critical constraints in vehicles with strict space and weight limitations. Furthermore, the need for containment systems to ensure passenger safety in the event of mechanical failure increases overall system complexity and integration challenges.

Safety considerations represent another major constraint. The presence of high-speed rotating masses introduces risks associated with structural failure, vibrations, and gyroscopic effects. These factors complicate vehicle certification processes, particularly in commercial applications where safety standards are especially stringent and tightly regulated.

In addition, the relatively low specific energy of flywheels significantly limits their use as standalone storage systems, as passenger EVs typically require high energy capacity to ensure sufficient driving range. As a result, FESSs generally need to be combined with other storage technologies to meet autonomy requirements (see Section 9.7.3).

Finally, economic and market-related factors also play a relevant role. FESSs typically involve higher installation costs than alternative storage systems, and the lack of a clear competitive advantage over established technologies results in limited user acceptance and reduced market adoption in real-world conditions.

9.7 Hybridization and multi-storage architectures

9.7.1 Battery + supercapacitor systems

Hybrid energy storage systems composed of SCs and batteries represent one of the most widespread and technically mature solutions in electric mobility. Their operating principle relies on the functional complementarity between both technologies: batteries provide relatively high specific energy, making them suitable for long-term energy storage and range extension, whereas SCs deliver high specific power, fast transient response, and excellent cycle durability, which makes them well suited for operating profiles characterized by frequent high-power transients.

Within this framework, SCs operate as a support system for the battery with the objective of reducing electrochemical stress and mitigating degradation mechanisms associated with high current events. In this sense, SCs act as a power buffer, absorbing and supplying energy during dynamic events such as rapid acceleration and regenerative braking. This buffering function prevents the battery from being exposed to sharp power peaks, contributing to improved lifetime, enhanced thermal behavior, and more stable operating conditions over extended usage periods.

However, the integration of both storage technologies introduces additional system complexity, particularly in terms of control architecture and power electronics. Bidirectional DC/DC converters, along with advanced monitoring and control systems, are typically required to manage power flow between the storage elements. In this context, control strategies based on frequency decoupling and power allocation are essential to fully exploit the advantages of each storage system, ensuring efficient and coordinated operation under varying driving conditions.

The application domain of these hybrid systems spans from light-duty passenger vehicles to heavy-duty transport, although their operational objectives differ. In passenger vehicles, the primary goal is to enhance energy efficiency and dynamic response. In heavy-duty applications, the integration of SCs and batteries aims to manage more demanding load profiles, reducing their impact on the battery system and improving overall system robustness and reliability under intensive duty cycles [5].

9.7.2 FC + supercapacitor systems

The integration of FCs and SCs in a hybrid energy storage system constitutes a promising solution in selected applications. This approach is primarily motivated by the limited dynamic response of FCs. While FCs offer high system-level energy density associated with hydrogen storage, their transient response is relatively slow, which constrains their ability to follow highly dynamic load profiles. In this context, SCs provide an effective complement, enhancing overall system performance.

Within this architecture, the SC operates as a support element. During high-power transient events, the SC acts as a power buffer, decoupling the FC from sharp power demands. Thus, during rapid acceleration phases, the SC is capable of delivering power with very fast response. Similarly, during regenerative braking, the SC can absorb high power levels efficiently, improving the recovery of braking energy and increasing overall system efficiency.

This hybridization provides two main advantages. First, it enables a more stable operating regime for the FC, closer to its optimal efficiency point, thereby reducing degradation mechanisms associated with frequent and severe transients [9]. Second, the integration of SCs allows a reduction in the required FC sizing, since peak power demands are handled by the SC, which generally leads to lower space and mass requirements at the system level.

These hybrid architectures have found application in heavy-duty transport segments such as buses and trucks, where operating profiles include frequent start-stop cycles. In such scenarios, the combination of FCs and SCs improves overall efficiency, reduces hydrogen consumption, and enhances vehicle dynamic performance. However, the integration of both systems introduces additional challenges related to energy management control, system cost, and overall complexity, which must be carefully addressed to ensure practical viability.

9.7.3 Flywheel-assisted systems

Flywheel-assisted energy storage systems have found limited application in electric mobility, primarily restricted to highly specialized use cases. In this context, FESSs provide strong performance in scenarios characterized by frequent and high-power demand peaks. Under such conditions, the high power density combined with the virtually unlimited cycle life of flywheels makes FESS-based solutions competitive with alternative storage technologies.

In this sense, the role of FESSs within a hybrid storage architecture is analogous to that of SCs, namely to provide support during transient events with high power demand. By absorbing and delivering short-duration power bursts, FESSs enable the decoupling of these transients from other storage technologies, thereby reducing their degradation and extending their operational lifetime. However, the superior durability of flywheels may provide an additional advantage in applications involving extremely high cycling frequencies or highly intermittent load profiles, where even SCs may experience non-negligible aging over time.

Flywheel-assisted configurations have been explored mainly in heavy-duty transport applications with frequent start-stop operation, such as industrial fleets, buses, and certain rail systems [10]. Furthermore, FESSs have been considered for grid-interactive roles, particularly in supporting charging infrastructure by mitigating power peaks and stabilizing demand profiles.

Despite these advantages, the deployment of FESS-based systems in electric mobility remains constrained by several factors, including mechanical complexity, safety requirements associated with high-speed rotating components, and integration limitations in terms of volume and mass. These constraints significantly hinder widespread adoption and ultimately favor more mature and easier-to-integrate technologies such as SCs in most mobility applications.

9.8 Emerging alternative storage technologies

9.8.1 SMES

The application of SMES in electric mobility has remained largely confined to experimental and research domains [11]. Their operating principle is based on the direct storage of energy in the form of a magnetic field generated by a current circulating through a superconducting coil. To achieve this, SMES systems typically operate at cryogenic temperatures in order to reach and maintain the superconducting state.

This operating principle provides two key characteristics: an exceptional capability for rapid power absorption and delivery over very short time scales, and high efficiency due to the nearly zero internal resistance. These features make SMES suitable for applications requiring fast transient support, voltage stabilization, or compensation of power fluctuations.

However, these advantages come at the cost of stringent operating conditions, particularly the requirement to maintain cryogenic temperatures. This introduces additional energy consumption associated with cooling, increased system complexity, and enhanced safety requirements. Furthermore, superconducting materials and cryogenic systems impose constraints in terms of cost and scalability.

As a result, the deployment of SMES in electric mobility has been limited mainly to wayside applications, where they can be used as grid support systems or to assist high-power charging infrastructure, rather than as onboard energy storage solutions.

9.8.2 Structural energy storage and multifunctional materials

Structural energy storage represents one of the most innovative approaches to energy storage in mobility applications. Unlike conventional storage technologies, it is not implemented as an independent subsystem, but instead relies on advanced materials capable of simultaneously fulfilling structural and energy storage functions [12].

This approach is based on the use of materials such as carbon fiber reinforced polymers, which can act as structural electrodes. In these configurations, energy storage and mechanical support are integrated, leading to concepts such as structural batteries or structural SCs. This functional integration provides a clear advantage in terms of mass reduction, as redundancies between structural and energy storage components are minimized, effectively embedding the storage function within the vehicle structure.

Despite their potential, structural energy storage systems still require significant development to improve performance. Current materials remain far from the levels of specific energy and power achieved by conventional technologies, which has limited their application primarily to experimental and prototyping stages.

9.8.3 Why most alternatives remain marginal

The storage technologies reviewed in Section 9.8 currently exhibit very limited deployment in electric mobility due to a combination of technical, economic, and operational constraints.

First, the specific cost of these technologies remains substantially higher than that of more mature solutions such as batteries, which constitutes a major barrier to large-scale implementation. Indeed, some of the materials involved, especially in structural energy storage, remain expensive and lack industrial maturity, thereby limiting manufacturability and large-scale scalability. In addition, their integration into vehicles generally requires advanced architectures, highly specific operating conditions, or structural redesign of the vehicle itself. These requirements increase engineering complexity, development effort, and overall system risk.

Significant performance limitations also remain. In particular, specific energy remains considerably lower than that of established storage technologies, which restricts the functional role of these emerging concepts to very specialized applications.

Moreover, certain technologies such as SMES require highly specific operating conditions, most notably cryogenic environments to maintain superconducting behavior. These requirements introduce additional energy consumption, auxiliary subsystems, and more demanding safety constraints, which further reduce practical suitability for onboard mobility applications.

Taken together, these limitations currently constitute a substantial barrier to the commercial deployment of SMES and structural energy storage in mainstream electric mobility. As a result, conventional technologies such as batteries, FCs, and SCs continue to dominate the broadest range of applications, while emerging alternatives remain largely confined to experimental studies, prototype validation, or highly specialized niche applications where their specific functional advantages may justify their complexity.

9.9 Economic, environmental, and strategic considerations

9.9.1 *Material availability and sustainability*

The analysis of materials employed in alternative energy storage systems (SCs and FESSs) represents, as in other storage technologies, a key factor in assessing their viability within the electric mobility sector. The impact associated with material selection is not uniform and varies significantly depending on the underlying technology.

In this context, SCs present a competitive advantage compared to storage technologies reliant on critical metals, such as batteries, whose dependence on geographically concentrated resources introduces supply uncertainty and potential stress within the supply chain (see Chapter 7). SC manufacturing is primarily based on carbon-derived materials, including activated carbon, carbon nanotubes, and graphene. These materials generally exhibit higher relative abundance and reduced exposure to geopolitical constraints, contributing to improved supply chain resilience.

By contrast, flywheel energy storage systems typically rely on high-strength composite materials, such as carbon fiber or glass fiber. In this case, the main challenges are associated with manufacturing complexity and energy-intensive production processes rather than raw material scarcity. Consequently, improvements in manufacturing efficiency, the development of alternative materials, and enhanced recyclability will be critical factors in strengthening the long-term sustainability and industrial relevance of these technologies within electric mobility systems.

9.9.2 *Lifecycle, recycling, and safety*

In terms of durability, SCs exhibit a high cycle life, generally exceeding that of batteries, which reduces replacement frequency and consequently the environmental impact associated with manufacturing. In the case of flywheel energy storage systems, this advantage can be even more pronounced, as they rely on mechanical components with limited degradation under controlled operating conditions.

However, recycling remains a relevant challenge. The recovery of carbon-based materials used in SCs is technically feasible, but current processes still face limitations in terms of efficiency and economic viability. Flywheel systems, on the other hand, are based on high-strength composite materials that require careful handling and specialized treatment at end of life.

From a safety perspective, both technologies offer improved thermal stability compared to electrochemical systems, avoiding issues such as thermal runaway. Nevertheless, they introduce additional challenges, including high operating voltages in SCs and mechanical risks associated with high-speed rotating components in flywheel systems.

9.9.3 Strategic industrial outlook

Within the electric mobility sector, energy storage systems based on SCs and FESSs exhibit a clearly application-oriented deployment rather than a widespread market adoption. Their primary use cases correspond to operational profiles requiring high power absorption and delivery capability, fast dynamic response, and high cycle durability. Consequently, SCs and FESSs have been predominantly implemented in specific niches such as railway systems and selected industrial applications.

In addition, these technologies are increasingly being deployed as grid-support storage systems. Their high power response makes them particularly suitable for applications such as charging support and systems integrating renewable generation. Furthermore, they can play a relevant role in improving the operational efficiency of heavy-duty and urban transport fleets.

From a strategic perspective, ongoing research is focused on improving integration with other energy storage technologies, enhancing material performance, and reducing system costs. In this context, the future development of alternative storage systems within the mobility sector is expected to depend on their ability to provide complementary functionalities in highly specialized applications alongside conventional energy storage technologies.

9.10 Conclusions and future outlook

9.10.1 Most promising roles in future EV ecosystems

Given their characteristics and inherent limitations, energy storage technologies based on SCs and flywheel systems are expected to play a complementary role alongside more mature solutions such as batteries and hydrogen, rather than displacing them from their established application domains. In this context, SCs and FESSs are more likely to be integrated into hybrid architectures or deployed in highly specific niches, particularly in applications requiring high cycle durability and fast power response.

SCs are expected to continue their predominant use in sectors such as urban public transport, railway systems, and fast-charging infrastructure, where ultra-fast charging capability and long operational lifetime are critical performance factors. In contrast, flywheel systems are likely to maintain a more limited presence, with applications focused on scenarios where durability, rapid dynamic response, and reduced sensitivity to thermal conditions are particularly advantageous. Their use in stationary energy recovery systems and specialized industrial environments is expected to remain their primary development niche, especially where high-frequency cycling is required.

Furthermore, these alternative technologies will continue to gain relevance within hybrid storage architectures, where SCs and FESSs operate in combination with batteries or hydrogen-based systems. In this regard, the development of coordinated energy management strategies across multiple storage technologies will play a key role in enabling efficient system-level operation, particularly under highly dynamic load conditions and variable energy demand profiles.

Finally, several factors will influence the future development and deployment of SCs and FESSs, including advances in material science, increasing regulatory pressure related to sustainability, and progressive industrial specialization. Economic considerations, system integration complexity, and infrastructure compatibility will also play a decisive role in their adoption. These drivers are expected to shape a diversified energy storage ecosystem in which multiple technologies coexist, each optimized for specific operational requirements and mission profiles.

9.10.2 Comparative assessment of all major energy storage technologies in electric mobility

The analyses presented in Chapters 7, 8 and 9 indicate a heterogeneous technological landscape in the field of energy storage systems for electric mobility applications, with multiple technologies exhibiting distinct characteristics and performance metrics. In order to provide a structured comparative overview of the available options, Table 9.1 summarizes the most relevant technology groups, including their primary advantages, key limitations, and dominant application niches. As shown in this table, the balance between strengths and weaknesses across technologies supports a scenario of coexistence, in which each solution is suited to specific operational requirements. In this context, hybrid architectures combining different storage technologies are expected to play an increasingly important role in future mobility systems.

Batteries and hydrogen-based storage systems are both characterized by high specific energy, which provides a clear advantage in applications requiring long driving range under strict mass and volume constraints. However, these technologies differ significantly in terms of charging or refuelling dynamics and overall system efficiency. Batteries exhibit high round-trip efficiency due to direct electrochemical energy conversion, involving a limited number of transformation stages. This feature has enabled their widespread adoption across nearly all electric mobility sectors. Nevertheless, hydrogen systems offer faster refuelling and higher gravimetric energy density, which provides an advantage in specific applications such as long-distance heavy-duty transport or high-utilization fleets, where downtime and mass limitations are critical constraints.

Table 9.1 Comparative overview of major energy storage technologies in electric mobility.

Technology	Main Advantage	Main Disadvantage	Typical Niche Application
Batteries	High specific energy	Limited cycle life and charging speed	Passenger EVs and general-purpose mobility
Hydrogen	Long range and fast refuelling	Low efficiency and complex infrastructure	Heavy-duty and long-range transport
SCs	Very high power density	Low specific energy	Regenerative braking and peak power support
FESSs	High cycle durability	Mechanical complexity and safety constraints	Rail systems and stationary energy recovery
Other technologies	Extreme power capabilities	Low maturity and high cost	Specialized industrial or experimental systems

In contrast, SCs and FESSs outperform both batteries and hydrogen systems in terms of power density, dynamic response, and cycle durability. Their ability to absorb and deliver power rapidly, combined with high thermal stability, makes them particularly suitable for high-frequency cycling applications. However, neither SCs nor FESSs are suitable for providing driving range due to their low specific energy and, in the case of SCs, relatively high self-discharge rates. These limitations restrict their use as primary energy storage systems. As a result, the main application domain of SCs and flywheel systems corresponds to specific niches where their performance characteristics provide a competitive advantage. Typical use cases include support for high-power transient events such as acceleration phases or take-off operations, as well as energy recovery through regenerative braking. In addition, their use as grid-support systems or as buffers in fast-charging infrastructure is becoming increasingly relevant within the evolving electric mobility ecosystem.

Finally, other emerging technologies such as SMES and structural energy storage systems are currently at an early stage of development and deployment in electric mobility. Although these technologies present promising features, such as high efficiency or the potential for structural integration and weight reduction, their level of maturity remains limited. Consequently, their application in EVs is currently restricted to research activities, experimental developments, and prototype validation.

References

[1] M. Tostado-Véliz, P. Arévalo, F. Jurado, An optimization framework for planning wayside and on-board hybrid storage systems for tramway applications, Journal of Energy Storage

43 (2021) 103207, https://doi.org/10.1016/j.est.2021.103207.
[2] Z. Ayaganov, V. Pavlenko, S.F.B. Haque, A. Tanybayeva, J. Ferraris, A. Zakhidov, Z. Mansurov, Z. Bakenov, A. Ng, A comprehensive study on effect of carbon nanomaterials as conductive additives in edlcs, Journal of Energy Storage 78 (2024) 110035, https://doi.org/10.1016/j.est.2023.110035.
[3] M. Stokić, B. Dimitrijević, Model for electrification of urban public transport lines with supercapacitor buses: a case study of Belgrade, Applied Energy 377 (2025) 124670, https://doi.org/10.1016/j.apenergy.2024.124670.
[4] K. Dissanayake, D. Kularatna-Abeywardana, A review of supercapacitors: materials, technology, challenges, and renewable energy applications, Journal of Energy Storage 96 (2024) 112563, https://doi.org/10.1016/j.est.2024.112563.
[5] C.V.M. Gopi, R. Ramesh, Review of battery-supercapacitor hybrid energy storage systems for electric vehicles, Results in Engineering 24 (2024) 103598, https://doi.org/10.1016/j.rineng.2024.103598.
[6] P. Rajnarayanan, K. Kathiravan, 6 – supercapacitors as distributed energy storage systems for ev charging infrastructure, in: S. Palanisamy, S. Chenniappan, S. Padmanaban (Eds.), Distributed Energy Storage Systems for Digital Power Systems, Elsevier, 2025, pp. 149–166.
[7] X. Li, C. Mittelstedt, A. Binder, A review of critical issues in the design of lightweight flywheel rotors with composite materials, E&I. Elektrotechnik und Informationstechnik 139 (2022) 204–221, https://doi.org/10.1007/s00502-022-01005-4.
[8] S. Si, B. Yang, B. Gao, X. Hou, B. Zhao, T. Zhang, Energy management strategy of flywheel hybrid electric vehicle based on driving condition recognition, Applied Thermal Engineering 272 (2025) 126445, https://doi.org/10.1016/j.applthermaleng.2025.126445.
[9] E. Oksuztepe, M. Yildirim, Pem fuel cell and supercapacitor hybrid power system for four in-wheel switched reluctance motors drive ev using geographic information system, International Journal of Hydrogen Energy 75 (2024) 74–87, https://doi.org/10.1016/j.ijhydene.2023.12.207.
[10] B. Sun, B. Li, J. Xing, X. Yu, M. Xie, Z. Hu, Analysis of the influence of electric flywheel and electromechanical flywheel on electric vehicle economy, Energy 295 (2024) 131069, https://doi.org/10.1016/j.energy.2024.131069.
[11] E.-L. Molina-Ibáñez, E. Rosales-Asensio, C. Pérez-Molina, F.M. Pérez, A. Colmenar-Santos, Analysis on the electric vehicle with a hybrid storage system and the use of superconducting magnetic energy storage (smes), Energy Reports 7 (2021) 854–873, https://doi.org/10.1016/j.egyr.2021.07.055, Technologies and Materials for Renewable Energy, Environment and Sustainability.
[12] H. Zhou, H. Li, L. Li, T. Liu, G. Chen, Y. Zhu, L. Zhou, H. Huang, Structural composite energy storage devices — a review, Materials Today Energy 24 (2022) 100924, https://doi.org/10.1016/j.mtener.2021.100924.

Vehicle-to-everything (V2X) 10

10.1 Concept and fundamentals of V2X

10.1.1 Definition and scope of V2X

The V2X concept refers to the capability of EVs to exchange energy, and potentially information, with different external agents within the energy system to which they are connected. In contrast to the conventional paradigm, where the EV behaves as a passive load with the sole objective of absorbing energy for charging purposes, under the V2X paradigm the EV becomes an active element within the system, enabling both the absorption and delivery of electrical power.

The V2X paradigm encompasses different configurations, which are analyzed in detail in Section 10.3, but can be initially classified according to the type of agent with which the EV exchanges energy. Based on this criterion, the configurations included within the V2X concept can be summarized as follows:

- Interaction with electrical infrastructures (distribution or transmission networks).
- Energy exchange with local environments (homes or buildings).
- Direct energy transfer between vehicles or to specific loads.
- Integration within distributed energy systems.

In all cases, the EV transitions from a passive element to an active component within the system, capable of exchanging energy with other agents. This behavior is aligned with the current paradigm of increasingly decentralized and distributed energy systems [1]. It is important to distinguish between two types of exchanges associated with V2X:

- **Energy exchange:** Involving the physical transfer of electrical power between the vehicle and the external system.
- **Service provision:** Where the vehicle contributes to the operation of the electrical system without requiring a continuous net energy transfer, as in regulation or support services.

This distinction is fundamental, since not all V2X configurations necessarily imply a direct energy exchange. Instead, the EV can be understood as a distributed energy storage element whose bidirectional capability can be exploited to provide services related to grid stability, efficiency, and overall system management.

10.1.2 EV as an energy asset

Traditionally, the EV has been viewed from the perspective of the electrical system as a load, intended to absorb energy solely to satisfy its charging requirements. However, the V2X paradigm requires a broader interpretation of the role of the EV within the system, in which it assumes a more complex function. In this context, bidirectional charging capability enables the EV to be considered a multifunctional agent, capable

Electric Mobility Systems. https://doi.org/10.1016/B978-0-44-349397-3.00018-5

of providing different services. From this perspective, the EV can be simultaneously characterized as:

- **Electrical load:** The EV maintains its conventional role when it draws energy from the grid to charge its onboard storage system.
- **Energy storage system:** Due to the presence of an onboard battery system capable of storing and delivering energy in a controlled manner.
- **Distributed Energy Resource (DER):** As it can actively participate in the operation of the electrical system.

The integration of the EV as a DER within the system is of particular relevance, especially given the increasing trend towards distributed energy systems in which small-scale generators, typically based on renewable sources, are deployed close to end users [2]. Within this context, the EV can be regarded as a flexible distributed element, assuming an active role in maintaining the balance between generation and demand.

In addition, the EV can adopt the role of a prosumer, that is, an agent capable of both consuming and supplying energy. In this way, the EV can contribute to increasing the distributed generation available in the grid when operating conditions allow. The interaction of the EV with the energy system can take place at different levels:

- Interaction with the electrical grid, participating in both power absorption and injection.
- Interaction with local loads, such as homes or buildings, contributing to their energy management.
- Integration within distributed systems, such as microgrids or environments with renewable generation, including energy communities [3].

In summary, the V2X paradigm requires EVs to be understood as multifunctional agents with a role that extends beyond that of a simple electrical load. The capability to fully exploit the onboard battery system positions the EV as a key element in the management of the energy system in which it is integrated.

10.1.3 Unidirectional vs bidirectional operation

Therefore, the V2X paradigm enables the definition of two operational models characterized by the direction of the energy flow. In this context, a distinction can be made between unidirectional and bidirectional operation. This classification establishes the fundamental framework for the operation and management of the EV within the energy system.

The unidirectional scheme corresponds to the conventional model, in which energy flows only in one direction, from the grid to the vehicle. Under this approach, the EV is regarded as a load, with the sole capability of absorbing energy to satisfy its charging requirements. This scheme is characterized by:

- Power flow in a single direction (from grid to vehicle).
- Relatively simple charging infrastructure.
- Absence of active interaction with the electrical system.

- Limitation to energy consumption functions.

In contrast, the bidirectional scheme allows energy to flow both to and from the vehicle, constituting the operational basis of the different V2X configurations. In this case, the EV is characterized as an active agent within the system, capable of participating in its operation. In general terms, the bidirectional scheme can be described as follows:

- Bidirectional power flow (from vehicle to grid and from grid to vehicle).
- Active participation in the energy system.
- Capability to provide services beyond energy consumption.
- Increased complexity in control and infrastructure.

It is important to distinguish between these two schemes, as this classification has significant implications for both infrastructure requirements and system management. While unidirectional operation relies on relatively simple infrastructure based on conventional chargers, bidirectional systems require more advanced converters, more sophisticated control systems, and additional communication mechanisms (see Section 10.2). Furthermore, although bidirectional operation enables several advantages from the grid perspective (see Section 10.5.1), the proper management of bidirectional power flows in EVs introduces additional challenges and complexities in system control (see Section 10.5.4).

10.2 Enabling technologies

10.2.1 Power electronics for bidirectional operation

10.2.1.1 Bidirectional converter topologies

In V2X applications, a set of devices is required to accommodate and manage bidirectional power flows within a vehicular system where both DC and AC domains coexist. The conversion and control of these power flows are performed by power electronic converters, which can be interpreted as elements capable of reconstructing electrical energy in order to adapt it to each domain, thus acting as interfaces between subsystems.

Independently of the specific architecture employed, several converters are common to most EV topologies (see Fig. 10.1). These converters constitute the key element that enables energy exchange, acting as interfaces between the electrical grid, the DC bus, and the onboard energy storage system. In this context, two main functional blocks can be identified:

- **Bidirectional AC/DC converters**: Typically implemented as the OBC in most EV architectures (see Chapter 6). These converters adapt energy from the electrical system, which is typically in AC form, into DC suitable for the DC bus, while also enabling the reverse operation under V2X conditions.
- **Bidirectional DC/DC converters**: Responsible for the interface between the DC bus and the battery system. Their main function is to regulate voltage and current levels, ensuring proper and controlled energy exchange with the storage system.

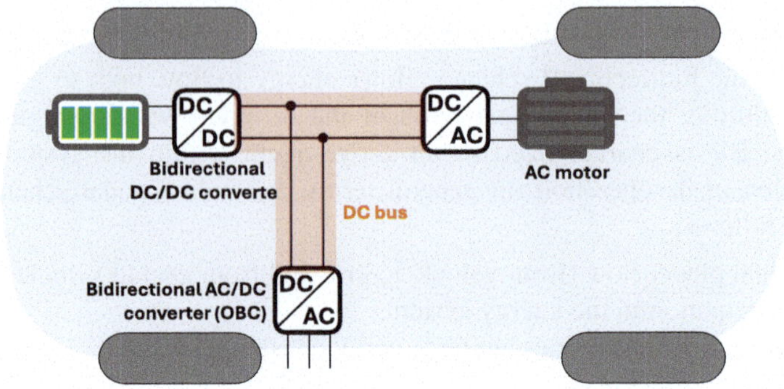

Figure 10.1 Simplified schematic of an EV powertrain highlighting the main bidirectional power electronic converters, including the AC/DC onboard charger (OBC), the DC/DC converter, and their connection to the DC bus and electric motor.

These converters can be implemented using different topological configurations, including full-bridge converters, multilevel converters, as well as isolated and non-isolated structures. The selection of a given configuration depends on factors such as power level, power quality requirements, efficiency, and safety constraints. A detailed analysis of these configurations is beyond the scope of this chapter and can be found in specialized literature such as [4].

Finally, bidirectional operation enables the EV to behave as a distributed energy source under certain conditions. This mode of operation imposes additional requirements, including grid synchronization, power factor control, and compliance with power quality standards [5]. These requirements are essential to ensure proper integration within the electrical system and directly influence the design and operation of EV converters.

10.2.1.2 Efficiency and loss mechanisms

The operation of the power electronic converters described in Section 10.2.1.1 leads to a set of unavoidable energy losses. These losses occur at each conversion stage and directly affect the overall system efficiency, making their joint evaluation necessary. In this context, it is useful to classify them according to their physical origin:

- **Switching losses**: These occur during the turn-on and turn-off transitions of semiconductor devices that constitute the core of power converters. Such losses depend on the switching frequency and on the dynamic characteristics of the devices employed.
- **Conduction losses**: Caused by the internal resistance of electronic devices when current flows through them. These losses are proportional to the square of the current (see Chapter 6).
- **Passive losses**: Associated with components such as inductors and capacitors, including resistive effects, magnetic core losses, and parasitic phenomena.
- **Auxiliary losses**: Related to control, sensing, and thermal management systems. These losses typically have a lower impact on overall efficiency compared to the previous mechanisms.

A critical aspect associated with these losses is the management of the generated heat. This thermal energy must be properly dissipated to prevent premature component degradation and to ensure safe operating conditions. In this context, the use of active cooling systems may be required, which in turn contributes to the overall energy balance of the system.

The analysis of energy losses becomes particularly relevant under prolonged operating conditions. In such scenarios, V2X operation may increase total losses due to the higher amount of energy flowing through the vehicle. Consequently, these losses not only reduce efficiency but may also affect the economic viability of the operation.

10.2.2 Communication infrastructure

10.2.2.1 Communication protocols

The V2X paradigm requires not only energy exchange between the vehicle and the system, but also a superimposed communication layer. This layer enables the exchange of information between the vehicle and the system or other agents, which is essential to coordinate energy flows and ensure safe, efficient, and controlled operation.

Within this framework, communication protocols play a central role, as they define the set of rules that structure and transmit information among the different system elements. The information exchanged in V2X environments typically includes:

- **Vehicle state information**: Including parameters such as the SoC, which describe the current operating condition of the vehicle.
- **Control signals**: Commands sent to the vehicle to modify its behavior, such as initiating or stopping charging or discharging processes, or adjusting exchanged power.
- **System information**: Data relevant to power system operation, including electricity prices, congestion indicators, or ancillary service requirements.

Among the most widely standardized protocols for communication between vehicles and charging infrastructure are ISO 15118,[1] which defines communication between the vehicle and the grid in road vehicles, and the Open Charge Point Protocol (OCPP),[2] which enables communication between charging stations and management systems, facilitating integration into control and aggregation platforms. Each protocol has particular features which are summarized in Table 10.1

Overall, the communication layer constitutes a fundamental element for enabling advanced V2X functionalities by ensuring effective coordination between energy and information flows. In particular, the correct implementation of communication protocols allows the system to react dynamically to external signals, adapt operating conditions in real time, and support coordinated operation across multiple agents. As a result, communication infrastructure becomes a key enabler for scalable and reliable V2X deployment.

[1] ISO 15118 standard.
[2] Open Charge Point Protocol (OCPP).

Table 10.1 Comparison between ISO 15118 and OCPP information exchange protocols.

Aspect	ISO 15118	OCPP
Communication scope	Vehicle–charger interface	Charger–backend interface
Primary function	Energy and control exchange	Infrastructure monitoring and management
Application level	Device-level communication	System-level communication
V2X support	Native support	Indirect support

10.2.2.2 Interoperability and data exchange

In systems where multiple agents interact (EVs, charging points, grid operators, etc.), interoperability and data exchange become fundamental requirements for the effective operation and management of V2X. In this context, interoperability can be understood as the ability of different agents to exchange information coherently regardless of manufacturer or specific implementation.

One of the main challenges affecting interoperability in V2X platforms is technological heterogeneity. Within a single V2X environment, devices and vehicles from different manufacturers must exchange information. Each implementation may rely on distinct architectures, protocols, and data formats, complicating direct integration. A key enabler to ensure effective information exchange under such conditions is the use of standardized communication protocols (see Section 10.2.2.1), which support functional compatibility across systems and reduce integration complexity in multi-vendor environments.

Another critical aspect is latency, defined as the delay between sending a data request or control signal and receiving the corresponding response. Certain V2X functionalities, such as grid service provision, require very short response times. Therefore, the communication infrastructure must ensure that latency levels remain within acceptable limits to avoid degrading system performance and compromising time-sensitive control actions.

Finally, in platforms involving multiple independent agents, ensuring data security is essential. This aspect directly influences the trust of users and operators, ultimately affecting participation. In this regard, robust mechanisms for authentication, encryption, and protection against unauthorized access must be implemented to prevent the exposure of sensitive information during data exchange [3,6].

10.2.3 Control architectures

10.2.3.1 Local control strategies

The implementation of V2X platforms requires the deployment of control strategies that govern power flows in terms of both magnitude and direction (towards or from the vehicle). In this context, local control strategies are executed independently by each vehicle, without coordination with the surrounding environment. As a result, these strategies rely exclusively on local information associated with the vehicle itself and its direct interface with the grid, neglecting the potential presence of other vehicles connected to the same network as well as interactions with external systems or operators. Within this local control paradigm, several approaches can be identified in the literature:

- **Rule-based control:** This approach represents the simplest and one of the most widely adopted strategies in commercial tools [7]. It is based on the definition of predefined logical conditions that trigger charging or discharging actions. Its main advantage lies in its ease of implementation, enabling deployment across low-cost platforms. However, its reliance on heuristic rules and its limited adaptability to dynamic environments hinder the achievement of optimal solutions from an economic or energy perspective.
- **Local optimization-based control:** In this case, control signals are obtained as the solution of an optimization problem that aims to optimize a given objective, typically defined through an economic or energy-related function [8]. In contrast to rule-based control, this approach enables the computation of optimal solutions within the local scope of the system. Nevertheless, it requires more advanced computational platforms for implementation and may involve a higher computational burden.
- **Artificial intelligence-based control:** This approach employs machine learning techniques to infer operational patterns and adapt system behavior based on either historical data or real-time information [9]. It provides a high degree of adaptability and can approximate optimal solutions in complex environments. However, its performance strongly depends on the quality and representativeness of the underlying model.

Local control strategies constitute a fundamental basis for the implementation of V2X platforms. However, by neglecting interactions with other agents within the system, these strategies may inherently lead to sub-optimal solutions from an aggregated system perspective.

10.2.3.2 Centralized and aggregated control

Centralized control strategies aim to coordinate multiple EVs (or other participating agents) in order to improve aggregated system behavior. As shown in Fig. 10.2, in contrast to local control strategies, centralized approaches rely on the coordination of different EVs through a higher-level entity, commonly referred to as an aggregator or system operator [3]. This entity is responsible for collecting information from all EVs connected to the system and generating global control actions that ultimately seek to optimize the overall system performance.

Within a centralized framework, EVs no longer operate as independent entities, but instead share their energy conversion and storage resources, which become integrated into a common aggregated asset. This aggregation enables the coordinated control of a fleet of vehicles as a single energy entity. Such an approach facilitates the

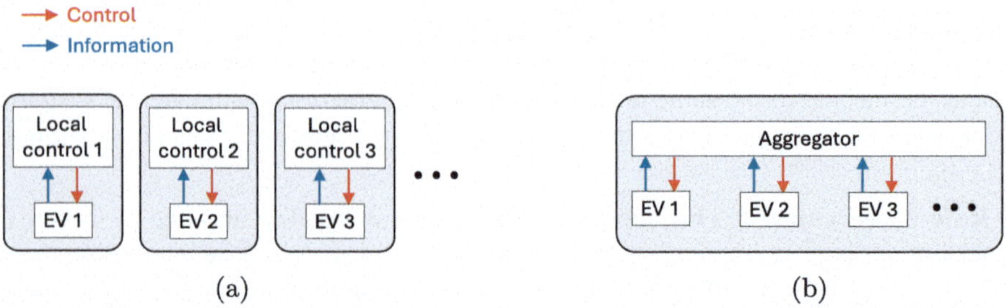

Figure 10.2 Comparison of V2X control architectures: (a) local control, where each EV operates independently based on local information; (b) centralized or aggregated control, where an aggregator coordinates multiple EVs through information exchange and supervisory control signals.

participation of EV fleets in large-scale applications, including grid services and electricity markets [10].

The main advantage of this approach lies in its ability to optimize energy resources at a higher level within a given system segment, enabling a collective utilization of distributed assets that can lead to improved aggregated performance. An additional advantage is its scalability, as it allows the simultaneous coordination of a large number of vehicles connected to the same network.

However, this approach introduces several challenges, including a stronger dependence on communication infrastructures, which reinforces the need to establish standardized communication protocols (see Section 10.2.2.1). Furthermore, centralized control strategies must ensure that all participating agents remain satisfied with the control decisions, preventing disengagement from the shared control framework.

10.3 V2X operating modes

This section presents the main V2X operating modes, each defined by the destination of the exchanged energy and the corresponding role of the EV within the energy system. These modes determine how EVs interact with different elements such as residential installations, the electrical grid, other vehicles, or standalone loads, thereby shaping both their technical requirements and operational capabilities. The diversity of V2X modes reflects the versatility of EVs as distributed energy resources, enabling applications that range from local energy management to system-level services.

A concise comparison of the most relevant operating modes is provided in Table 10.2. This table summarizes their main characteristics in terms of energy destination, key advantage, and principal limitation, offering a structured overview that facilitates understanding of their respective roles within the broader V2X framework. The following subsections examine each mode in detail, addressing their operating principles, applications, and technical constraints.

Table 10.2 Summary of V2X operating modes, including energy destination, main advantage, and main limitation.

Mode	Energy Destination	Main Advantage	Main Limitation
Vehicle-to-Home (V2H)	Residential loads	Enables local energy management and self-consumption	Limited power levels and dependency on user-defined constraints
Vehicle-to-Grid (V2G)	Electrical grid	Provides grid services at system level	Requires advanced coordination
Vehicle-to-Vehicle (V2V)	Other EVs	Allows direct energy transfer without relying on grid infrastructure	Low scalability and reduced efficiency
Vehicle-to-Building (V2B)	Building loads	Enhances demand flexibility	Integration complexity
Vehicle-to-Load (V2L)	Standalone loads	Offers high flexibility and mobility	Limited power capacity

10.3.1 Vehicle-to-home (V2H)

10.3.1.1 Operating principles

The V2H operating mode refers to the direct transfer of electrical energy from the vehicle to the residential installation to which it is connected. This mode of operation is typically associated with households equipped with a bidirectional charging point [11], enabling the presence of a local domestic energy source capable of fully or partially supplying household demand. In this context, the EV complements other local generation systems that may be installed, such as rooftop PV installations.

From an operational perspective, the V2H power flow is typically activated in response to specific signals or events that indicate favorable conditions for vehicle discharge. The most commonly used signals for enabling this bidirectional operation are the following:

- The presence of sufficient instantaneous demand, which is often a necessary condition. This ensures that the energy discharged from the vehicle is directly used to supply household loads, providing an immediate functional purpose.
- A minimum SoC threshold is defined to guarantee the subsequent availability of the vehicle. When the SoC of the EV exceeds this threshold, the onboard battery can be discharged down to the predefined limit.
- External signals such as high electricity prices, which incentivize the use of stored energy instead of importing energy from the grid.
- Specific grid-related events, including constraints or stress conditions, which may require a reduction in external power consumption.

- The absence or insufficiency of other local generation sources such as PV systems. In such cases, a priority scheme is typically implemented, where PV generation is utilized first up to its available limit, and the EV supplies the remaining demand if necessary.

As observed, SoC management is the key parameter in defining an appropriate V2H strategy. Control approaches based on SoC not only ensure a minimum energy reserve for vehicle operation, but also contribute to mitigating battery degradation by avoiding excessive discharge cycles.

10.3.1.2 Residential applications

The V2H operating mode enables a local energy source that improves residential energy management through the effective use of the energy stored in the vehicle. This functionality allows the stored energy to be exploited in several typical applications, among which the following are particularly relevant:

- **Self-consumption**: This is one of the most relevant applications of V2H, as this operating mode enables a local energy source capable of fully or partially supplying household demand. As a result, the energy imported from the grid can be significantly reduced, improving overall energy efficiency at the residential level and reducing dependence on external supply.
- **Renewable energy utilization**: The availability of a local storage system allows excess PV generation during periods of low demand to be stored for later use. This capability increases the utilization factor of renewable energy and reduces curtailment, contributing to a more efficient integration of distributed generation within the residential system.
- **Peak demand reduction**: V2H enables the implementation of peak shaving strategies through the controlled discharge of the onboard battery, limiting the power peaks that must be imported from the grid. This approach can provide economic benefits under power-dependent tariffs and contributes to smoothing the residential load profile, thereby supporting grid operation and reducing infrastructure stress.
- **Backup supply**: By enabling V2H operation, the EV can act as a local generator that is activated during grid outages. This functionality enhances the resilience of individual households and may positively impact the overall reliability of the electrical system at the community level.

10.3.1.3 Technical constraints

In V2H operation mode, several technical constraints must be considered, as they impose functional limits primarily on the power and energy that can be discharged. These limitations arise from both intrinsic technical constraints of the equipment and operational limits intentionally defined to ensure appropriate operating conditions and long-term system reliability.

The most relevant constraint is associated with the power limits of the bidirectional charging equipment, the onboard vehicle converters, and the electrical installation.

In residential environments, these constraints typically restrict the charging and discharging power to a few kilowatts [3], due to the standard design thresholds of domestic charging systems. As a result, the EV may not be able to fully supply the household demand under all operating conditions, requiring coordination with the grid supply or other local energy sources.

Secondly, a minimum SoC threshold is usually imposed to ensure that the energy stored in the vehicle remains sufficient for its primary transportation function. This threshold is typically defined by the user and is strongly influenced by the habitual usage profile of the EV. Therefore, the minimum SoC must be selected as a trade-off that enables effective energy utilization in the household while preserving vehicle availability.

Finally, additional constraints may arise from battery operating conditions. These constraints are intended to mitigate battery degradation and typically include limits on the number of charge-discharge cycles, operating temperature, DoD, and current rate. Such restrictions directly limit the usable energy capacity and, consequently, the amount of energy that can be delivered during V2H operation.

10.3.2 Vehicle-to-grid (V2G)

10.3.2.1 Grid services

In contrast to V2H operation, where the EV interacts directly with a residential installation, in V2G mode the vehicle exchanges energy directly with the electrical grid. In this context, the EV operates as a DER at system level rather than as a local asset. This mode of operation is typically associated with public bidirectional charging points that provide a direct interface with the grid. Due to this interaction, V2G enables the vehicle to provide a range of grid services. Relevant examples of these services include the following:

- **Grid stabilization**: EVs can actively contribute to grid stability by providing ancillary services such as voltage support and frequency regulation [12]. In this scenario, EVs may receive control signals from the grid operator requesting controlled charging or discharging of the onboard battery to support system operation. The fast response capability of power electronic converters allows EVs to participate in frequency control actions, improving overall system reliability, dynamic performance, and operational security under varying load conditions and contingencies.
- **Peak shaving**: This functionality is conceptually similar to that provided in V2H mode (see Section 10.3.1.2), but applied at grid level rather than at a single installation. In this case, peak demand reduction is performed in response to system-level conditions, independently of the local demand at the connection point. These strategies contribute to smoothing the aggregate demand profile and limiting power gradients, which enhances grid efficiency, reduces stress on infrastructure, and mitigates the need for peaking generation units [13].
- **Generation-demand balancing**: As flexible energy assets capable of operating over short time intervals, EVs with V2G capability can contribute to maintaining

the balance between generation and demand. This is achieved by either injecting power into the grid or absorbing excess energy, depending on system requirements and operating conditions, particularly in systems with high renewable energy penetration.

10.3.2.2 Integration with renewable energy

EVs with V2G capability can significantly contribute to the integration of renewable energy into the electrical grid. As distributed and flexible assets capable of both absorbing and supplying energy almost instantaneously, EVs can dynamically adapt their charge-discharge profiles to match renewable generation patterns [14]. This capability is particularly valuable in systems with high penetration of intermittent renewable sources such as PV and wind power, where variability and uncertainty must be actively managed.

The inherent variability of these energy sources introduces challenges in maintaining the balance between generation and demand, requiring flexible resources that can respond to rapidly changing system conditions. In this context, EVs can operate either as controllable loads or as energy suppliers at specific time intervals. They can store excess renewable energy during periods of low demand and high generation, and subsequently return this energy to the grid when renewable output decreases or demand increases. This operational flexibility not only supports generation-demand balancing but also enhances renewable energy utilization and overall system efficiency.

Another key advantage of V2G-enabled EVs lies in their distributed nature. Many renewable energy sources, particularly small-scale installations, are geographically dispersed across the grid. The presence of distributed storage resources such as EVs enables local absorption of renewable generation, reducing the need for long-distance energy transmission and consequently minimizing associated losses. This contributes to a more efficient and resilient energy system, while also alleviating congestion in distribution networks.

Finally, EVs can act as backup storage for renewable generation, supporting its integration even during periods of reduced output caused by sudden cloud coverage or low wind conditions. In such scenarios, the energy stored in EVs can help maintain generation commitments, facilitating the participation of renewable assets in electricity markets and in financial mechanisms such as power purchase agreements [15]. In addition, this capability enhances system reliability and provides additional operational flexibility under uncertain conditions.

10.3.2.3 Aggregation concepts

The V2G operating mode reaches its full potential when multiple EVs are managed in a coordinated manner through aggregation schemes. This paradigm arises from the intrinsically limited capacity of individual EVs in terms of both energy and power. In this context, the coordination of a large number of vehicles enables the system to reach scales that are relevant for the effective provision of grid services. Within this framework, the role of the aggregator emerges as the entity responsible for coordinating, monitoring, and optimizing the behavior of a fleet of vehicles.

The aggregator acts as an intermediary entity between the vehicles and the electrical grid, with the objective of consolidating the individual characteristics of each EV so that the system or the market perceives the fleet as a single energy resource. To achieve this, the aggregator must process information from the EVs such as SoC, temporal availability, user preferences, and operational constraints. Based on this information, the aggregator defines individual charge-discharge strategies that allow system requirements to be satisfied without compromising the primary functionality of the vehicles, typically following a centralized control approach (see Section 10.2.3.2).

This aggregation of resources enables the scalability of EVs as grid assets, allowing their evolution from local resources to system-level agents. As the number of participating EVs increases, the aggregated capacity in terms of available energy and power also grows, enhancing the capability to participate in ancillary services and energy markets [10].

However, the coordination of a large number of EVs introduces several operational challenges. The heterogeneity of vehicles, uncertainty in their availability, and dependence on user behavior increase the complexity of control. In addition, the management of aggregated systems requires robust communication infrastructures capable of handling large volumes of data. Finally, privacy concerns and potential conflicts among users may arise, requiring advanced control and communication strategies [3].

10.3.3 Vehicle-to-vehicle (V2V)

10.3.3.1 Operating concept

The V2V concept refers to the direct transfer of electrical energy between two EVs, without requiring the energy to flow through the electrical grid or any intermediate installation. Within this scheme, one vehicle operates as an energy source by exporting power, while the other behaves as a load, absorbing the energy delivered by the source EV. This direct interaction enables localized energy exchange with minimal reliance on external infrastructure.

From an operational perspective, this energy transfer typically requires intermediate power electronic converters to adapt voltage and current levels from the source vehicle to values compatible with the receiving vehicle. The V2V mode also involves the exchange of key operational variables between both EVs, such as SoC, power limits, and charging constraints. In addition, dedicated communication systems are required to coordinate the process, ensuring that the transfer is executed under safe and optimal conditions for both vehicles, while respecting electrical and thermal constraints.

In terms of applications, V2V operation is particularly relevant in scenarios with limited access to charging infrastructure. In such cases, an EV with insufficient SoC can receive enough energy from another vehicle to reach a nearby charging point. This mode is also applicable in fleet management contexts, where it can support energy balancing among vehicles, improving overall operational flexibility and resource utilization.

10.3.3.2 Practical limitations

In practice, the V2V mode faces significant technical limitations that constrain its large-scale deployment. These limitations mainly arise from technical and energy-related constraints associated with the infrastructure required to enable direct energy transfer between vehicles. They can be summarized in the following key aspects:

- **Efficiency**: Unlike other operating modes such as V2H or V2G, V2V may require multiple energy conversion stages in order to adapt voltage and current levels to those required by the EV acting as a load. This process can involve several power electronic converters, which introduce additional losses and reduce the overall system efficiency (see Section 10.2.1.2). As a result, the net transferred energy may be significantly lower than the initially available energy.
- **Infrastructure**: The infrastructure required for reliable V2V energy transfer is not yet widely deployed. The need for compatible physical interfaces and communication systems between vehicles highlights the importance of standardized protocols. However, such standards may not be sufficiently mature or universally adopted, representing a major barrier to interoperability and large-scale implementation.
- **Scalability**: While other modes such as V2G enable aggregation of multiple EVs, increasing the overall system capacity, V2V is inherently limited to pairwise interactions between vehicles. This limitation is further exacerbated by the requirement of physical proximity and simultaneous availability of both EVs, introducing operational uncertainty and restricting applicability in broader system-level scenarios.

10.3.4 Other modes

10.3.4.1 Vehicle-to-building (V2B)

The V2B operating mode extends the bidirectional energy flow paradigm to the building domain. Its main difference with respect to the V2H mode lies in the integration of tertiary or commercial buildings, where the EV operates as a DER within installations characterized by higher and more complex demand profiles than residential environments. In such contexts, the vehicle must typically be integrated into pre-existing energy management frameworks.

In this regard, the functions of the EV in a V2B scheme are conceptually similar to those in V2H operation, namely the enhancement of self-consumption, peak demand reduction, and the optimization of electricity costs under dynamic pricing schemes. However, in V2B platforms the EV must coordinate with a more heterogeneous portfolio of resources compared to V2H, including PV generators, stationary storage systems, and critical or non-critical loads, among others [16].

Another distinguishing feature of V2B is its higher scalability. In buildings equipped with collective parking facilities, fleet-level aggregation schemes can be implemented, enabling the coordinated operation of multiple EVs. This aggregation capability increases the overall flexibility and capacity of the bidirectional energy exchange system, reinforcing the role of EVs as active elements within advanced building energy management architectures.

10.3.4.2 Vehicle-to-load (V2L)

The V2L operating mode represents the most direct and decentralized paradigm for the utilization of the energy stored in onboard battery systems of EVs. In this scheme, the EV effectively operates as a mobile power source, supplying energy directly to an external load without requiring additional infrastructure or advanced energy management strategies.

From an operational perspective, V2L is the simplest mode to implement, as it does not rely on advanced communication infrastructures or coordination with other energy resources. In this context, the activation of the energy flow from the vehicle is primarily driven by user commands, where the user determines when and how the available energy is utilized, subject to basic constraints such as the SoC and the maximum output power of the system.

Consequently, V2L is particularly suitable for applications requiring portable energy supply in locations with limited or no access to the electrical grid. Representative use cases include outdoor activities, emergency operations, field work, and temporary power supply during grid outages. However, the V2L scheme is inherently constrained by the physical limitations of the EV, including restricted power capability and potential battery degradation under intensive or prolonged use.

10.4 Applications and use cases

10.4.1 Integration in microgrids

A microgrid can be defined as a set of interconnected loads and DERs, with clearly defined operational boundaries, that act as a single energy entity at the system level [17]. In this context, the integration of EVs into microgrids represents one of the most paradigmatic use cases of V2X implementation, as it enables the active participation of EVs in energy systems with a certain degree of autonomy. Within this framework, EVs can operate as DERs, exporting energy to the microgrid during specific periods and thereby contributing to the balance between generation and demand. Additionally, their modular and distributed nature facilitates scalable deployment across different microgrid configurations.

One of the most distinctive characteristics of a microgrid is its ability to operate in isolated mode, meaning that it can supply local demand without support from an external grid. Under these conditions, EVs can play a fundamental role by providing the energy stored in their batteries to meet local demand [18]. This capability becomes particularly relevant in microgrids with high penetration of intermittent renewable energy sources, such as PV or wind, where variability and uncertainty in generation may compromise system stability and power quality.

From an operational perspective, the effective implementation of V2X platforms in microgrid environments requires local control strategies capable of coordinating the different resources connected to the system, including DERs, stationary storage systems, and loads. Depending on system complexity and operational requirements, more advanced approaches such as decentralized management mechanisms may be required, particularly in scenarios involving a large number of distributed assets.

In addition to supporting the generation-demand balance, EVs can also provide small-scale ancillary services, such as frequency support and voltage regulation. These capabilities enhance the operational performance of the microgrid and ultimately contribute to improving its overall resilience, reliability, and adaptability under dynamic operating conditions, especially during disturbances or rapid demand fluctuations.

10.4.2 Renewable energy support

In power systems with high penetration of renewable energy sources, significant energy curtailment may occur due to the temporal mismatch between generation and demand. This phenomenon is particularly common in solar-based systems, where peak generation typically takes place during midday hours. If demand is not sufficiently high during these periods, the generated solar energy must be curtailed or dissipated. In this context, energy storage systems have gained relevance as key enablers to capture excess renewable generation and shift its use to periods of higher demand, thereby increasing overall system efficiency.

Within this framework, EVs can operate as flexible loads, absorbing surplus renewable energy through controlled charging processes. As a result, renewable energy support has become one of the most prominent application domains of V2X strategies [19]. In addition, EVs can complement other energy management and storage solutions, such as hydrogen production chains or stationary storage systems, contributing to a more efficient utilization of renewable resources.

Furthermore, the geographically distributed nature of charging infrastructure makes EVs a highly decentralized and spatially dispersed storage resource. This characteristic reduces the need for long-distance energy transport thus reducing transmission losses, and enables localized support for distributed renewable generation, particularly in low-voltage or medium-voltage networks, where congestion issues may arise under high renewable penetration.

However, the effective utilization of EVs in this context requires well-coordinated control mechanisms between renewable generation and vehicle operation. Aggregation schemes can play a key role, not only by increasing the effective capacity of the EV fleet, but also by simplifying control strategies and improving scalability. In this way, EVs are emerging as a critical asset in the decarbonization of both the transportation sector and the broader power system, supporting higher integration of renewable energy sources and enhancing overall system flexibility, reliability, and operational efficiency under variable generation conditions.

10.4.3 Energy resilience and backup

In scenarios where ensuring continuity of supply is a critical requirement, V2X schemes can play a relevant role. EVs can effectively operate as backup generators during supply interruptions, thereby enhancing overall system resilience [20]. Under normal operating conditions, EVs remain connected to charging points, absorbing energy from the grid and storing it in onboard battery systems. However, reverse power flow can be enabled when required. In the event of contingencies, EVs can therefore supply energy to critical loads, operating as distributed backup sources within the system.

These schemes can be implemented at different levels. At the most basic level, EVs can provide backup to specific loads through V2L configurations. More advanced implementations include coordinated backup at building level through V2H or V2B platforms, where multiple loads can be supported in an organized manner. At a broader scale, V2G schemes can extend resilience capabilities to the grid level, contributing to the continuity of supply across larger network segments. As a result, EVs have gained relevance as support systems in critical infrastructures, such as hospitals, data centers, or emergency facilities, where uninterrupted power supply is essential. Similarly, they can play an important role in energy systems with limited or unreliable grid access, acting as flexible and mobile energy sources that enhance energy accessibility.

The use of EVs as backup systems requires the implementation of control strategies that ensure availability when needed. For instance, control algorithms may enforce a minimum SoC to guarantee a predefined backup duration. In addition, prioritization mechanisms may be required to supply critical loads over non-essential ones during emergency operation modes. Nevertheless, this application presents inherent limitations, particularly in terms of operating time, which is directly constrained by battery capacity, and available power, which is limited by the rated power of charging interfaces. These constraints must be considered when designing reliable backup strategies.

10.5 Advantages and limitations

10.5.1 Technical advantages

The V2X operating mode introduces a set of technical advantages that position EVs as a relevant agent in the evolution of the power system. These advantages encompass both operational aspects, contributing to increased flexibility and improved system operation, as well as benefits from an energy efficiency perspective. As a summary, the main advantages provided by the V2X mode can be outlined as follows:

- **Operational flexibility**: The V2X framework expands the operational range of EVs, enabling them to transition from passive loads to active energy sources when required. This capability facilitates real-time adaptation to variations in demand, electricity prices, or generation availability, thereby improving overall system efficiency and enabling more dynamic grid operation under uncertain conditions.
- **Efficiency**: The proper management of EVs as system assets allows controlled charge and discharge cycles, enabling temporal energy shifting. EVs can absorb excess energy during low-demand periods and release it during peak demand,

reducing curtailment and improving the utilization of renewable resources while enhancing overall energy balancing capabilities.

- **System support**: EVs can actively participate in the provision of ancillary services such as frequency support, voltage regulation, and peak demand mitigation. These capabilities contribute to enhancing system stability, security, and resilience under variable operating conditions, particularly in systems with high renewable penetration.
- **Decentralization**: The connection interfaces of EVs are inherently distributed across the power system. This provides EV fleets with a decentralized nature, reducing dependence on centralized infrastructures and enabling more resilient system architectures with improved fault tolerance.
- **Scalability**: V2X platforms naturally increase their capabilities as the EV fleet expands. This reduces the need for large, discrete investments, which in the case of traditional grid assets are often required upfront and frequently supported by public entities, allowing progressive system evolution.
- **Integration with distributed generation**: The distributed nature of EVs enables effective coordination with local generation resources such as small-scale renewable systems. This coordination improves local energy management and reduces congestion in distribution networks, enhancing overall system efficiency.
- **Reduction of network losses**: EVs integrated within V2X frameworks promote local consumption and generation, thereby reducing energy losses associated with long-distance power transmission, particularly in constrained or highly loaded networks, and improving grid utilization.
- **Fast response capability**: Due to the characteristics of their power electronic interfaces, EVs can respond rapidly to control signals. This makes them particularly suitable for applications requiring high-speed response, such as ancillary services or transient system events, improving dynamic system performance.

10.5.2 Technical limitations

Despite the advantages described in Section 10.5.1, the deployment of V2X platforms presents several technical limitations that hinder their large-scale implementation. These limitations are mainly associated with the operation of onboard energy storage systems, the efficiency of the energy conversion process, and the overall complexity of system integration. A summary of the main limitations of the V2X operating mode is provided below:

- **Battery degradation**: The V2X operating mode involves an increased number of charge and discharge cycles, which accelerates the degradation of battery energy storage systems (see Chapter 7). The repeated activation of bidirectional energy flows can reduce the lifetime of EV batteries, depending on factors such as DoD, charge rates, and thermal conditions. This degradation introduces both technical and economic implications that may significantly affect system viability.
- **Energy losses**: Bidirectional energy transfer requires multiple conversion stages, typically involving AC/DC and DC/DC converters (see Section 10.2.1). The

operation of these converters introduces energy losses (see Section 10.2.1.2), leading to a reduction in overall system efficiency compared to conventional unidirectional operation. Furthermore, the heat generated by these losses imposes additional constraints on thermal management, system design, and operational safety.

- **Technical complexity**: The implementation of V2X schemes requires the integration of advanced power electronic devices, robust and reliable communication infrastructures, and sophisticated control systems. This level of complexity increases design, installation, and operational requirements, as well as associated costs and maintenance needs, potentially limiting widespread adoption.
- **Operational limitations**: EV usage patterns introduce uncertainty due to their variable availability, making them less predictable energy assets compared to stationary storage systems. This variability complicates system planning and reduces the reliability of energy services provided by EVs, particularly in applications requiring guaranteed availability.
- **Power constraints**: The maximum power that an EV can deliver is limited by both the charging interface and the onboard power electronics. These constraints may restrict participation in certain applications, especially those requiring high power levels or sustained energy delivery over extended periods.
- **Interoperability and standardization**: The coexistence of multiple manufacturers, communication protocols, and system architectures complicates the homogeneous integration of V2X systems. In this context, standardization efforts and the adoption of common communication protocols are critical factors influencing the scalability and deployment of the technology (see Section 10.2.2.2).
- **Impact on SoC management**: Participation in V2X services may compromise the SoC of the vehicle to levels that do not guarantee its primary mobility function. This necessitates the implementation of additional control strategies to enforce minimum SoC thresholds, thereby increasing system complexity and operational constraints.

10.5.3 Economic considerations

The acquisition and installation costs associated with the infrastructure required for the deployment of V2X technologies remain significantly higher than those associated with conventional unidirectional charging systems. As shown in Table 10.3, the costs related to bidirectional systems are considerably greater than those required for a typical unidirectional charger, particularly regarding charger acquisition and additional expenses such as permits and administrative procedures. This difference is mainly attributed to the fact that bidirectional chargers require more advanced power electronics, protection systems, communication capabilities, and control functionalities, all of which substantially increase the initial installation cost.

Due to these elevated upfront costs, the economic return associated with the implementation of V2X technologies must be carefully evaluated from the user perspective. In this context, several factors must be considered, including electricity

Table 10.3 Representative comparison of average acquisition and installation costs associated with residential unidirectional and bidirectional EV chargers, according to data from [21,22]. (All values given in US dollars).

Cost Component	Unidirectional	Bidirectional
Charger	550	3100
Labor & Installation	1350	2550
Panel upgrade	2250	4250
Others	125	1225

prices, energy compensation mechanisms, participation in ancillary service markets, and vehicle usage patterns. These aspects are critical and may strongly influence the overall economic feasibility of the infrastructure associated with a V2X platform.

In addition to infrastructure-related costs, the uncertainty associated with battery degradation caused by the increased utilization of onboard energy storage systems must also be considered. This degradation cost is difficult to quantify accurately, particularly because it is associated with long-term aging phenomena that strongly depend on environmental conditions, charging and discharging cycles, and user operating patterns (see Chapter 7).

10.5.4 Operational challenges

Beyond the technical limitations (Section 10.5.2) and economic constraints (Section 10.5.3), V2X operation presents several operational challenges that may condition its implementation in certain scenarios. These challenges arise from different aspects such as coordination and communication requirements, as well as user-related considerations, and can be summarized as follows:

- **Increased coordination complexity:** Bidirectional EV operation requires continuous information exchange in order to properly coordinate vehicles with other EVs, charging infrastructure, grid operators, and energy management systems. Furthermore, V2X operation demands a high degree of synchronization among the involved agents, particularly in scenarios with a large number of simultaneously connected EVs. As system scale increases, maintaining stable and coordinated operation becomes progressively more challenging.
- **Communication requirements:** Due to the increased coordination demands, V2X systems require more complex communication infrastructures than unidirectional charging systems. In addition, the need for coordination among heterogeneous agents introduces additional challenges related to technological interoperability, which may require the adoption of standardized communication protocols (see Section 10.2.2.1).
- **User dependency:** Unlike stationary energy storage systems, EVs present highly variable and uncertain availability, strongly dependent on user mobility

patterns. Consequently, the energy available at different times of the day may differ substantially from the estimations performed by different agents such as system operators [10]. This variability introduces uncertainty and an unavoidable operational risk associated with V2X platforms.

- **Mobility requirements:** V2X operation may conflict with user mobility needs. If appropriate control mechanisms are not implemented, onboard batteries may reach energy levels below those required for transportation purposes. This factor must be carefully considered during the development of bidirectional systems, since poor user experience may become a significant barrier to the adoption of V2X technologies.

10.6 Regulatory and market framework

10.6.1 Regulatory overview

Since bidirectional EV operation involves not only energy consumption but also direct energy injection into the electrical grid, the technical and regulatory requirements associated with bidirectional systems must be more stringent than those applicable to conventional unidirectional charging systems. In bidirectional operation, aspects related to grid connection, power quality, and electrical protection must be properly addressed, together with the communication mechanisms established among vehicles, chargers, and system operators. In general, the regulatory framework associated with V2X operation must ensure the following operational conditions for the electrical system:

- Operational safety.
- Interoperability among agents.
- System stability.

One of the main regulatory challenges associated with V2X operation is the lack of harmonization among different regions. Currently, the development status of specific regulatory frameworks varies significantly across countries. Furthermore, in those regions where dedicated regulations already exist, their objectives and underlying principles differ considerably (see Section 10.6.3). This regulatory heterogeneity affects interoperability among manufacturers, platforms, and energy operators, as well as communication protocols and certification requirements for bidirectional chargers and energy management systems.

In many cases, EVs are still considered purely as electrical loads within existing regulatory frameworks, thereby limiting the large-scale integration of V2G functionalities. Consequently, regulatory frameworks must evolve toward the effective integration of EVs into electricity markets. In this regard, dedicated regulations are required to define operational responsibilities, economic compensation mechanisms, and access conditions to energy and ancillary service markets. The development of such regulatory frameworks may create new business opportunities for V2X platforms, thus strongly influencing their economic feasibility (see Section 10.6.2).

Finally, V2X systems must comply with several technical requirements intended to guarantee system stability during grid injection operation. Therefore, regulations must address technical aspects related to anti-islanding protection, grid synchronization, power limitations, and cybersecurity.

10.6.2 Market structures and business models

The economic viability of V2X platforms strongly depends on their capability to actively participate in different electricity markets. Therefore, the development of market structures capable of integrating EVs as active energy resources within the electrical system is essential for the large-scale deployment of V2X technologies.

In this context, EVs provide significant operational flexibility and may participate in electricity markets through different mechanisms, supplying services such as frequency regulation, load balancing, peak shaving, or support for renewable energy integration. However, the relatively limited energy and power capacity of individual EVs complicates their direct participation in electricity markets [10]. Consequently, the role of the aggregator becomes particularly relevant, since aggregators are capable of coordinating large EV fleets and managing their collective participation in energy and ancillary service markets.

Beyond enabling the participation of EV fleets in electricity markets, aggregators also contribute to increasing the available aggregated energy capacity, reducing operational uncertainty, and facilitating the interaction among users, grid operators, and electricity markets [23]. In practice, aggregators act as intermediary entities capable of transforming distributed and highly variable EV resources into manageable and commercially viable energy assets.

Nevertheless, the economic feasibility of V2X platforms operating under market-based schemes still depends on several factors, including regulatory conditions, electricity prices, infrastructure costs, and the adoption level of bidirectional charging technologies. As a consequence, many initiatives related to EV participation in electricity markets remain in pilot phases or early commercialization stages [24].

10.6.3 Barriers to deployment

In summary, V2X operation may provide significant potential advantages in terms of energy management and operational flexibility (see Section 10.5.1). However, the development and deployment of these technologies remain at a relatively early stage and still occupy a marginal position compared with conventional unidirectional charging systems. This situation is mainly caused by several technological, regulatory, and economic limitations affecting different agents within the electrical system, including users, manufacturers, infrastructure operators, and energy market participants.

One of the main barriers to the implementation of V2X technologies is the high degree of heterogeneity among manufacturers, regions, and technological platforms. This heterogeneity results in substantial differences regarding communication protocols, certification requirements, and control architectures employed in bidirectional

Table 10.4 Representative comparison of V2X regulatory and standardization approaches in different regions.

Region	Main Standards / Regulations	Current Status	Main Focus
European Union	AFIR, ISO 15118, IEC 61851	Advanced regulatory deployment	Interoperability and smart charging integration
United States	IEEE 1547, ISO 15118, OCPP	Partial deployment and state-level implementation	Grid integration and DER coordination
Japan	CHAdeMO, ISO 15118	Early V2G adoption	Bidirectional charging and resilience
China	GB/T standards, pilot V2G regulations	Rapid expansion phase	Large-scale infrastructure deployment

systems. Consequently, the current operational and regulatory framework complicates interoperability among vehicles, chargers, and grid operators, thereby limiting the coordinated integration of V2X platforms into large-scale electrical systems.

Another important barrier concerns the regulatory framework itself. As shown in Table 10.4, where different V2X regulatory frameworks across representative regions are compared, significant differences exist regarding both the definition and objectives of the associated regulations. In addition to this regulatory heterogeneity, the degree of development of V2X regulations also varies considerably among regions. While some countries already possess relatively advanced regulatory frameworks, in other regions the regulations associated with V2X operation remain in early development stages or are simply nonexistent.

Finally, economic and operational factors further complicate the deployment of V2X platforms. In particular, the elevated costs associated with the required infrastructure (see Section 10.5.3), together with the increasing complexity related to communication, operation, and coordination requirements, constitute major barriers to the development and adoption of bidirectional charging systems.

10.7 Case study: rule-based V2X operation

10.7.1 Control strategy

To illustrate the operational capabilities of V2X technologies, this section presents a case study representing a simplified V2H platform. The considered system is composed of an EV, a residential load with demand profile P_{load}, a PV generation system with output power P_{PV}, and a grid connection. In order to control the bidirectional energy flow associated with the EV battery, a rule-based control algorithm has been implemented. The general operation of the proposed algorithm is summarized in Fig. 10.3, whereas the `MATLAB`® code is provided in Section 10.7.4.

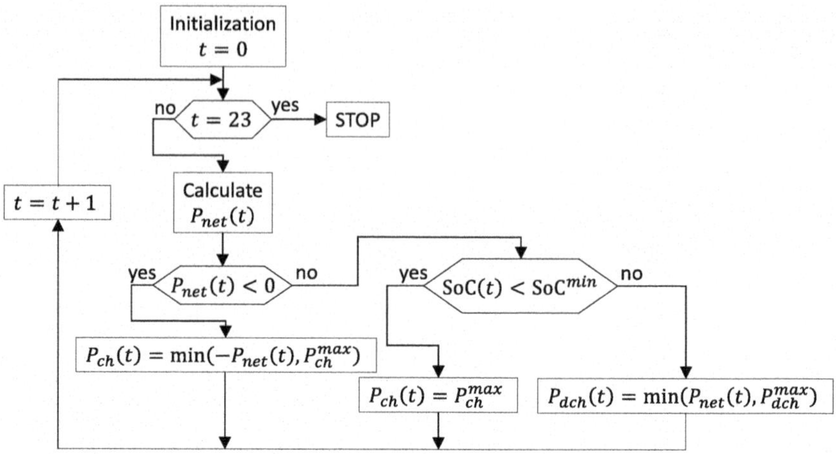

Figure 10.3 Flowchart of the heuristic control strategy considered in the case study.

The proposed strategy is based on the calculation of the net residential demand at each time instant as:

$$P_{net}(t) = P_{load}(t) - P_{PV}(t) \tag{10.1}$$

In this way, positive values of P_{net} indicate an energy deficit in the residential system, whereas negative values represent a PV generation surplus that may be used to charge the EV battery.

Depending on the value of P_{net}, the algorithm applies simple heuristic rules to determine whether the EV should operate in charging or discharging mode. In particular, at each time instant t, the total exchanged energy is calculated as:

$$\Delta E(t) = P_{ch}(t)\eta_{ch} - \frac{P_{dch}(t)}{\eta_{dch}} \tag{10.2}$$

where P_{ch} and P_{dch} correspond to charging and discharging powers, respectively, while η_{ch} and η_{dch} represent charging and discharging efficiencies.

Finally, the battery SoC evolution is updated according to:

$$\mathrm{SoC}(t+1) = \mathrm{SoC}(t) + \frac{\Delta E(t)}{E^{max}} \tag{10.3}$$

while continuously satisfying the operational constraints defined for SOC_{min} and SOC_{max}.

10.7.2 Case description

The considered case study assumes a simulation horizon of 24 hours divided into time intervals of 1 hour. A single EV connected to the residential installation is considered,

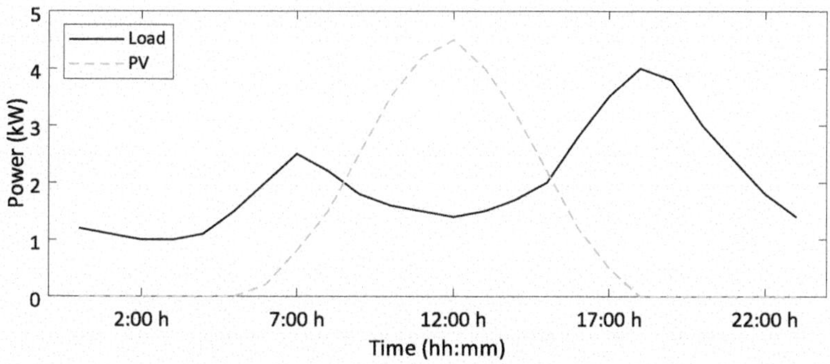

Figure 10.4 Home demand and PV generation of the case study.

with a total battery capacity of 60 kWh and an initial SoC of 60%. In addition, lower and upper operational limits are imposed on the battery state of charge, corresponding to 50% and 90%, respectively. These constraints ensure an operation compatible with technical limitations and minimum mobility requirements.

The EV battery system can operate in both charging and discharging modes through a bidirectional charger with a maximum power of 7 kW. Charging and discharging processes are assumed to occur with a constant efficiency of 95%.

The temporal profiles corresponding to residential demand and PV generation are shown in Fig. 10.4. As can be observed, the residential load profile presents maximum demand levels during evening and night hours, whereas PV generation is concentrated around the central hours of the day. Consequently, the selected profiles provide representative operating conditions to analyze the interaction between renewable generation, residential demand, and bidirectional EV operation within a V2H framework.

Furthermore, the proposed model assumes deterministic load and PV generation profiles, neglecting forecasting errors, communication delays, and stochastic variations associated with user behavior or environmental conditions.

10.7.3 Results and discussion

This section analyzes the results obtained by applying the control strategy described in Section 10.7.1 under the operating conditions presented in Section 10.7.2.

First, Fig. 10.5 shows the instantaneous charging and discharging power profiles for two operating conditions: one with the V2H functionality enabled and another corresponding to conventional unidirectional charging operation. In this figure, negative power values indicate battery discharging operation. As can be observed, in both scenarios the EV charges during the central hours of the day, coinciding with the period of maximum PV generation. However, when the V2H mode is enabled, the vehicle additionally discharges energy during specific periods, particularly during early morning and evening hours.

This operating strategy reduces the dependence on grid energy, as illustrated in Fig. 10.6, which presents the grid power exchange under both operating conditions.

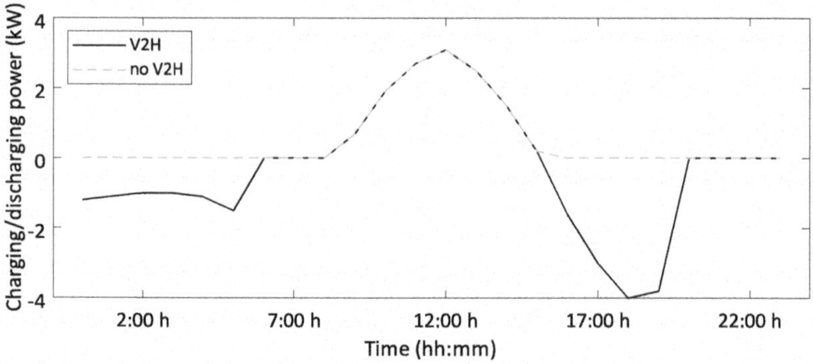

Figure 10.5 Charging and discharging power profiles for V2H and non-V2H operation modes. Negative power values indicate battery discharging operation.

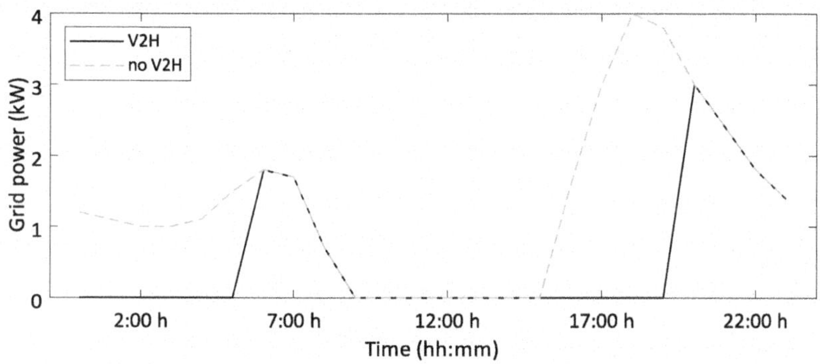

Figure 10.6 Grid power exchange under V2H and non-V2H operating conditions.

The obtained results demonstrate that the V2H mode decreases the energy imported from the electrical grid, especially during the periods in which the EV operates in discharging mode. Consequently, the bidirectional operation improves local utilization of PV generation and contributes to peak demand reduction.

Finally, Fig. 10.7 shows the temporal evolution of the battery SoC. In both operating scenarios, the SoC increases during the central hours of the day due to PV surplus utilization. Nevertheless, when the V2H mode is activated, the SoC decreases during discharging intervals, resulting in lower stored energy levels at the end of the simulation horizon. This result highlights the importance of carefully controlling battery energy levels during bidirectional operation in order to avoid interference with user mobility requirements.

10.7.4 MATLAB code of the rule-based V2H control

The MATLAB implementation corresponding to the proposed rule-based V2H strategy is presented below. The code includes the definition of system parameters, residential demand and PV generation profiles, battery operational constraints, and the heuristic

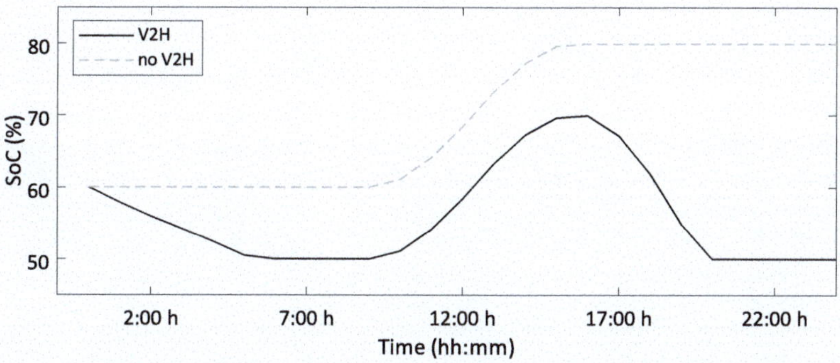

Figure 10.7 Battery SoC evolution for V2H and non-V2H operating conditions.

control rules governing EV charging and discharging operation. In addition, the algorithm computes the temporal evolution of the battery SoC and the power exchanged with the electrical grid during the simulation horizon.

```

%% Simulation parameters

dt = 1; % Time step (h)
T = 24; % Simulation horizon (h)
t = 0:dt:T-1;

V2H = 0; % V2H mode (if = 1, V2H mode is activated)

%% EV parameters

Ebat = 60; % Battery capacity (kWh)

SOC_init = 0.60; % Initial SOC (pu)
SOC_min  = 0.50; % Minimum SOC (pu)
SOC_max  = 0.90; % Maximum SOC (pu)

Pcharge_max    = 7; % Maximum charging power (kW)
Pdischarge_max = 7; % Maximum discharging power (kW)

eta_charge    = 0.95; % Charging efficiency (pu)
eta_discharge = 0.95; % Discharging efficiency (pu)

%% Residential load profile (kW)

Pload = [ ...
    1.2 1.1 1.0 1.0 1.1 1.5 ...
    2.0 2.5 2.2 1.8 1.6 1.5 ...
    1.4 1.5 1.7 2.0 2.8 3.5 ...
    4.0 3.8 3.0 2.4 1.8 1.4 ];
```

```
%% PV generation profile (kW)

PPV = [ ...
    0 0 0 0 0 0 ...
    0.2 0.8 1.5 2.5 3.5 4.2 ...
    4.5 4.0 3.2 2.2 1.2 0.5 ...
    0 0 0 0 0 0 ];

%% Initialization

Results.SOC = zeros(1,T+1);
Results.SOC(1) = SOC_init; % Initial SoC

Results.Pcharge    = zeros(1,T);
Results.Pdischarge = zeros(1,T);

Results.Pgrid = zeros(1,T);

%% Rule-based control

for k = 1:T

    % Net residential demand
    Pnet = Pload(k) - PPV(k);

    %% Charging mode

    % Case 1: PV surplus available
    if PPV(k) > Pload(k)

        Psurplus = PPV(k) - Pload(k);

        if Results.SOC(k) < SOC_max

            Results.Pcharge(k) = ...
                min(Psurplus, Pcharge_max);

        end

    % Case 2: SOC below minimum threshold
    elseif Results.SOC(k) < SOC_min

        Results.Pcharge(k) = Pcharge_max;

    end

    %% Discharging mode (V2H)
```

```
    if V2H == 1

        if Pload(k) > PPV(k) && ...
           Results.SOC(k) > SOC_min

            Results.Pdischarge(k) = ...
                min(Pnet, Pdischarge_max);

        end

    end

    %% Prevent simultaneous charge/discharge

    if Results.Pcharge(k) > 0 && ...
       Results.Pdischarge(k) > 0

        Results.Pdischarge(k) = 0;

    end

    %% SOC update

    DeltaE = ...
        eta_charge * Results.Pcharge(k) * dt ...
        - Results.Pdischarge(k) * dt ...
        / eta_discharge;

    Results.SOC(k+1) = ...
        Results.SOC(k) + DeltaE / Ebat;

    % SOC limits
    Results.SOC(k+1) = ...
        max(SOC_min, ...
        min(Results.SOC(k+1), SOC_max));

    %% Grid power

    Results.Pgrid(k) = ...
        Pload(k) ...
        - PPV(k) ...
        - Results.Pdischarge(k) ...
        + Results.Pcharge(k);

end
```

References

[1] Z. Salmani Khankahdani, M.S. Ghazizadeh, V. Vahidinasab, A decentralized optimization framework for enhancing resilience in reconfigurable smart grids: leveraging mobile battery units, Journal of Energy Storage 155 (2026) 121413, https://doi.org/10.1016/j.est.2026.121413.

[2] M. Adham, S. Keene, R.B. Bass, Distributed energy resources: a systematic literature review, Energy Reports 13 (2025) 1980–1999, https://doi.org/10.1016/j.egyr.2025.01.026.

[3] M. Tostado-Véliz, A. Borghetti, P. Siano, F. Jurado, Privacy-preserving energy management in local energy communities with evs – an enhanced benders-like solution strategy, Sustainable Cities and Society 132 (2025) 106823, https://doi.org/10.1016/j.scs.2025.106823.

[4] M.H. Rashid (Ed.), Handbook of Power Electronics in Autonomous and Electric Vehicles, 1st edition, Academic Press, 2024, eBook ISBN: 9780323950985.

[5] M.K. Khan, K. Kauhaniemi, H. Laaksonen, M.A. Hassan, Review of recent developments in grid codes: focus on compliance testing and grid-forming inverter-based resources, Renewable and Sustainable Energy Reviews 227 (2026) 116509, https://doi.org/10.1016/j.rser.2025.116509.

[6] V. Dvorkin, F. Fioretto, P. Van Hentenryck, P. Pinson, J. Kazempour, Differentially private optimal power flow for distribution grids, IEEE Transactions on Power Systems 36 (3) (2021) 2186–2196, https://doi.org/10.1109/TPWRS.2020.3031314.

[7] O. Izquierdo-Monge, A.Z. Velasco Bonilla, M. Lafuente-Cacho, P. Peña-Carro, Ángel Hernández-Jiménez, Heuristic method for electric vehicle charging in a Spanish microgrid: leveraging renewable energy surplus, Journal of Power Sources 629 (2025) 235945, https://doi.org/10.1016/j.jpowsour.2024.235945.

[8] M. Tostado-Véliz, S. Kamel, H.M. Hasanien, R.A. Turky, F. Jurado, Optimal energy management of cooperative energy communities considering flexible demand, storage and vehicle-to-grid under uncertainties, Sustainable Cities and Society 84 (2022) 104019, https://doi.org/10.1016/j.scs.2022.104019.

[9] M. Elnady, S. Ozana, Artificial intelligence techniques in vehicle-to-grid (v2g) systems: a review, comparative study, and model evaluation, Journal of Energy Storage 135 (2025) 118155, https://doi.org/10.1016/j.est.2025.118155.

[10] M. Tostado-Véliz, H.M. Hasanien, J. Carpio, F. Jurado, Risk-aware strategies for optimal participation of parking lots in day-ahead electricity markets, Energy 322 (2025) 135406, https://doi.org/10.1016/j.energy.2025.135406.

[11] S. Tian, G. Razeghi, S. Samuelsen, System-level demonstration of vehicle-to-home integration with a mass-market ev, Energy Conversion and Management 348 (2026) 120629, https://doi.org/10.1016/j.enconman.2025.120629.

[12] K. Sevdari, L. Calearo, P.B. Andersen, M. Marinelli, Ancillary services and electric vehicles: an overview from charging clusters and chargers technology perspectives, Renewable and Sustainable Energy Reviews 167 (2022) 112666, https://doi.org/10.1016/j.rser.2022.112666.

[13] S. Yun, J. Woo, K. Kwak, Unlocking peak shaving: how ev driver heterogeneity shapes v2g potential, Energy 329 (2025) 136773, https://doi.org/10.1016/j.energy.2025.136773.

[14] M. Sandström, P. Huang, C. Bales, E. Dotzauer, Combined hosting capacity analysis for electric vehicles and rooftop photovoltaic systems – impacts on a Swedish residential power grid, Energy 351 (2026) 140805, https://doi.org/10.1016/j.energy.2026.140805.

[15] D. Zhao, Z. Zhou, D. Boff, A. Fernandez, Hybrid power purchase agreements for flexible 24/7 energy delivery – a comprehensive review of current practices and research pathways, Applied Energy 410 (2026) 127509, https://doi.org/10.1016/j.apenergy.2026.127509.

[16] K.-Y. Lo, J.H. Yeoh, I.-Y.L. Hsieh, Towards nearly zero-energy buildings: smart energy management of vehicle-to-building (v2b) strategy and renewable energy sources, Sustainable Cities and Society 99 (2023) 104941, https://doi.org/10.1016/j.scs.2023.104941.
[17] D.T. Ton, M.A. Smith, The U.S. Department of Energy's microgrid initiative, The Electricity Journal 25 (8) (2012) 84–94, https://doi.org/10.1016/j.tej.2012.09.013.
[18] J.I. Schmidt, M.M. Wilber, T. Schwoerer, T. Bodony, B. Sanderson, L. Albertson, M. Bergan, Off the grid and on the fence: unpacking electric vehicle adoption barriers in isolated microgrids, Energy Research & Social Science 127 (2025) 104281, https://doi.org/10.1016/j.erss.2025.104281.
[19] T. Rawat, J. Singh, V.C. Pandey, S. Sharma, Planning renewables in distribution system with electric vehicles to improve hosting capacity and energy losses: two-stage stochastic optimization framework, Sustainable Energy, Grids and Networks 39 (2024) 101445, https://doi.org/10.1016/j.segan.2024.101445.
[20] P. Huang, A. Kidanemariam, L.-H. Bjornsson, Transforming electric vehicles into mobile power sources: technical performance evaluation of the electric vehicle based virtual electricity network (even) solution for improving the power supply resilience, Sustainable Cities and Society 128 (2025) 106477, https://doi.org/10.1016/j.scs.2025.106477.
[21] EVQuoter, Ev charger installation cost guide 2025, https://evquoter.com/ev-charger-installation-cost-guide-2025, 2025. (Accessed 7 May 2026).
[22] LatestCost, Bidirectional ev charger cost guide 2026, https://latestcost.com/bidirectional-ev-charger-cost/, 2025. (Accessed 7 May 2026).
[23] F. De Caro, G. Graber, V. Calderaro, A. Vaccaro, V. Galdi, Exploring economic profit of ev aggregators in supplying flexibility services through v2g, Sustainable Energy, Grids and Networks 43 (2025) 101851, https://doi.org/10.1016/j.segan.2025.101851.
[24] X. Lei, J. Zhong, Y. Chen, Z. Shao, L. Jian, Grid integration of electric vehicles within electricity and carbon markets: a comprehensive overview, eTransportation 25 (2025) 100435, https://doi.org/10.1016/j.etran.2025.100435.

Smart charging strategies

11

11.1 Introduction to smart charging

11.1.1 Definition and role of smart charging

With the accelerated growth in the number of EVs in operation (see Chapter 3 for the specific case of ECs), EV charging has become a major challenge from the perspective of power system operation. Conventionally, EVs start charging at the maximum available power immediately after being connected to the charging infrastructure and continue operating under this condition until the required energy demand is satisfied. This uncontrolled charging approach is simple from both operational and implementation perspectives. However, it can generate significant technical problems, particularly in scenarios with high EV penetration levels (see Section 11.1.2).

As a counterpart to uncontrolled charging, the concept of smart charging has emerged. In general terms, smart charging can be understood as the set of technologies and control strategies intended to optimally manage the EV charging process while simultaneously considering user requirements, grid conditions, and economic or operational criteria. To achieve this objective, smart charging systems must incorporate monitoring, communication, and dynamic control capabilities that are not available in conventional charging systems.

Naturally, the smart charging paradigm is closely associated with the development of smart grids, the digitalization of energy systems, and demand-side management strategies [1]. Advanced communication infrastructures and real-time data acquisition systems enable charging stations, aggregators, and grid operators to exchange operational information continuously, thereby facilitating coordinated charging management.

Furthermore, smart charging facilitates the integration of EVs with other DERs connected to the power system, such as renewable generation technologies. In this context, smart charging strategies can properly manage charging schedules by exploiting periods with more favorable renewable generation conditions, reducing grid stress and improving the overall flexibility and efficiency of the electrical infrastructure.

11.1.2 Challenges of uncontrolled EV charging

Although uncontrolled charging presents certain operational advantages due to its simplicity, its large-scale implementation introduces several important technical and economic challenges. Some of the most relevant issues are summarized below:

- **Peak demand increase:** The temporal coincidence of uncontrolled charging processes in a large number of EVs can produce significant demand peaks (see Fig. 11.1), particularly when fast charging infrastructures are employed (see Chapter 6). This problem becomes more severe when charging periods coincide

Electric Mobility Systems. https://doi.org/10.1016/B978-0-44-349397-3.00019-7

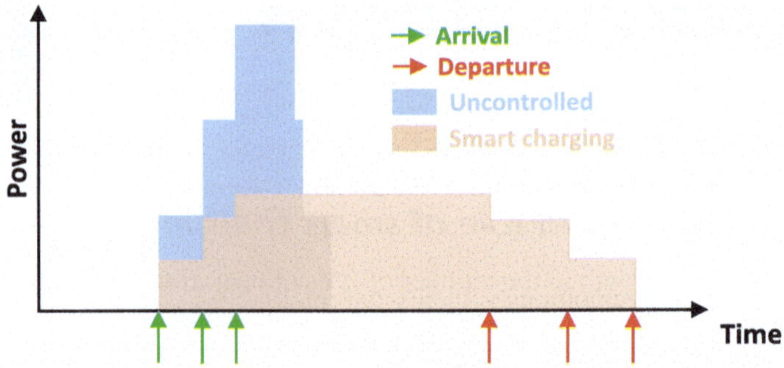

Figure 11.1 Comparison between uncontrolled charging and smart charging strategies for EVs. In uncontrolled charging, EVs start charging immediately after connection and typically operate at maximum available power until the required energy demand is satisfied. In contrast, smart charging dynamically adjusts the charging power profile according to grid conditions, operational constraints, electricity prices, or renewable energy availability, thereby reducing peak demand and improving overall grid flexibility.

with high residential or commercial demand intervals, potentially requiring reinforcement of grid infrastructure elements such as transformers or distribution lines.

- **Distribution network congestion:** In addition to temporal simultaneity, the geographical concentration of charging stations can generate severe network operation problems, including line congestion and voltage instability phenomena. These issues can be particularly critical in low-voltage distribution networks, which in most cases were not originally designed to support the high power demand associated with the charging of large EV fleets.
- **Economic impacts:** The appearance of high charging demand peaks can also increase electricity prices in wholesale electricity markets. In most electricity market structures, energy prices are determined by the intersection between supply and demand curves (see Chapter 6). Consequently, significant increases in electricity demand may require the operation of less economical generation units, thereby increasing the overall electricity price.
- **Renewable integration challenges:** Uncontrolled charging can also complicate the integration of renewable energy sources, particularly highly intermittent technologies such as PV and wind power generation. In many cases, EV charging demand peaks may occur during periods with limited renewable generation availability, reducing renewable energy utilization and increasing dependence on conventional generation technologies.

11.1.3 General architecture of smart charging systems

A smart charging system is structured around an architecture capable of integrating power, communication, and control elements in order to optimally manage the charging process of multiple EVs. Although particular implementations may differ

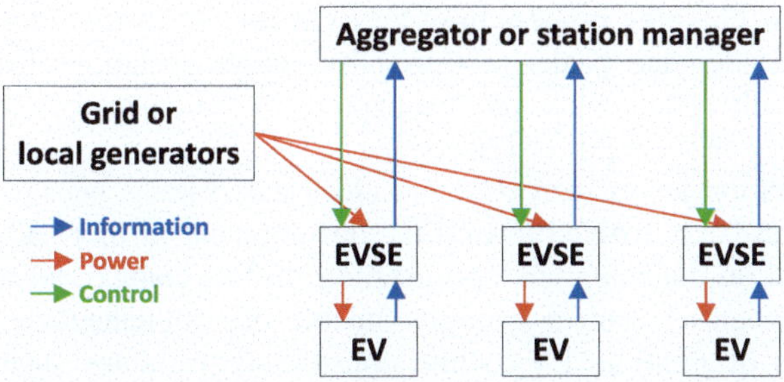

Figure 11.2 General architecture of a smart charging system. Information flows enable the exchange of operational and status data between EVs, EVSEs, and the aggregator or station manager. Control signals are used by the management entity to coordinate and regulate charging operation according to grid conditions, operational constraints, or optimization objectives. Power flows represent the transfer of electrical energy from the grid or local generators to the EVSEs and, subsequently, to the EVs.

depending on the application or charging infrastructure, smart charging architectures are generally organized around the interaction between the EV, the EV Supply Equipment (EVSE), and different management systems.

Within this architecture, several information, control, and power flows are established among the different system elements, as illustrated in Fig. 11.2. At the lowest hierarchical level, EVs and EVSEs exchange information related to battery SoC, available charging power, charging limitations, and estimated connection times. This information enables the dynamic control of the charging process while simultaneously considering grid operational constraints and user requirements.

When multiple EVSEs are connected within the same charging facility, a higher-level management layer is commonly incorporated, usually referred to as an aggregator or station manager [2]. These entities are responsible for supervising and coordinating the operation of the different charging points, assigning charging power references, executing optimization strategies, and ensuring compliance with operational or grid-related constraints.

To support all these functionalities, smart charging systems must incorporate advanced monitoring, communication, and control infrastructures, typically supported through cloud-based platforms, standardized communication protocols, and Internet of Things (IoT) technologies. These capabilities enable real-time data acquisition and coordinated energy management across the charging infrastructure.

11.2 State of the art and literature review

11.2.1 Evolution of smart charging research

The concept of smart charging has received significant attention during the last decades, with research evolving in parallel with the development of electric mobility and smart grid technologies. Early EV charging strategies were mainly based on simple models and basic load management approaches, whose primary objective was to control the charging process in order to avoid overloads in local distribution networks or limit demand peaks produced by charging stations [3]. These early approaches typically relied on simple management mechanisms based on predefined charging schedules or static power limitations.

As EV penetration levels increased and electrical systems became progressively more complex, research evolved toward the development of more advanced control approaches based on mathematical optimization. In this context, advanced charging coordination strategies began to emerge, capable of simultaneously managing the charging process of multiple EVs while considering variables such as electricity prices, grid constraints, user preferences, or renewable generation availability [4]. As a result, numerous optimization strategies appeared, targeting objectives such as peak shaving [5], valley filling [6], or operational cost minimization [4].

In parallel, the development of communication technologies, IoT systems, and cloud-based platforms enabled the emergence of highly digitalized and interconnected smart charging architectures [7]. Furthermore, new strategies focused on the optimal integration of EV charging with DERs and the implementation of demand-side management schemes were progressively introduced [8]. In recent years, research has increasingly shifted toward more advanced approaches based on Artificial Intelligence (AI), predictive control, machine learning, and multi-agent systems [9].

11.2.2 Main smart charging objectives

Smart charging strategies can be designed to optimize different technical, economic, and operational objectives. In most cases, these objectives aim to simultaneously optimize the EV charging process from the perspective of the electrical grid, charging operators, and end users. Among the objectives most commonly considered within the smart charging paradigm, the following can be highlighted:

- **Peak shaving:** This has been one of the most extensively studied objectives in smart charging strategies [10]. These approaches seek to temporally redistribute charging power in order to avoid the demand peaks typically associated with uncontrolled charging (see Fig. 11.1). Through this objective, it is possible to reduce stress on grid components such as transformers or power lines, while potentially decreasing electricity prices in wholesale energy markets.
- **Valley filling:** This strategy consists of shifting part of the charging demand toward periods of lower aggregated electricity consumption. In this way, a flatter demand profile can be achieved, thereby avoiding large power gradients that may require the operation of expensive generation units or lead to grid stability issues [11].
- **Economic objectives:** Many smart charging strategies are mainly focused on minimizing operational costs. In this context, dynamic electricity pricing schemes and variable charging tariffs become particularly relevant, combining periods of high electricity prices with periods of lower energy costs [12].
- **Renewable integration:** These strategies aim to shift charging demand toward periods with higher expected renewable energy production. Consequently, two

Table 11.1 Comparison between centralized and decentralized smart charging approaches.

Approach	Advantages	Disadvantages
Centralized	Global system visibility Easier implementation of global optimization strategies Improved coordination among EVs and EVSEs	Limited scalability High computational burden Strong dependence on communication systems
Decentralized	Improved scalability Higher robustness against communication failures Faster local decision-making	More difficult global coordination Potential synchronization issues Possible suboptimal global performance

main objectives can be achieved: improving the utilization of renewable energy sources while simultaneously reducing charging energy costs [13].

- **User comfort:** Other implementations incorporate objectives related to user comfort, considering aspects such as battery degradation reduction, departure time compliance, minimum SoC requirements, or other individual user preferences [14].

11.2.3 Centralized and decentralized approaches

Depending on how control and coordination decisions are structured within the charging system, smart charging strategies can generally be classified into centralized and decentralized approaches, following a classification similar to the control architectures discussed in Chapter 10. Each approach presents different advantages and limitations, which are summarized in Table 11.1.

Centralized approaches are characterized by the presence of a central entity, commonly referred to as an aggregator or station manager, responsible for collecting information from the different EVs and EVSEs connected to the infrastructure [15]. This central entity processes the acquired information by executing optimization algorithms from which optimal charging control strategies are derived and subsequently transmitted to the different charging points. Such an approach enables global system optimization, since the controller has complete visibility over all connected EVs, thereby facilitating the implementation of advanced coordination and optimization strategies. However, centralized architectures may present scalability limitations, high computational burden, strong dependence on communication infrastructure, as well as privacy and cybersecurity concerns associated with information exchange.

In contrast, decentralized schemes distribute control decisions among EVs, EVSEs, or local controllers connected to the charging infrastructure [16]. Under this paradigm, each agent performs individual decision-making processes based on locally available information or shared system signals. Decentralized approaches generally exhibit higher robustness against communication failures and improved scalability as the number of connected agents increases. Nevertheless, achieving global coordination becomes more challenging, and issues related to synchronization, convergence, or overall system stability may arise.

Hybrid strategies combining characteristics from both centralized and decentralized schemes have also been proposed in the literature [17]. These architectures typically incorporate intermediate management layers based on local aggregators, aiming to balance optimization performance, operational flexibility, scalability, and computational complexity.

11.2.4 Optimization and AI methods

Within smart charging optimization strategies, several methodological categories and fundamental approaches can be identified. This classification criterion not only provides a global overview of the existing methodologies, but also facilitates the selection of appropriate techniques depending on the specific optimization problem under consideration. Based on this criterion, smart charging optimization methodologies can be broadly classified as follows:

- **Mathematical programming:** These approaches employ explicit mathematical formulations of the optimization problem and rely on exact solvers for its resolution [12]. Within this category, different formulations can be distinguished, including linear, quadratic, mixed-integer, and nonlinear programming, depending on the type of variables and objective functions involved. These methods are widely used due to their high accuracy and ability to explicitly incorporate operational and network constraints.
- **Uncertainty modeling:** Smart charging strategies are subject to significant uncertainty associated with non-predictable user behavior, including arrival times, charging duration, or initial SoC, among others. Consequently, different formulations have been developed to model uncertainty and support risk-aware decision-making processes, including stochastic programming [18] and robust optimization [15].
- **AI-based methods:** These techniques have gained considerable relevance in recent years due to the rapid development of AI tools and methodologies. Such approaches employ learning-based models capable of reproducing pseudo-optimal operational patterns, typically inferred from historical datasets. In this context, reinforcement learning models [19], evolutionary algorithms [20], and predictive models [21] have attracted significant attention within smart charging applications.

11.2.5 Standards, protocols, and industrial deployment

The practical implementation of smart charging strategies requires the deployment of communication and interoperability infrastructures capable of coordinating the different agents involved in the charging ecosystem, including EVs, EVSEs, and external entities such as grid operators. To facilitate information exchange among these heterogeneous agents, several standards and communication protocols have been developed to ensure compatibility, security, and scalability across devices and platforms from different manufacturers.

One of the most widely adopted communication protocols is OCPP. This protocol is commonly employed for communication between charging stations and central management systems, enabling functionalities such as remote monitoring, operational control, firmware updates, and the implementation of dynamic charging strategies. Other relevant protocols include ISO 15118, which enables advanced functionalities such as automatic authentication, bidirectional energy information exchange, and integration of V2X services.

In parallel with communication protocols, international standards such as IEC regulations[1] define technical requirements associated with charging equipment, including electrical safety, interoperability, and operational requirements related to charging infrastructure deployment. Together, these standards and protocols constitute the technological foundation required for the large-scale implementation of interoperable, secure, and intelligent smart charging systems.

11.3 Fundamentals of EV charging optimization

11.3.1 Optimization problem formulation

The mathematical formulation of the optimization problem associated with smart charging constitutes one of the fundamental elements of its implementation. In general terms, the objective of this optimization problem is to determine how the charging process of one or multiple EVs should be managed in order to satisfy specific operational objectives while guaranteeing compliance with technical, economic, and energy-related constraints. Although different mathematical formulations can be employed depending on the application, most approaches share several common elements, including decision variables, objective functions, and operational constraints. A generic smart charging optimization problem can be formulated using a mixed-integer linear programming model, as shown in (11.1).

$$\min_{\mathbf{x},\mathbf{y}} \quad \mathbf{c}^\top \mathbf{x} + \mathbf{d}^\top \mathbf{y} \tag{11.1a}$$

$$\text{s.t.} \quad \mathbf{A}\mathbf{x} + \mathbf{B}\mathbf{y} \le \mathbf{b} \tag{11.1b}$$

$$\mathbf{A}_{\text{eq}}\mathbf{x} + \mathbf{B}_{\text{eq}}\mathbf{y} = \mathbf{b}_{\text{eq}} \tag{11.1c}$$

$$\mathbf{x}^{\min} \le \mathbf{x} \le \mathbf{x}^{\max} \tag{11.1d}$$

$$\mathbf{y}^{\min} \le \mathbf{y} \le \mathbf{y}^{\max} \tag{11.1e}$$

$$\mathbf{x} \in \mathbb{R}^n, \quad \mathbf{y} \in \mathbb{Z}^m \tag{11.1f}$$

where **x** represents the vector of continuous decision variables, typically associated with charging power levels, energy exchange, or charging durations, while **y** denotes the vector of integer or binary variables related to discrete operational decisions such

[1] https://iec.ch/homepage.

as charger activation states or EV assignment variables. The vectors **c** and **d** contain the coefficients of the objective function (11.1a), whereas matrices **A**, **B**, $\mathbf{A}_{\text{eq}}$, and $\mathbf{B}_{\text{eq}}$ define the inequality (11.1b) and equality (11.1c) constraints associated with operational, electrical, or scheduling limitations. The vectors $\mathbf{x}^{\text{min}}$ and $\mathbf{x}^{\text{max}}$ establish the lower and upper bounds of the continuous variables in (11.1d), usually associated with charger power ratings, battery charging limits, or grid capacity restrictions. Similarly, $\mathbf{y}^{\text{min}}$ and $\mathbf{y}^{\text{max}}$ define the admissible range of integer decision variables in (11.1e). Finally, domains of variables are given by (11.1f).

Another important aspect in the formulation and solution of smart charging optimization problems is the discretization of the optimization horizon. In most implementations, the scheduling horizon is divided into fixed-duration time intervals, although variable-step approaches can also be employed depending on the characteristics of the problem [22]. For each time interval, the optimization problem computes the optimal setpoints associated with variables **x** and **y**, which ultimately determine the resulting charging strategy.

11.3.2 Objective functions in smart charging

The objective function mathematically represents the criterion intended to be optimized within the charging management problem. Accordingly, the objective function is formulated as a scalar function whose value must be minimized (or maximized) while satisfying a set of operational constraints. In general, the objective function is expressed as a function of the decision variables of the optimization problem, namely the controllable quantities associated with the charging process.

As an illustrative example, one of the most widely adopted objective functions corresponds to the minimization of charging costs. For a scheduling horizon divided into $\mathcal{T}$ time intervals, the objective function may be expressed as:

$$\min \sum_{t=1}^{\mathcal{T}} c(t) \sum_{i=1}^{N} P_{ch}(i,t) \tag{11.2}$$

where $c(t)$ represents the electricity price at time interval t, N is the total number of EVs and $P_{ch}(i,t)$ denotes the charging power assigned to vehicle i during time interval t.

Alternative objective functions may be formulated depending on the specific purpose of the smart charging strategy, including objectives such as peak shaving, valley fillin,g or user preference satisfaction (see Section 11.2.2). In some situations, it may be desirable to optimize several criteria simultaneously. Such formulations belong to the family of multiobjective optimization problems, for which different solution methodologies have been proposed [23]. One of the most common approaches corresponds to weighted-sum scalarization functions. Assuming that three independent objective functions, namely F_1, F_2, and F_3, must be minimized simultaneously, the combined objective function can be formulated as:

$$\min(\alpha_1 F_1 + \alpha_2 F_2 + \alpha_3 F_3) \tag{11.3}$$

where the α coefficients represent weighting factors used to scale the relative importance of each objective.

An important aspect concerns the linearity and continuity of the objective function, since these characteristics define the type of optimization problem and strongly influence the solution methodology. Linear objective functions typically lead to linear programming problems, whereas the inclusion of binary variables or nonlinear terms, such as products of variables, may require branch-and-bound algorithms or heuristic optimization techniques [24].

11.3.3 Operational and grid constraints

Any optimization problem related to smart charging must incorporate a set of constraints that guarantee the feasibility of the obtained solutions. These constraints ultimately define the admissible solution space and may include technical, economic, or user-comfort criteria, among others. Within the smart charging paradigm, several typical constraints are commonly considered:

- **Equipment constraints:** These constraints aim to model the technical limitations of charging equipment, energy storage systems and other electrical infrastructure components. For instance, each charging point operates within minimum and maximum power limits, meaning that the charging power assigned to each EV must remain within these boundaries throughout the scheduling horizon. Another representative example concerns the onboard battery SoC, which is constrained by the battery capacity itself as well as by potential DoD limitations intended to reduce accelerated cell degradation (see Chapter 7).
- **Temporal constraints:** Smart charging problems frequently include restrictions associated with EV arrival and departure times. Consequently, the optimization problem cannot allocate charging power outside the time interval during which the EV remains parked and physically connected to the charging infrastructure. Additionally, constraints are usually incorporated to ensure that a minimum SoC level is achieved before vehicle departure.
- **Grid constraints:** In some cases, additional constraints related to the electrical network are included. The most common examples involve power capacity limitations of electrical infrastructure components such as transformers and distribution lines. However, grid stability criteria may also be considered, including voltage deviation limits or distribution network congestion restrictions.

Similarly to the objective function formulation (see Section 11.3.2), the mathematical formulation of the constraints strongly influences the adopted solution methodology. Therefore, careful problem formulation is required, since this aspect directly affects both the quality of the obtained solutions and the computational cost associated with solving the optimization problem.

11.3.4 Forecasting and uncertainty management

Smart charging strategies typically provide solutions for scheduling horizons ranging from several hours to multiple days or even longer-term operation. Consequently, the proper execution of the associated optimization problem requires the prediction of several model parameters. For this purpose, a wide variety of forecasting tools can be incorporated into the optimization framework as auxiliary routines [25].

Among the most relevant parameters requiring prediction are the charging power demand as well as EV arrival and departure times, since these variables strongly determine the charging power allocation at each time interval. Another important parameter corresponds to the available renewable generation, particularly in systems integrating highly intermittent energy sources such as PV or wind generation. In scenarios involving dynamic electricity prices linked to electricity markets, such as real-time pricing tariffs (see Section 11.5), electricity price evolution must also be forecasted.

Despite the use of sophisticated forecasting techniques, the estimated parameters inevitably contain uncertainty components associated with modeling errors, weather variability or unpredictable user behavior. As a consequence, this uncertainty must be properly quantified and managed within the optimization problem itself. Several approaches have been proposed for this purpose, including stochastic programming, robust optimization, and predictive control methodologies (see Section 11.2.4).

11.4 Coordinated charging strategies

11.4.1 Concept of coordinated charging

Coordinated charging of EVs consists of an advanced optimization strategy in which the charging process of multiple vehicles is managed in a coordinated manner within a shared charging infrastructure. The main objective of these strategies is to determine charging power setpoints for each EV in order to optimize a predefined criterion while simultaneously considering:

- User charging requirements.
- Electrical grid constraints.
- Charging infrastructure capacity.
- Operational objectives of the management system.

The implementation of coordinated charging strategies requires advanced monitoring, communication and control systems capable of enabling information exchange between EVs, charging infrastructure and management entities. Among the most relevant parameters considered during the optimization process are:

- Battery SoC.
- Estimated arrival and departure times.
- Maximum available charging power.
- Electricity tariffs.
- Distribution network limitations.

Based on this information, individual charging profiles are optimized over a predefined scheduling horizon in order to satisfy user charging requirements while simultaneously improving the overall operational performance of the charging infrastructure and the electrical network. The main objectives of these optimization strategies are:

- Improve charging infrastructure utilization.
- Increase operational flexibility.
- Facilitate renewable energy integration.
- Reduce operational costs.
- Smooth demand profiles.
- Improve overall grid stability.

11.4.2 Centralized coordinated charging

Centralized coordinated charging constitutes one of the most widely adopted smart charging strategies, mainly due to its relatively straightforward implementation and its capability to make decisions from a global system perspective. In this approach, all decisions regarding the charging profiles of individual EVs are determined by a central management entity. This entity, commonly referred to as an aggregator or station manager, simultaneously coordinates the operation of multiple chargers and connected EVs within the same charging infrastructure based on information collected from each EVSE.

The implementation of a centralized control strategy requires a hierarchical infrastructure such as the one illustrated in Fig. 11.2. Within this structure, the central coordinator acts as the main decision-making unit, transmitting control signals to the different EVSEs, which subsequently execute the established charging profiles. Since this strategy follows a centralized control architecture, its main advantages and limitations are directly associated with the characteristics of centralized control schemes (see Section 11.2.3), including:

- Availability of a global system view, enabling simultaneous supervision of all available energy resources and charging units.
- Simultaneous coordination of multiple EVs, facilitating joint optimization of the charging demand and reducing congestion situations.
- High operational and economic efficiency due to the use of global optimization algorithms.
- Simplified integration of renewable energy sources and shared energy storage systems within the management strategy.
- Homogeneous implementation of control policies and operational criteria across the entire charging infrastructure.
- Centralized monitoring capabilities, simplifying supervision, maintenance, and fault detection procedures.

However, centralized charging strategies also present several important limitations, mainly associated with scalability constraints and communication vulnerabilities:

- Significant increase in computational complexity as the number of EVs grows.
- Strong dependence on communication infrastructure and continuous operation of the central controller.

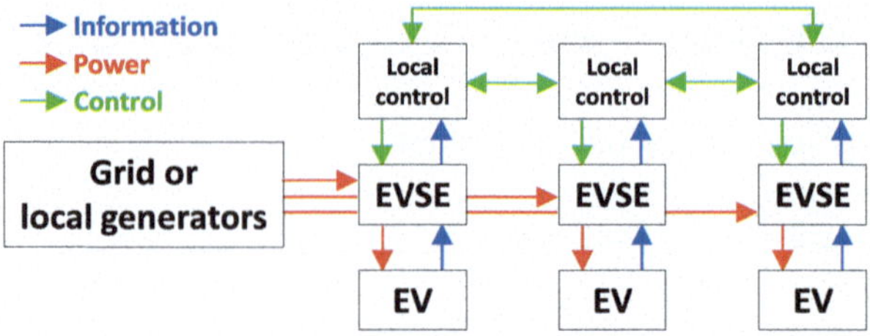

Figure 11.3 Example of a decentralized coordinated charging architecture in which multiple local controllers independently manage the charging process of different EVSEs while exchanging limited information signals with the electrical grid and other system entities.

- Potential scalability limitations in large-scale charging networks.
- Sensitivity to communication latencies in real-time applications.
- Vulnerability to central controller failures and cybersecurity attacks.

Due to these characteristics, centralized coordinated charging currently represents the standard approach in many smart charging applications. Nevertheless, scalability limitations and privacy concerns motivate the development of alternative decentralized solutions for systems involving large EV populations or scenarios where sensitive data exchange constitutes a major implementation barrier.

11.4.3 Decentralized coordinated charging

Decentralized coordinated charging emerges as an alternative to the centralized charging approach presented in Section 11.4.2, particularly in large-scale charging infrastructures or in systems where privacy and scalability constitute critical implementation factors. Unlike centralized charging schemes, charging profile decisions for individual EVs are not determined by a single central entity. Instead, the decision-making process is distributed among multiple local controllers or intelligent agents with partial decision-making capabilities, following an architecture similar to the one illustrated in Fig. 11.3.

Within a decentralized architecture, each EV, EVSE or group of chargers can incorporate a local controller capable of making decisions over its own control domain based on partial system information. These controllers only manage the infrastructure belonging to their corresponding operational scope, although they can obtain limited global information through distributed data exchange with other controllers. As a result, global optimization is achieved through iterative coordination processes between distributed agents (see Section 11.4.4.2). The main advantages of decentralized coordinated charging include:

- High scalability in charging infrastructures involving large EV populations.
- Reduced computational burden associated with a single central controller.
- Lower dependence on a unique control entity.
- Increased robustness against local failures or communication disruptions.

- Improved privacy preservation by reducing the exchange of sensitive information between entities.

In addition to these intrinsic advantages, decentralized coordinated charging strategies provide greater implementation flexibility, which becomes particularly relevant in geographically distributed systems, microgrids, or heterogeneous charging networks where multiple operators simultaneously participate in the energy management process.

Nevertheless, decentralized charging architectures also face several important limitations and implementation challenges:

- The absence of a complete global system view can complicate the achievement of globally optimal solutions.
- Iterative coordination processes between agents require additional synchronization and communication mechanisms, potentially introducing convergence, stability, or latency issues.

Overall, decentralized coordinated charging represents an attractive alternative to centralized approaches in scenarios where scalability, privacy or communication constraints limit centralized implementation feasibility. However, decentralized charging strategies still constitute an active research field within smart charging applications, and their state of the art remains less mature than centralized solutions, limiting the availability of standardized large-scale implementations.

11.4.4 Optimization techniques for coordinated charging

11.4.4.1 Mathematical solvers for coordinated charging

Coordinated charging strategies are commonly formulated as optimization problems aimed at determining optimal charging profiles in order to optimize a specific management criterion while guaranteeing solution feasibility under multiple operational constraints (see Section 11.3.1). The resolution of these optimization problems requires the use of specialized solvers. The selection of a particular solver depends on several factors, including the mathematical formulation of the problem, its complexity and the overall system size.

In mathematical formulations where both the objective function and the constraints can be represented through linear relationships, conventional linear programming solvers can be employed, including interior-point algorithms [26]. These solvers are computationally efficient and guarantee the identification of the global optimum. However, they are only directly applicable to continuous optimization problems. Consequently, when binary or integer decision variables are incorporated into the formulation, more advanced optimization techniques such as branch-and-bound algorithms may be required [24].

When the mathematical formulation becomes nonlinear, alternative classes of solvers must be adopted depending on the type of nonlinearity involved. In this context, specific solvers exist for quadratic objective functions, conic constraints, or general nonlinear optimization problems [27]. In highly complex optimization scenarios where exact mathematical formulations become difficult to solve in real time, heuristic and metaheuristic methods can also be employed. Among the most widely adopted approaches are genetic algorithms and particle swarm optimization techniques, which are capable of obtaining near-optimal solutions with lower computational requirements [28].

11.4.4.2 Algorithms for decentralized coordinated charging

Decentralized coordinated charging schemes typically require the use of specific algorithms capable of solving optimization problems in a distributed manner while maintaining coordination between multiple agents or local controllers. Unlike centralized approaches, these algorithms must guarantee convergence toward globally consistent solutions while relying only on partial information exchanges between distributed entities.

One of the most widely adopted algorithms in distributed optimization is the Alternating Direction Method of Multipliers (ADMM) [29]. This algorithm decomposes the global optimization problem into multiple local subproblems that can be solved independently by each agent, subsequently incorporating iterative coordination mechanisms to ensure global solution consistency. As a result, ADMM presents a naturally scalable structure capable of maintaining computational costs under control even in large-scale charging infrastructures involving numerous local controllers.

Although ADMM enables efficient distributed global optimization, it still requires information exchange between controllers, which may compromise user privacy. As an alternative, specific decomposition techniques such as the Benders decomposition algorithm have been employed in distributed optimization problems with privacy-preserving capabilities in other related applications [30]. However, the implementation of Benders decomposition generally requires some form of centralized entity capable of collecting information and coordinating global decision-making processes.

In order to preserve a fully decentralized architecture, differential privacy techniques can also be employed [31]. These approaches enable the exchange of aggregated or intentionally perturbed information between agents without revealing sensitive data associated with users, consumption profiles, or internal EV states. Consequently, system privacy can be significantly improved without completely compromising the global coordination capabilities of the decentralized charging strategy.

11.4.5 Practical challenges and limitations

Despite the operational advantages and flexibility offered by coordinated charging strategies, their practical implementation in real-world infrastructures remains limited due to several technical and operational challenges. These limitations mainly arise from the inherent uncertainty associated with both user behavior and the dynamic conditions of the electrical system. In general terms, the following aspects represent the main challenges currently faced by coordinated charging implementations:

- **Parameter forecasting:** Optimization problems governing coordinated charging strategies are typically solved over finite prediction horizons. Under these conditions, accurate estimation of several parameters becomes essential. Although advanced forecasting techniques are currently available (see Section 11.3.4), deviations between predicted and actual parameter realizations remain unavoidable. Such discrepancies may significantly degrade optimization performance, leading to suboptimal charging profiles or operational constraint violations.
- **Uncertainty management:** EV charging processes are strongly influenced by user behavior and external boundary conditions, such as renewable energy availability, both of which are subject to considerable uncertainty. Consequently, advanced optimization and control approaches capable of handling uncertainty and operational risk must be incorporated into smart charging frameworks (see Section 11.3.4).
- **Computational complexity:** As the number of EVs connected to the same charging infrastructure increases, the dimensionality and computational burden of the optimization problem grow considerably. This situation often requires the implementation of advanced optimization methods, which introduce additional challenges related to scalability, convergence, and real-time execution (see Section 11.4.3).
- **Infrastructure dependency:** Coordinated charging strategies rely heavily on digital communication and control infrastructures. As a result, these systems become vulnerable to communication failures, synchronization problems, and cybersecurity threats. In addition, interoperability issues may arise due to the coexistence of multiple protocols, manufacturers, and charging platforms (see Section 11.2.5).

11.4.6 Impact on grid operation and infrastructure

Coordinated charging strategies have a direct impact on power system operation, providing multiple operational and energy-related benefits. Among the most relevant advantages, the following aspects can be highlighted:

- **Peak demand limitation:** As discussed in Section 11.1.2, uncontrolled EV charging may produce significant demand peaks, creating major operational challenges for power systems. In this context, coordinated charging strategies enable the simultaneous management of multiple EV charging processes, substantially reducing peak power consumption and smoothing aggregated load profiles.
- **Infrastructure utilization:** Coordinated charging strategies improve the management of charging power profiles, resulting in more efficient utilization of electrical infrastructure. This capability helps prevent transformer overloading, feeder congestion, and excessive stress on distribution assets, potentially extending equipment operational lifetime.
- **System efficiency:** Coordinated EV charging enables the temporal shifting of large demand blocks toward periods with higher renewable energy availability or more

favorable operating conditions. Consequently, these strategies facilitate renewable energy integration while improving overall system efficiency and sustainability.

- **Grid stability:** The implementation of coordinated charging strategies allows the incorporation of stability, security and resilience constraints directly into charging management frameworks. As a result, large EV fleets may contribute to ancillary service provision, voltage regulation, and frequency support, ultimately improving overall grid stability and reducing the probability of supply interruptions.

11.5 Dynamic pricing strategies

11.5.1 Fundamentals of dynamic pricing

In conventional charging systems, users typically pay a fixed price based either on the total energy delivered or on the total connection time of the vehicle to the charging infrastructure. However, this fixed-pricing strategy does not provide sufficient incentives to modify charging habits. In this context, dynamic charging pricing strategies have emerged as an alternative to fixed tariffs, aiming to exploit the intrinsic flexibility of EV charging processes in favor of system operation [15].

The primary objective of dynamic charging pricing is to provide economic signals to EV users that influence charging behavior, enabling the modification of consumption profiles in a manner beneficial for the system. Consequently, dynamic charging prices can provide several advantages for both the electrical system and urban transportation infrastructures, among which the following can be highlighted:

- Improved renewable integration by shifting part of the charging demand toward periods with higher renewable energy availability.
- Enhanced grid stability and operational performance by penalizing charging demand during periods of high network loading or limited generation availability.
- Improved urban traffic distribution by redirecting charging demand toward areas with lower vehicle concentration levels (see Section 11.5.4).

Conceptually, dynamic charging prices maintain a close relationship with the wide variety of dynamic tariffs currently implemented in retail electricity markets and conventionally offered by electricity retailers. As a result, these pricing strategies present a direct connection with demand response programs deployed in smart grids, potentially incorporating information derived from wholesale electricity prices, network congestion conditions, or real-time renewable energy availability.

11.5.2 TOU and real-time pricing (RTP)

Different dynamic pricing strategies can be implemented depending primarily on how charging prices are determined and which operational criteria are considered when updating tariffs over time. In this context, two of the most widely adopted approaches are TOU and RTP tariffs. Although both strategies share a similar objective, namely influencing charging behavior through dynamic economic signals, they differ significantly in terms of pricing dynamism and temporal granularity.

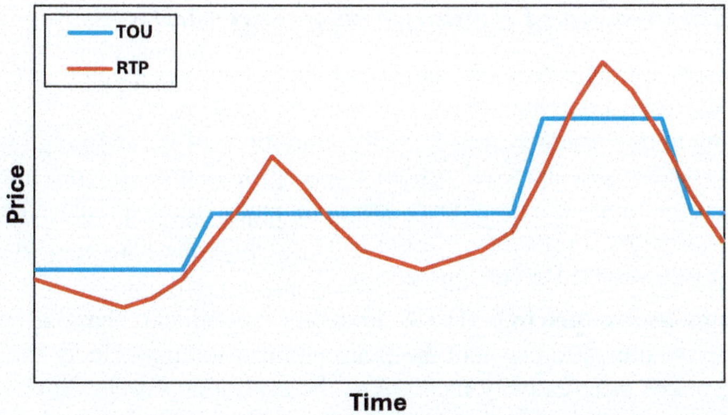

Figure 11.4 Comparison between TOU and RTP electricity tariff profiles over a 24-hour operating horizon. TOU tariffs employ predefined price intervals with fixed peak and off-peak periods, whereas RTP schemes continuously adapt electricity prices according to real-time market and grid operating conditions.

TOU tariffs are characterized by predefined fixed pricing periods, typically differentiating between peak periods with high electricity prices, off-peak periods with reduced tariffs, and intermediate price intervals. Under this pricing mechanism, users know electricity prices in advance, considerably simplifying charging scheduling and operational planning. Typically, TOU schemes are designed to shift charging demand away from periods less favorable for system operation, encouraging energy consumption during off-peak intervals that usually coincide with periods of lower demand or higher renewable penetration, where operating conditions are more favorable. As a result, TOU tariffs present a relatively simple implementation and a high degree of user acceptance due to their predictable and stable structure.

In contrast, RTP tariffs dynamically adjust electricity prices in real time without relying on predefined pricing schedules, instead responding directly to actual electrical system conditions. These mechanisms can present higher temporal granularity than TOU tariffs, which conventionally employ hourly prices, potentially updating tariffs every few minutes. RTP strategies typically use economic signals derived from wholesale electricity markets or operational signals provided directly by grid operators. Consequently, users receive economic signals that more accurately reflect the real operational cost of the electrical system, enabling significantly more flexible and efficient demand-side management. However, the real-time nature of RTP updates and the associated price volatility introduces additional uncertainty, making charging scheduling considerably more complex.

Fig. 11.4 illustrates representative examples of TOU and RTP tariffs. As shown in the figure, while the illustrated TOU tariff presents three clearly differentiated pricing periods, the RTP tariff exhibits a considerably more variable profile in which electricity prices continuously adapt to changing system operating conditions.

11.5.3 Station-level dynamic charging tariffs

Similarly to conventional electricity tariffs, which normally respond to grid operating conditions, dynamic charging prices can also be implemented at charging-station level, adapting to the instantaneous operating conditions of the charging facility itself rather than to the broader electrical system. Consequently, these pricing strategies aim to simultaneously improve station operational efficiency, user experience and operator economic profitability. In practice, station-level dynamic pricing mechanisms can be implemented following different approaches:

- **Occupancy-aware pricing:** These strategies implement charging tariffs that depend on the utilization level of the charging infrastructure. Under this approach, charging prices progressively increase as the station approaches high occupancy levels. Consequently, this mechanism discourages temporary use of highly congested facilities, promoting the redistribution of charging demand toward alternative infrastructures or less congested time periods. Conversely, during periods of low occupancy, reduced charging prices can be offered to attract additional users, thus improving asset utilization and reducing infrastructure underutilization. Additionally, these mechanisms may contribute to reducing waiting times and improving the overall quality of service perceived by EV users.
- **Renewable-aware pricing:** This type of pricing strategy is commonly employed in charging stations integrating local renewable generation systems [15]. In this case, charging prices are dynamically adapted according to the local renewable energy availability. Lower charging prices are typically offered during periods with high renewable generation, incentivizing EV charging when renewable resources are abundant, while higher prices may be applied during periods with limited renewable availability. As a result, local renewable resource utilization is improved, dependence on the electrical grid is reduced, and local self-consumption is promoted. Furthermore, these approaches can contribute to mitigating renewable curtailment and improving local energy management flexibility.
- **Station profitability-oriented pricing:** Other station-level pricing strategies follow primarily economic criteria. In these approaches, charging prices are dynamically adjusted to maximize charging station profitability, minimize operational energy costs or avoid economic penalties associated with grid operation, such as demand charges caused by high peak power consumption. These mechanisms enable charging operators to actively optimize infrastructure economic performance while maintaining operational flexibility.

11.5.4 Network-level pricing mechanisms

Network-level dynamic charging prices aim to extend the concept of economic demand management beyond the charging infrastructure itself by incorporating information associated with external systems such as the electrical grid or transportation network. Unlike station-level pricing mechanisms (see Section 11.5.3), network-level tariffs utilize information collected from multiple infrastructures, seeking to simultaneously optimize energy, operational and urban mobility variables. Consequently, these pricing strategies transform charging infrastructures into active tools for both electrical grid and transportation network management. These mechanisms currently constitute one of the most active research areas in smart charging systems, although several major approaches can be identified:

- **Locational electricity network pricing:** These strategies establish different charging prices at different grid nodes according to the actual energy cost associated with each location. Consequently, prices may reflect real-time electrical network conditions such as line congestion, transformer loading or local generation availability. In essence, higher prices are applied at nodes where electricity supply costs are higher, directly influencing the charging cost of stations connected to those network locations while encouraging the use of infrastructures connected to more favorable grid nodes. As a result, these tariffs present a strong electrical grid operation perspective and can be employed to mitigate operational problems such as distribution network congestion.
- **Transportation network-aware pricing:** These mechanisms follow a philosophy similar to the previous approaches, establishing different electricity prices depending on the charging location. However, these pricing mechanisms are guided by transportation network indicators rather than electrical grid conditions [32]. Their objective is to mitigate transportation-related problems through localized economic signals. Under this framework, higher charging prices may be applied at charging infrastructures located near highly congested traffic areas or regions with limited parking availability, encouraging EV users to utilize alternative charging stations located in more favorable traffic conditions. Consequently, these mechanisms can simultaneously influence mobility patterns and electrical demand profiles, enabling integrated approaches between smart grids and smart cities.
- **Ancillary service-oriented pricing:** Finally, some strategies specifically focus on incentivizing the provision of grid-support services such as frequency regulation or voltage support. These mechanisms typically respond to signals transmitted by system operators, aiming to promote the provision of V2G services (see Chapter 10) in order to improve electrical system security, operational flexibility, and resilience during specific operating conditions or emergency situations.

11.5.5 Economic and behavioral impacts

Dynamic charging pricing strategies produce significant economic and behavioral impacts on EV users. One of the most important effects is associated with charging demand elasticity, which describes the ability of users to modify their charging habits in response to variations in electricity prices. In general, users with higher temporal flexibility tend to shift charging sessions toward periods with lower electricity costs, contributing to peak demand reduction and improving infrastructure utilization. However, not all users exhibit the same adaptation capability, particularly those requiring immediate charging availability or operating under restricted schedules.

In parallel, the implementation of dynamic pricing mechanisms may introduce fairness concerns, since some users, especially those with limited charging flexibility, can be disproportionately affected by high charging prices during congested periods. Consequently, poorly designed pricing strategies may generate unequal economic impacts among different user groups.

Furthermore, dynamic pricing strategies can significantly influence the selection of charging locations. Price differences among charging infrastructures may encourage EV migration from one charging station to another, modifying infrastructure utilization patterns and altering the spatial distribution of traffic flows. Although this behavior may alleviate local congestion in certain areas, excessive charging migration can create new operational imbalances within the charging network.

11.5.6 Limitations and regulatory challenges

The implementation of dynamic charging pricing strategies presents several important limitations and challenges that must be carefully considered, among which the following can be highlighted:

- **Uncertainty:** The application of RTP-based tariffs (see Section 11.5.2) introduces uncertainty that must be properly managed. Since charging prices vary according to electricity market conditions or network operation states, charging planning becomes more complex and less predictable for EV users.
- **Complexity:** Smart charging tariffs may become considerably more complex than conventional electricity pricing schemes. The continuous monitoring of variable prices, time intervals, and congestion signals increases both communication requirements and the complexity of energy management platforms.
- **Regulatory framework:** Many countries still lack regulatory frameworks specifically adapted to dynamic tariff models for EV charging infrastructures.
- **User acceptance:** Excessively complex or unpredictable charging tariffs may generate user dissatisfaction and negatively affect the large-scale adoption of smart charging systems.

11.6 Integration with renewable energy and energy storage

11.6.1 Renewable-based charging infrastructures

The integration of charging infrastructures with local renewable generation units constitutes one of the most widespread and effective solutions in charging facilities implementing smart charging strategies. In general, the integration of renewable energy sources into EV charging infrastructures provides several strategic advantages:

- Reduced dependence on the electrical grid.

- Reduction of emissions associated with EV charging processes.
- Lower operational costs through local energy self-consumption.
- Improved local energy stability and peak demand reduction.
- Enhanced overall sustainability of charging infrastructures.

In this context, PV generators constitute the most widely adopted solution [15], mainly due to their reduced visual and acoustic impact, high technological maturity, and relatively low installation requirements. In many cases, PV systems can be installed directly on existing rooftops or parking canopies within the charging infrastructure itself, improving land utilization and reducing additional installation costs.

On the other hand, small-scale wind generators may also represent an attractive solution in certain scenarios, particularly in isolated installations or high-power charging facilities located in regions with favorable wind resources [33]. Although their deployment is considerably less common than PV systems, wind-assisted charging infrastructures can complement renewable generation profiles and improve local renewable energy utilization.

Nevertheless, the integration of renewable generation introduces additional operational challenges associated with the inherent intermittency and variability of renewable energy sources. Renewable generation profiles introduce uncertainty into smart charging strategies, requiring the implementation of forecasting methods and advanced energy management techniques to properly coordinate EV charging demand with renewable energy availability (see Section 11.3.4).

11.6.2 Integration with battery energy storage systems

Although less common than renewable integration, the installation of stationary energy storage systems in charging infrastructures is progressively gaining relevance and may represent a technically and economically viable investment in certain scenarios [34]. In most applications, stationary energy storage systems are based on electrochemical batteries due to their technological maturity, high efficiency, and suitable long-term storage capabilities (see Chapter 7). Nevertheless, alternative technologies such as hydrogen-based storage systems (see Chapter 8) or other storage technologies discussed in Chapter 9 may also be considered depending on the specific application requirements.

The combined integration of renewable generation and stationary energy storage systems leads to complex and highly flexible architectures such as the one illustrated in Fig. 11.5. These infrastructures enable multiple bidirectional power flows among renewable generators, storage systems, EV charging stations, and the electrical grid, potentially allowing local energy injection into the distribution network.

In general, the integration of stationary energy storage systems provides several important operational advantages:

- **Improved charging management:** Local energy storage significantly increases system flexibility, enabling more effective charging management strategies. One of the main benefits is peak shaving capability, where the storage system supplies

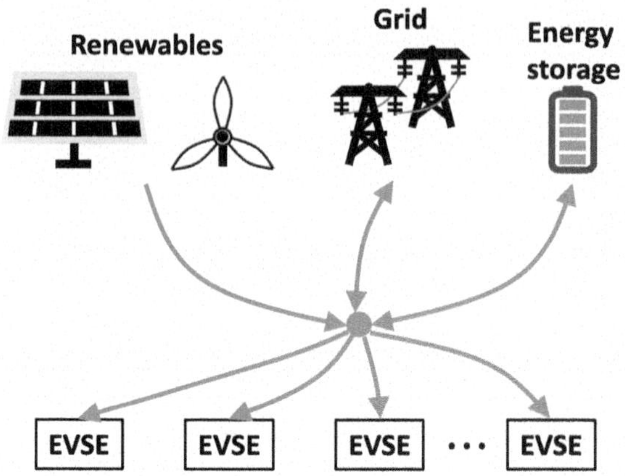

Figure 11.5 General architecture of a smart charging infrastructure integrating renewable generation, stationary energy storage systems, grid connection, and multiple EVSE units.

temporary charging demand peaks, reducing the power requested from the electrical grid and mitigating transformer or distribution line overloading.

- **Energy arbitrage:** Stationary storage systems allow energy to be stored during favorable operating conditions, such as periods with low electricity prices or high renewable generation. The stored energy can later be used to decouple EV charging demand from the grid during expensive periods or even be injected into the grid for economic profit, improving both infrastructure efficiency and operational profitability.
- **Provision of ancillary services:** Certain storage technologies, particularly SC-based systems, exhibit operational characteristics suitable for ancillary grid services. These systems may contribute to applications requiring fast dynamic response, such as frequency support, short-term power balancing, or local voltage stabilization.

11.6.3 Interaction with V2G and microgrids

The integration of V2G functionalities requires a substantial adaptation of conventional smart charging strategies, where only unidirectional power flow (EV charging) is considered. In this context, the mathematical optimization models, solution techniques, control algorithms, and overall energy management architecture must evolve from approaches focused exclusively on charging scheduling toward more advanced frameworks capable of simultaneously coordinating charging and discharging processes.

Specifically, the incorporation of V2G functionalities introduces new decision variables associated with onboard energy storage discharge operations. Similarly, the formulation of the objective function may also change, for example by incorporating economic revenues associated with energy exportation to the electrical grid. In parallel, the operational constraints of the optimization problem must accurately model V2G capabilities, including maximum discharge power limits, battery SOC constraints, and non-simultaneity conditions between charging and discharging processes.

In essence, V2G platforms transform EVs into portable power generation and storage units (see Chapter 10), allowing vehicles to operate as active agents within modern electrical systems. This functionality presents major implications in systems with limited local generation capacity or weak grid connections, such as isolated or partially islanded microgrids. In these scenarios, EVs may temporarily operate as distributed energy storage systems while connected to the charging infrastructure, contributing additional storage capacity, improving local energy balancing, and enhancing the overall efficiency, operational flexibility, and resilience of the energy system.

11.7 Communication, control and cybersecurity

11.7.1 Communication and control infrastructure

The communication and control infrastructure constitutes the backbone of the entire ecosystem surrounding the implementation of smart charging strategies. The deployment of an adequate communication infrastructure is essential to efficiently coordinate energy flows, EVSE operation, and the interaction among the different agents involved in the charging system.

Modern smart charging architectures have progressively evolved through the incorporation of advanced digitalization technologies. As a consequence, communication systems have transitioned from conventional data acquisition and monitoring platforms toward more sophisticated frameworks capable of actively contributing to the operation and optimization of smart charging strategies. In this context, three technologies are becoming increasingly relevant in EV charging communication infrastructures:

- **IoT:** These technologies enable real-time monitoring of operational variables within charging infrastructures, including EVSE status, energy demand, station occupancy, power quality, and renewable generation availability, among others. Consequently, these technologies provide a distributed data acquisition framework that improves infrastructure supervision and enhances smart charging optimization capabilities [35].
- **Cloud computing:** Cloud-based platforms provide large-scale data storage, centralized infrastructure supervision, and high computational capacity. These capabilities enable the execution of advanced functions such as demand forecasting, historical operational analysis, and large-scale energy management optimization.
- **Edge computing:** Edge-based systems execute local control routines close to the physical charging infrastructure, enabling rapid responses to load variations, operational faults, or congestion signals. This approach reduces communication latency and decreases dependence on remote centralized servers, improving the robustness of real-time smart charging applications.

11.7.2 Data management and monitoring

The operation of algorithms and control routines associated with smart charging strategies depends directly on the availability of reliable and updated operational information. This information is essential for dynamically adapting smart charging strategies in response to operational changes, grid conditions, or unexpected events occurring within the charging infrastructure.

Modern monitoring and data acquisition systems may incorporate smart sensors, advanced metering devices, and distributed digital platforms capable of collecting information with high temporal resolution. These technologies enable continuous supervision of relevant operational variables such as charging demand, EVSE status, infrastructure occupancy, renewable generation availability, or local grid conditions.

The deployment of these devices generates large volumes of operational data that can be exploited through predictive analytics techniques. In this context, data-driven approaches may be employed for applications such as charging demand forecasting, renewable generation estimation, user behavior analysis, or early detection of operational anomalies.

Finally, the deployment of charging management platforms enables the automation of a significant part of smart charging system operation while facilitating the integration of advanced energy management algorithms in large-scale charging infrastructures. These platforms centralize infrastructure supervision and coordination, providing operators with monitoring, control, and optimization capabilities within a unified digital environment.

11.7.3 Cybersecurity and privacy challenges

The increasing digitalization of the infrastructure associated with smart charging applications has introduced several challenges and risks related to cybersecurity and data privacy. In this context, the communication, monitoring and control infrastructure of charging systems is exposed to three major categories of cyberattacks:

- **False data injection attacks:** In this type of cyberattack, a malicious agent manipulates critical variables employed by smart charging algorithms, such as demand profiles or electricity prices, potentially leading to incorrect or suboptimal operational decisions.
- **Spoofing attacks:** These attacks allow malicious entities to impersonate legitimate users or devices in order to gain unauthorized access to the charging management system.
- **Denial-of-service attacks:** The objective of these cyberattacks is to saturate communication channels or management platforms, thereby degrading or interrupting the normal operation of the charging infrastructure.

To mitigate the risks associated with these cyberattacks, charging management systems incorporate multiple security layers based on authentication mechanisms, encryption protocols, and digital certificates to guarantee the integrity and confidentiality of the exchanged information. Furthermore, advanced optimization and management techniques can also be employed to enhance the privacy level of the information exchanged among system entities (see Section 11.4.4.2).

11.8 Future trends and emerging approaches

11.8.1 AI and predictive charging

Although several AI-based techniques have already been successfully developed and applied in the smart charging literature [19], this field continues to represent one of the research areas with the greatest future development potential. In this context, further advances are expected in AI-based methods capable of managing highly dynamic and uncertain environments through advanced learning, prediction, and real-time adaptation capabilities.

One of the areas with the highest potential corresponds to the application of machine learning tools for the prediction of key parameters involved in smart charging strategies, such as renewable generation, electricity prices, and charging demand. As discussed in Section 11.3.4, the prediction accuracy of these parameters is critical for the proper operation of the optimization models governing smart charging applications. Consequently, the future evolution of smart charging strategies will likely be strongly associated with the integration of advanced AI-based forecasting techniques capable of improving operational efficiency and decision-making processes.

The application of reinforcement learning techniques constitutes another important development area within the smart charging paradigm [36]. These approaches are capable of learning typical consumption patterns and deriving optimal management strategies through continuous interaction with the surrounding environment. As a result, charging decisions can be dynamically adapted according to grid conditions, economic signals, and energy availability, improving both operational flexibility and overall system performance.

11.8.2 Autonomous charging and self-managed energy systems

The automation of charging infrastructures constitutes another emerging research direction with a prominent role within the smart charging paradigm. In this context, the development of autonomous charging systems capable of operating in a fully or partially autonomous manner is expected, integrating advanced monitoring, optimization and real-time control functionalities. These automated infrastructures may incorporate self-optimization capabilities able to execute smart charging routines in order to dynamically adapt power allocation, charging schedules, and energy management strategies according to grid conditions, renewable energy availability, and user demand.

The integration of autonomous vehicles and intelligent fleets represents one of the main driving forces behind this paradigm. In particular, future charging stations may be capable of automatically coordinating reservation procedures, vehicle connection processes, and energy scheduling operations without human intervention. Within this context, technologies such as robotic charging systems and wireless charging may act as additional enablers for the complete automation of charging operations, improving operational flexibility, reducing manual intervention requirements, and facilitating the deployment of highly automated smart charging infrastructures.

11.8.3 Vehicle automation and Mobility-as-a-Service (MaaS) integration

The smart charging paradigm must also adapt to the emergence and development of new mobility models. In this context, MaaS has emerged as a transportation paradigm in which different mobility services are integrated into a single digital platform accessible to users. Consequently, MaaS requires advanced energy management and optimization platforms capable of simultaneously integrating mobility, logistics, and electrical operation in real time.

As a result, future smart charging strategies should be capable of holistically integrating not only energy-related variables, but also routing functions, fleet management, and dynamic charging scheduling. Such capabilities are particularly relevant for applications involving autonomous vehicles, shared mobility services, and electrified commercial fleets, where operational availability and charging coordination constitute critical factors.

Furthermore, energy management systems may automatically coordinate charging processes at depots and urban charging stations, optimizing both energy consumption and infrastructure utilization. The combination of vehicle automation, real-time communication, and intelligent optimization is expected to significantly improve the overall efficiency, operational flexibility, and scalability of future intelligent electric mobility systems.

11.8.4 Emerging decentralized energy ecosystems

Smart charging strategies must also adapt to future scenarios in which EVs will evolve into flexible energy assets capable of exchanging energy and grid services with other system entities. These future energy ecosystems are expected to operate with a high degree of decentralization, where exchange technologies such as blockchain [37] may become naturally integrated within smart charging strategies.

Future smart charging frameworks will also need to interact with local energy trading and flexibility service platforms. In this context, it will be essential to incorporate functionalities associated with local energy markets and peer-to-peer trading schemes [38], through which EVs may exchange energy and services with other distributed energy assets such as prosumers. Furthermore, EVs may actively participate in collective energy management strategies, including energy arbitrage and participation in wholesale electricity markets.

Another emerging research direction corresponds to the development of grid-forming EVs capable of actively contributing to the stability and regulation of electrical systems with high renewable energy penetration. In parallel, the implementation of digital twins may enable the modeling and supervision of complex charging infrastructures, improving predictive capabilities, operational planning, and the overall resilience of future intelligent energy ecosystems.

11.9 Case study: smart charging management in a multi-EV charging facility

11.9.1 Optimization model

This case study evaluates an optimal smart charging strategy for a fleet of three EVs operating under a RTP mechanism. The objective is to minimize the total charging cost while satisfying battery, user, and grid operational constraints. The optimization problem is formulated as a mixed-integer linear programming model, where the charging power allocation is dynamically adjusted according to the temporal variation of electricity prices.

The formulation of the optimization problem is given below while its MATLAB® code is provided in Section 11.9.4.

$$\min_{P_{\mathrm{ch}}(i,t),\, x(i,t)} \quad \sum_{t=1}^{T} c(t) \left(\sum_{i=1}^{N} P_{\mathrm{ch}}(i,t) \right) \tag{11.4a}$$

$$\text{s.t.} \quad \mathrm{SoC}(i,t+1) = \mathrm{SoC}(i,t) + \frac{\eta_{\mathrm{ch}}(i)\, P_{\mathrm{ch}}(i,t)}{C(i)} \qquad \forall i, \forall t \tag{11.4b}$$

$$\mathrm{SoC}^{\min}(i) \le \mathrm{SoC}(i,t) \le \mathrm{SoC}^{\max}(i) \qquad \forall i, \forall t \tag{11.4c}$$

$$0 \le P_{\mathrm{ch}}(i,t) \le x(i,t)\, P_{\mathrm{ch}}^{\max}(i) \qquad \forall i, \forall t \tag{11.4d}$$

$$x(i,t) \in \{0, 1\} \qquad \forall i, \forall t \tag{11.4e}$$

$$\mathrm{SoC}\left(i, t^{\mathrm{arr}}(i)\right) = \mathrm{SoC}^{\mathrm{init}}(i) \qquad \forall i \tag{11.4f}$$

$$\mathrm{SoC}\left(i, t^{\mathrm{dep}}(i)\right) \ge \mathrm{SoC}^{\mathrm{req}}(i) \qquad \forall i \tag{11.4g}$$

$$P_{\mathrm{ch}}(i,t) = 0 \qquad \forall i, \forall t \notin \left[t^{\mathrm{arr}}(i), t^{\mathrm{dep}}(i)\right] \tag{11.4h}$$

$$\sum_{i=1}^{N} P^{\mathrm{ch}}(i,t) \le P_{\mathrm{grid}}^{\max} \qquad \forall t \tag{11.4i}$$

The objective function in (11.4a) minimizes the aggregated charging cost over the scheduling horizon by considering the RTP signal $c(t)$ and the charging power demanded by each EV. The battery energy evolution is modeled through the SoC

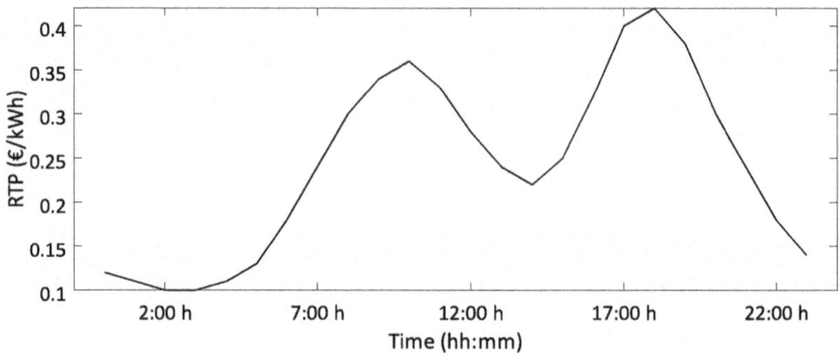

Figure 11.6 RTP tariff considered in the case study.

dynamics in (11.4b), where the charging efficiency and battery capacity determine the increment in stored energy at each time step.

Battery operational limits are imposed in (11.4c) to guarantee safe operation within the allowable SoC range. Charging power constraints are defined in (11.4d), where the binary variable $x\ (i, t)$ establishes the charging state of each EV. The binary nature of this variable is enforced in (11.4e). Initial SoC is fixed by (11.4f), whereas user mobility requirements are represented in (11.4g), ensuring that each EV reaches the required SoC before departure. Availability constraints are incorporated in (11.4h), while the aggregate grid power limitation is imposed in (11.4i) to avoid exceeding the maximum admissible feeder capacity.

11.9.2 Case data

The considered case study evaluates the coordinated charging of three EVs under a RTP scheme over a 24-hour scheduling horizon. The optimization problem is solved with an hourly resolution. The RTP signal employed in the simulations is illustrated in Fig. 11.6. As observed, electricity prices vary significantly throughout the day, with lower tariffs during off-peak periods and substantially higher prices during evening peak demand intervals. This pricing structure encourages load shifting toward economically favorable charging periods.

The main characteristics of the EV fleet are summarized in Table 11.2. The analyzed vehicles present same roundtrip efficiency (95%) but different battery capacities, charging power limits, and connection intervals, thereby introducing heterogeneous operational constraints into the optimization problem. The initial and required SoC values are defined in per-unit form in order to normalize the charging requirements among vehicles with different battery capacities. Furthermore, all EVs are assumed to operate within the same minimum and maximum SoC limits, while identical charging efficiencies are considered for simplicity.

Finally, the charging station is subjected to a maximum grid power constraint of 30 kW, which limits the aggregate charging demand that can be simultaneously supplied by the distribution grid.

Table 11.2 Main parameters of the EV fleet considered in the case study.

EV	C (kWh)	SoC^{init}	SoC^{req}	P_{ch}^{max} (kW)	T^{arr} (h)	T^{dep} (h)
1	55	0.25	0.90	7.4	6	22
2	70	0.30	0.90	11	7	23
3	50	0.20	0.85	7.4	8	24

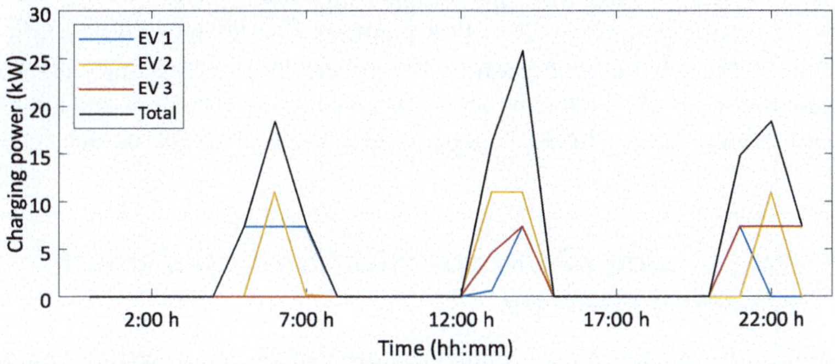

Figure 11.7 Charging power of EVs in the case study.

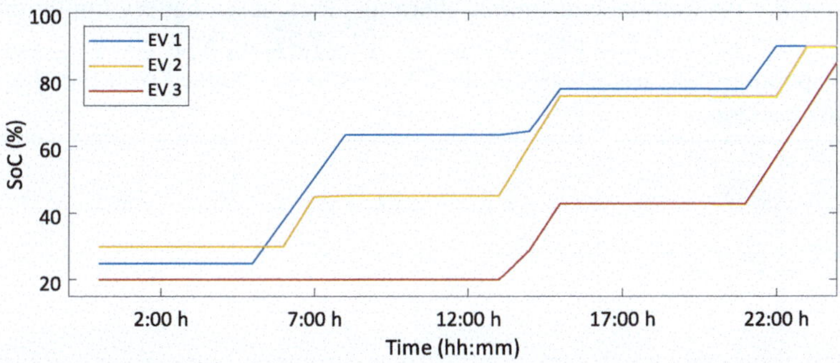

Figure 11.8 SoC of EVs in the case study.

11.9.3 Results and discussion

The optimized charging profiles obtained for the three EVs are presented in Fig. 11.7, while the corresponding state-of-charge trajectories are illustrated in Fig. 11.8. The results demonstrate the capability of the coordinated charging strategy to schedule the charging process according to both electricity price variations and system-level operational constraints.

As observed in Fig. 11.7, the charging demand is primarily concentrated during low-price periods of the RTP signal. In particular, the optimization algorithm prioritizes charging during nighttime and early morning intervals, where electricity prices remain significantly lower than during daytime peak periods. Conversely, charging activity is reduced or temporarily interrupted during hours characterized by high electricity prices, especially during the evening peak interval. This behavior confirms that the optimization framework effectively minimizes the overall charging cost while satisfying the energy requirements of all vehicles.

The aggregate charging power always remains below the grid power limit of 30 kW, thereby ensuring compliance with the infrastructure constraints. Furthermore, the charging process is distributed among the vehicles according to their arrival and departure times, battery capacities, and required final SoC values.

The SoC evolution shown in Fig. 11.8 confirms that all EVs successfully reach their required final SoC before departure. Moreover, the charging trajectories exhibit smooth and progressive behavior, avoiding abrupt charging variations and maintaining all vehicles within their admissible operational limits throughout the scheduling horizon.

11.9.4 MATLAB code of the optimization-based smart charging strategy

This section presents the MATLAB implementation of the optimization-based smart charging strategy analyzed throughout the case study below. The provided code formulates and solves the coordinated charging problem under a RTP framework while considering EV operational constraints, charging efficiencies, battery limitations, and grid power restrictions. The implementation allows the optimal charging schedules and corresponding SoC trajectories of the EV fleet to be obtained over the considered scheduling horizon.

```
%% Optimal Smart Charging of 3 EVs under RTP Pricing
% MATLAB Optimization Toolbox is required

clear;
clc;

%% Simulation Parameters

T = 24; % Time horizon (h)
N = 3; % Number of EVs
dt = 1; % Time step (h)

time = 1:T;

%% RTP Electricity Prices (euro/kWh)

c = [0.12 0.11 0.10 0.10 0.11 0.13 ...
     0.18 0.24 0.30 0.34 0.36 0.33 ...
     0.28 0.24 0.22 0.25 0.32 0.40 ...
     0.42 0.38 0.30 0.24 0.18 0.14]';

%% EV Parameters
```

```
C = [55; 70; 50]; % Battery capacity (kWh)

SoC_init = [0.25; 0.30; 0.20]; % Initial SoC (pu)
SoC_req  = [0.90; 0.90; 0.85]; % Required SoC (pu)

SoC_min = [0.20; 0.20; 0.20]; % Minimum SoC (pu)
SoC_max = [1.00; 1.00; 1.00]; % Maximum SoC (pu)

eta_ch = [0.95; 0.95; 0.95]; % Charging efficiency (pu)

Pmax = [7.4; 11; 7.4]; % Maximum charging power (kW)

Tarr = [6; 7; 8]; % Arrival times
Tdep = [22; 23; 24]; % Departure times

Pgrid_max = 30; % Grid power limit (kW)

%% Optimization Variables

Pch = optimvar('Pch',N,T,'LowerBound',0); % Charging power (kW)

x = optimvar('x',N,T,'Type','integer',...
             'LowerBound',0,'UpperBound',1); % Charging status

SoC = optimvar('SoC',N,T+1,...
               'LowerBound',0,'UpperBound',1); % SoC (pu)

%% Optimization Problem

prob = optimproblem('ObjectiveSense','minimize');

%% Objective Function

prob.Objective = dt*c'*sum(Pch,1)';

%% Constraints

% Initial SoC

prob.Constraints.SoC_init = SoC(:,1) == SoC_init;

% SoC Dynamics

eta_matrix = repmat(eta_ch./C,1,T);

prob.Constraints.SoC_dynamics = ...
    SoC(:,2:T+1) == ...
```

```
    SoC(:,1:T) + eta_matrix.*Pch*dt;

% SoC Limits

prob.Constraints.SoC_min = ...
    SoC >= repmat(SoC_min,1,T+1);

prob.Constraints.SoC_max = ...
    SoC <= repmat(SoC_max,1,T+1);

% Charging Power Limits

prob.Constraints.Power_limits = ...
    Pch <= repmat(Pmax,1,T).*x;

% Final SoC Requirement

final_soc_expr = optimconstr(N,1);

for i = 1:N

    final_soc_expr(i) = ...
        SoC(i,Tdep(i)+1) >= SoC_req(i);

end

prob.Constraints.Final_SoC = final_soc_expr;

% Availability Constraints

availability = ones(N,T);

for i = 1:N

    availability(i,1:Tarr(i)-1) = 0;
    availability(i,Tdep(i)+1:end) = 0;

end

prob.Constraints.Availability = ...
    Pch <= repmat(Pmax,1,T).*availability;

% Grid Power Constraint

prob.Constraints.Grid_limit = ...
    sum(Pch,1)' <= Pgrid_max;

%% Solve Optimization Problem
```

```
options = optimoptions('intlinprog','Display','iter');

[sol,fval,exitflag,output] = ...
    solve(prob,'Options',options);
```

References

[1] S. Mohanty, S. Panda, S.M. Parida, P.K. Rout, B.K. Sahu, M. Bajaj, H.M. Zawbaa, N.M. Kumar, S. Kamel, Demand side management of electric vehicles in smart grids: a survey on strategies, challenges, modeling, and optimization, Energy Reports 8 (2022) 12466–12490, https://doi.org/10.1016/j.egyr.2022.09.023.

[2] M. Tostado-Véliz, X. Jin, R. Bhakar, F. Jurado, Coordinated pricing mechanism for parking clusters considering interval-guided uncertainty-aware strategies, Applied Energy 355 (2024) 122373, https://doi.org/10.1016/j.apenergy.2023.122373.

[3] P.T. Staats, W.M. Grady, A. Arapostathis, R.S. Thallam, A procedure for derating a substation transformer in the presence of widespread electric vehicle battery charging, IEEE Transactions on Power Delivery 12 (4) (1997) 1562–1568, https://doi.org/10.1109/61.634176.

[4] E. Sortomme, M.A. El-Sharkawi, Optimal charging strategies for unidirectional vehicle-to-grid, IEEE Transactions on Smart Grid 2 (1) (2011) 131–138, https://doi.org/10.1109/TSG.2010.2090910.

[5] K. Clement-Nyns, E. Haesen, J. Driesen, The impact of charging plug-in hybrid electric vehicles on a residential distribution grid, IEEE Transactions on Power Systems 25 (1) (2010) 371–380, https://doi.org/10.1109/TPWRS.2009.2036481.

[6] L. Gan, U. Topcu, S. Low, Optimal decentralized protocol for electric vehicle charging, in: 2011 50th IEEE Conference on Decision and Control and European Control Conference, 2011, pp. 5798–5804.

[7] F. Mwasilu, J.J. Justo, E.-K. Kim, T.D. Do, J.-W. Jung, Electric vehicles and smart grid interaction: a review on vehicle to grid and renewable energy sources integration, Renewable and Sustainable Energy Reviews 34 (2014) 501–516, https://doi.org/10.1016/j.rser.2014.03.031.

[8] J.A.P. Lopes, F.J. Soares, P.M.R. Almeida, Integration of electric vehicles in the electric power system, Proceedings of the IEEE 99 (1) (2011) 168–183, https://doi.org/10.1109/JPROC.2010.2066250.

[9] S. Vandael, B. Claessens, M. Hommelberg, T. Holvoet, G. Deconinck, A scalable three-step approach for demand side management of plug-in hybrid vehicles, IEEE Transactions on Smart Grid 4 (2) (2013) 720–728, https://doi.org/10.1109/TSG.2012.2213847.

[10] J. Yu, Y. Zhang, Z. Yan, H. Ye, W. Fu, H. Deng, Z. Qian, A coordinated scheduling approach with electricity peak shaving capability for integrated office building-ev system, Energy and Buildings 363 (2026) 117547, https://doi.org/10.1016/j.enbuild.2026.117547.

[11] H. Gharavi, R.G. Ahangar, S. McLoone, A coordinated cvr-evdc approach for electric vehicle charging demand management, Applied Energy 392 (2025) 125916, https://doi.org/10.1016/j.apenergy.2025.125916.

[12] M. Tostado-Véliz, H.M. Hasanien, S. Kamel, R.A. Turky, F. Jurado, M. Elkadeem, Multiobjective home energy management systems in nearly-zero energy buildings under uncertainties considering vehicle-to-home: a novel lexicographic-based stochastic-

information gap decision theory approach, Electric Power Systems Research 214 (2023) 108946, https://doi.org/10.1016/j.epsr.2022.108946.
[13] T. Yuvaraj, R. Krishnamoorthy, S. Arun, M. Bajaj, V. Blazek, L. Prokop, Smart energy strategies for dynamic ev charging and renewable integration in distribution systems: a multi-objective optimization framework, Energy Strategy Reviews 63 (2026) 102051, https://doi.org/10.1016/j.esr.2026.102051.
[14] C. Gómez-Quiles, P. Paramo-Balsa, J. Riquelme-Santos, A. Gómez-Expósito, Minimum-time milp-based ev charging scheduling for prosumers, International Journal of Electrical Power & Energy Systems 172 (2025) 111293, https://doi.org/10.1016/j.ijepes.2025.111293.
[15] M. Tostado-Véliz, H.M. Hasanien, P. Arévalo, F. Jurado, Robust dynamic charging price in pv-assisted charging stations, Applied Energy 395 (2025) 126251, https://doi.org/10.1016/j.apenergy.2025.126251.
[16] Z. Zhou, H. Xu, Mean field game-based decentralized optimal charging control for large-scale of electric vehicles, IFAC-PapersOnLine 55 (15) (2022) 111–116, https://doi.org/10.1016/j.ifacol.2022.07.617, 6th IFAC Conference on Intelligent Control and Automation Sciences ICONS 2022.
[17] R. Wang, G. Xiao, P. Wang, Hybrid centralized-decentralized (hcd) charging control of electric vehicles, IEEE Transactions on Vehicular Technology 66 (8) (2017) 6728–6741, https://doi.org/10.1109/TVT.2017.2668443.
[18] B. Hashemi, M. Shahabi, P. Teimourzadeh-Baboli, Stochastic-based optimal charging strategy for plug-in electric vehicles aggregator under incentive and regulatory policies of dso, IEEE Transactions on Vehicular Technology 68 (4) (2019) 3234–3245, https://doi.org/10.1109/TVT.2019.2900931.
[19] Z. Zhao, C.K. Lee, X. Yan, H. Wang, Reinforcement learning for electric vehicle charging scheduling: a systematic review, Transportation Research. Part E, Logistics and Transportation Review 190 (2024) 103698, https://doi.org/10.1016/j.tre.2024.103698.
[20] A.M. Elkholy, A.N. Rozhkov, A.V. Badalyan, I.A. Cherdintsev, Adaptive genetic algorithms enhance ev charging infrastructure resilience through multi-constraint optimization of grid resources and traffic dynamics, Electric Power Systems Research 250 (2026) 112045, https://doi.org/10.1016/j.epsr.2025.112045.
[21] B. Hermans, S. Walker, J. Ludlage, L. Özkan, Model predictive control of vehicle charging stations in grid-connected microgrids: an implementation study, Applied Energy 368 (2024) 123210, https://doi.org/10.1016/j.apenergy.2024.123210.
[22] M. Tostado-Véliz, P. Arévalo, F. Jurado, An optimization framework for planning wayside and on-board hybrid storage systems for tramway applications, Journal of Energy Storage 43 (2021) 103207, https://doi.org/10.1016/j.est.2021.103207.
[23] M. Tostado-Véliz, S. Gurung, F. Jurado, Efficient solution of many-objective home energy management systems, International Journal of Electrical Power & Energy Systems 136 (2022) 107666, https://doi.org/10.1016/j.ijepes.2021.107666.
[24] J.C. Silva, N. Lopes, C. Alves, G. Putnik, L. Varela, L. Ferreira, M. Cruz-Cunha, Evaluation of solvers' performance for solving the flexible job-shop scheduling problem, Procedia Computer Science 219 (2023) 1043–1048, https://doi.org/10.1016/j.procs.2023.01.382, cENTERIS – International Conference on ENTERprise Information Systems / ProjMAN – International Conference on Project MANagement / HCist – International Conference on Health and Social Care Information Systems and Technologies 2022.
[25] R. Tawn, J. Browell, A review of very short-term wind and solar power forecasting, Renewable and Sustainable Energy Reviews 153 (2022) 111758, https://doi.org/10.1016/j.rser.2021.111758.

[26] J. Zhang, Y. He, M. Cui, Y. Lu, Primal dual interior point dynamic programming for coordinated charging of electric vehicles, Journal of Modern Power Systems and Clean Energy 5 (6) (2017) 1004–1015, https://doi.org/10.1007/s40565-016-0224-5.
[27] A. Wächter, L.T. Biegler, On the implementation of a primal-dual interior point filter line search algorithm for large-scale nonlinear programming, Mathematical Programming 106 (1) (2006) 25–57, https://doi.org/10.1007/s10107-004-0559-y.
[28] W. Yin, W. Liang, J. Ji, Study on charge and discharge control strategy of improved pso for ev, Energy 304 (2024) 132061, https://doi.org/10.1016/j.energy.2024.132061.
[29] L. Yang, Q. Zhang, X. Zhang, M. Tayyab, An enhanced admm-based distributed framework for coordinated wind power and electric vehicle scheduling optimization, Energy Strategy Reviews 65 (2026) 102235, https://doi.org/10.1016/j.esr.2026.102235.
[30] M. Tostado-Véliz, A. Borghetti, P. Siano, F. Jurado, Privacy-preserving energy management in local energy communities with evs – an enhanced benders-like solution strategy, Sustainable Cities and Society 132 (2025) 106823, https://doi.org/10.1016/j.scs.2025.106823.
[31] T. Wu, A. Scaglione, N. Ravi, S. Peisert, D. Arnold, Optimal control of differentially private ev charging: a scalable learning approach under uncertainty, IEEE Internet of Things Journal 12 (24) (2025) 55061–55075, https://doi.org/10.1109/JIOT.2025.3622650.
[32] J. Liu, G. Lin, S. Huang, Y. Zhou, C. Rehtanz, Y. Li, Collaborative ev routing and charging scheduling with power distribution and traffic networks interaction, IEEE Transactions on Power Systems 37 (5) (2022) 3923–3936, https://doi.org/10.1109/TPWRS.2022.3142256.
[33] L. He, Z. Wu, Advancing sustainable ev charging infrastructure: a hybrid solar-wind fast charging station with demand response, Renewable Energy 237 (2024) 121843, https://doi.org/10.1016/j.renene.2024.121843.
[34] D. Çelik, M.A. Khan, N. Khosravi, M. Waseem, H. Ahmed, A review of energy storage systems for facilitating large-scale ev charger integration in electric power grid, Journal of Energy Storage 112 (2025) 115496, https://doi.org/10.1016/j.est.2025.115496.
[35] T. Kunj, K. Pal, Role, application and challenges of iot in smart ev charging management: a review, Future Batteries 7 (2025) 100102, https://doi.org/10.1016/j.fub.2025.100102.
[36] Q. Xiao, R. Zhang, Y. Wang, P. Shi, X. Wang, B. Chen, C. Fan, G. Chen, A deep reinforcement learning based charging and discharging scheduling strategy for electric vehicles, Energy Reports 12 (2024) 4854–4863, https://doi.org/10.1016/j.egyr.2024.10.056.
[37] G.C. Okwuibe, Z. Li, T. Brenner, O. Langniss, A blockchain based electric vehicle smart charging system with flexibility$_{**}$ this project is sponsored by oli systems gmbh, Stuttgart, IFAC-PapersOnLine 53 (2) (2020) 13557–13561, https://doi.org/10.1016/j.ifacol.2020.12.800, 21st IFAC World Congress.
[38] Q. Li, N. Zhao, B. Martinez-Pastor, P. Carroll, Equitable rural electric vehicle charging through peer-to-peer curbside networks, Transportation Research. Part D, Transport and Environment 154 (2026) 105268, https://doi.org/10.1016/j.trd.2026.105268.

Nomenclature

Acronyms

AC	Alternate-Current
ACEA	European Automobile Manufacturers Association
ADMM	Alternating Direction Method of Multipliers
AI	Artificial Intelligence
BASA	Bilateral Aviation Safety Agreement
BEC	Battery Electric Car
BEMU	Battery Electric Multiple Units
BEV	Battery Electric Vehicle
BLDC	Brushless Direct-Current Motor
BMS	Battery Management System
BoP	Balance of Plant
BVLOS	Beyond Visual Line of Sight
CAN	Controller Area Network
CAPEX	Capital Expenditures
CARB	California Air Resources Board
CC-CV	Constant Current-Constant Voltage
CCS	Combined Charging System
CCUS	Carbon Capture, Utilization and Storage
DC	Direct-Current
DEP	Distributed Electric Propulsion
DER	Distributed Energy Resource
DoD	Depth-of-Discharge
DoH	Degree-of-Hybridization
EASA	European Union Aviation Safety Agency
EC	Electric Car
EDLC	Electric Double-Layer Capacitor
EREC	Extended-Range Electric Car
ESC	Electronic Speed Controller
EV	Electric Vehicle
EVSE	Electric Vehicle Supply Equipment
eVTOL	Electric Vertical Take-Off and Landing
FAA	Federal Aviation Administration
FC	Fuel-Cell
FCEC	Fuel-Cell Electric Car
FCU	Flight Control Unit
FESS	Flywheel Energy Storage System
FPV	First-Person View
GM	General Motors
GNSS	Global Navigation Satellite System
HEC	Hybrid Electric Car
HEV	Hybrid Electric Vehicle

HPC	High Power Charging
HRS	Hydrogen Refuelling Station
ICE	Internal Combustion Engine
IEA	International Energy Agency
IMU	Inertial Measurement Unit
IoT	Internet of Things
ISR	Intelligence, Surveillance, and Reconnaissance
KERS	Kinetic Energy Recovery System
LFP	Lithium-Iron Phosphate
LHV	Lower Heating Value
LMO	Lithium-Manganese Oxide
LOHC	Liquid Organic Hydrogen Carriers
MaaS	Mobility-as-a-Service
MOF	Metal-Organic Framework
MTOW	Maximum Take-Off Weight
NCA	Nickel-Cobalt-Aluminum
NiMH	Nickel-Metal Hydride
NMC	Nickel-Manganese-Cobalt
OBC	Onboard Charger
OCPP	Open Charge Point Protocol
OPEX	Operational Expenditures
PEM	Proton Exchange Membrane
PEMFC	Proton Exchange Membrane Fuel Cell
PHEC	Plugin Hybrid Electric Car
PID	Proportional-Integral-Derivative
PLC	Power Line Communication
PV	Photovoltaic
PWM	Pulse Width Modulation
RTP	Real-Time Pricing
SC	Supercapacitor
SEI	Solid Electrolyte Interphase
SMES	Superconducting Magnetic Energy Storage
SMR	Steam Methane Reforming
SoC	State-of-Charge
SoH	State-of-Health
TCO	Total Cost of Ownership
TOU	Time-of-Use
UAM	Urban Air Mobility
UAS	Unmanned Aircraft System
UAV	Unmanned Aerial Vehicle
UITP	International Association of Public Transport
V2B	Vehicle-to-Building
V2G	Vehicle-to-Grid
V2H	Vehicle-to-Home
V2L	Vehicle-to-Load
V2V	Vehicle-to-Vehicle
V2X	Vehicle-to-Everything
VLOS	Visual Line of Sight
VTOL	Vertical Take-Off & Landing

ZEV	Zero Emission Vehicle

Energy and Power

E	Energy (kWh)
P	Power (kW)
P_{avg}	Average power (kW)
P_{loss}	Power losses (kW)
t	Time (h)
I	Electric current (A)
V	Voltage (V)
R	Electrical resistance (Ω)

Storage-related parameters

C	Capacity (kWh)
SoC	State of Charge (% or p.u.)
SoH	State of Health
ΔG	Free Gibbs energy (J)
n	Number of electrons
F	Faraday constant (C/mol)
q	Charge (C)
C_{rate}	Charge/discharge rate (h^{-1})
η	Efficiency
T	Temperature (°C)
$\mathcal{C}$	Capacitance (F)
$\mathcal{I}$	Moment of inertia (kg·m^2)
ω	Angular velocity (s^{-1})

Optimization

c	Cost of electricity (€/kWh)
P_{load}	Load demand (kW)
P_{grid}	Grid supplied power (kW)
P_{PV}	PV generation (kW)
P_{ch}	Charging power (kW)
P_{dch}	Discharging power (kW)
η_{ch}	Charging efficiency (% or p.u.)
η_{dch}	Discharging efficiency (% or p.u.)
$\mathcal{T}$	Time horizon
α	Multi-objective weight
N	Number of EVs

Superscripts

max	Maximum value
min	Minimum value
req	Required
dep	Departure

Index

Printed and bound by CPI Group (UK) Ltd, Croydon, CR0 4YY
17/09/2026
02218533-0004